© 2007 by David P. Bayles

About the Author

EDWARD HUMES is the author of eight critically acclaimed narrative nonfiction books,
including the bestseller *Mississippi Mud, School
of Dreams,* and *Over Here.* He is the recipient of
the Pulitzer Prize and a PEN Award. He is
presently writer at large for *Los Angeles* magazine. For more information about Humes please
visit his Web site at www.edwardhumes.com.
He lives in California.

"*Monkey Girl* is a fascinatingly detailed record of the creationist war on science, one that links national movements to their impacts on the lives of individuals and the education of our children."   —*Skeptic* magazine

"Rich back stories of characters from both sides of the courtroom make self-evident the fundamental culture war brewing beneath America's surface, while a play-by-play account of the Dover trial—injected with both legal and scientific detail—adds real-time color to this comprehensive read."   —*Seed* magazine

"*Monkey Girl* is compelling and unsettling—and a Rosetta Stone for understanding the frightened and oftentimes frightening subculture of Darwin Deniers."   —Patt Morrison, *Los Angeles Times* columnist, NPR host, and author of *Rio LA*

"A compelling, page-turning narrative. . . . A must read for anyone who cares about science, education, and liberty."
—Michael Shermer, author of *Why Darwin Matters*

"A riveting account of the modern 'holy war' being waged by American fundamentalists in the schoolyard battle over Darwin's dangerous idea. Humes brings this modern American story to life through the richly drawn portrayals of the citizen-members of the Dover Township school board, as their battle over what our children should be taught moves from one small Pennsylvania town to the country at large."
—Lee M. Silver, professor of molecular biology and public policy, Princeton University, and author of *Challenging Nature*

"Unusually deft analysis . . . extensive behind-the-scenes interviews . . . well illuminated by this skilled writer. . . . Highly recommended read."
—Eugenie C. Scott, executive director, National Center for Science Education and author of *Evolution vs. Creationism*

MONKEY GIRL

EVOLUTION, EDUCATION, RELIGION, AND THE BATTLE FOR AMERICA'S SOUL

EDWARD HUMES

AN ecco BOOK

HARPER PERENNIAL

NEW YORK • LONDON • TORONTO • SYDNEY

HARPER ⬤ PERENNIAL

A hardcover edition of this book was published in 2007 by Ecco,
an imprint of HarperCollins Publishers.

FIRST HARPER PERENNIAL EDITION PUBLISHED 2008.

Designed by Fearn Cutler de Vicq

The Library of Congress has catalogued the hardcover edition as follows:

Humes, Edward.
   Monkey girl : evolution, education, religion, and the battle for America's soul /
Edward Humes.
      p. cm.
Includes index.
ISBN: 978-0-06-088548-9
ISBN-10: 0-06-088548-3
1. Evolution (Biology)—Study and teaching (Secondary)—Pennsylvania—Dover
(Township). 2. Creationism—Study and teaching (Secondary)—Pennsylvania—
Dover (Township). 3. Intelligent design (Teleology)—Study and teaching
(Secondary)—Pennsylvania—Dover (Township). I. Title.

QH362 H86 2007
231.7'6520973—dc22   2006050263

ISBN: 978-0-06-088549-6 (pbk.)

08 09 10 11 12 WBC/RRD 10 9 8 7 6 5 4 3 2

PREFACE
AND ACKNOWLEDGMENTS

As I finished writing *Monkey Girl*, it became abundantly clear that, though the drama portrayed in these pages has a powerful denouement, the underlying story never really ends—it evolves. There's really no other word for it. Americans are becoming more divided than ever on the question of evolution and the teaching of origins in our public schools, and while debate is an essential part of any healthy democracy, the evolution wars—and the conflict between faith and science that they represent—have been far more destructive than nourishing to our children and our Republic.

To the extent that this conflict thrives on too little knowledge rather than too much, it is my hope that *Monkey Girl* will help end misunderstandings about exactly what happened in Dover, Pennsylvania, and help dispel the larger myths about evolutionary theory, its relationship to religion, and the questions that science can and cannot answer. I ask readers of all perspectives to turn these pages with open minds. There is no greater waste or tragedy than a war based on falsehoods; if the evolution wars are to continue, let the combatants be armed with facts, not fiction.

I wish to acknowledge the invaluable assistance, patience, and kindness I received from those who shared their expertise and experiences in the course of my research for *Monkey Girl*:

In Pennsylvania: Mike Argento, Carol "Casey" Brown, Jeff Brown, William Buckingham, Robert Eschbach, Burt Humburg, Jeri Lee Jones, Bryan Rehm, Christy Rehm, Bertha Spahr, Steven Stough, and Laurie Lebo, whose news reporting for the *York Daily Record* on the complex Dover intelligent design case sets the standard for excellence.

In Kansas: Steve Abrams and Jack Krebs.

The attorneys: Steve Harvey, Richard Katzskee, Eric Rothschild, Richard Thompson, and Vic Walczak.

The experts: Michael Behe, Barbara Forest, Phillip Johnson, Nick Matzke, P. Z. Myers, and Kevin Padian.

I am particularly grateful to U.S. District Court Judge John E. Jones III for his exceptional openness and willingness to facilitate the efforts of all working journalists; and to the National Center for Science Education, for its extensive efforts to make available to the public a complete record of the Dover case.

Last, but certainly not least, I wish to thank Emily Takoudes for her excellent editing (and kind patience while awaiting something to edit); my publisher at Ecco, Dan Halpern, for believing in *Monkey Girl*; my incomparable literary agent, Susan Ginsburg of Writers House, for believing in me; and my beloved, Donna, and my wonderful, rascally children, Gaby and Eben, for enduring the Dreaded Deadline Days as *Monkey Girl* approached the finish line.

CONTENTS

Prologue: One Thing's for Sure,
We Didn't Come from Any Monkey   *xi*

PART I: ORIGINS

1 Balancing Act   3

2 What Lies Beneath   16

3 I'd Rather Take a Beating Than Back Down   35

4 Darwin's Nemesis   63

5 Class Acts   79

PART II: SURVIVAL OF THE FITTEST

6 Broken Watches   109

7 The Watchmaker Returns   128

8 The Waters of Kansas Part   146

9 What Will We Tell the Children?   161

PART III: UNNATURAL SELECTIONS

10 Send Lawyers, Geeks, and Money   181

11 Monkey Suit   201

12 Sword and Shield, Shock and Awe   228

13 Paleozoic Roadkill, Kentucky Fried Chicken,
     and Bad Frog Beer   254

14 Of Panders and People   279

15 Under the Microscope, Deer in the Headlights   297

16 Forty Days and Forty Nights   316

17 Breathtaking Inanity   328

Epilogue   339

Notes   353

Index   367

Prologue

ONE THING'S FOR SURE, WE DIDN'T COME FROM ANY MONKEY

The Reverend Jim Grove is a wiry and intense man, his eyes burning and birdlike as he takes the measure of each person entering the ninth-floor federal courtroom in downtown Harrisburg. He is among the first people whom visitors encounter as they arrive to watch the trial billed in the media as the second coming of the legendary Scopes Monkey Trial. "Yes," Grove says, "it's a monkey trial all right. And the evolutionists are the monkeys." He does not smile when he says this.

Now and then, whenever he feels the moment is right (generally whenever someone initiates eye contact), the reverend unfolds himself from one of the old wood and leather chairs in the courthouse hallway and offers a small green flyer. This is an invitation to an event in the modest confines of the Dover, Pennsylvania, fire hall entitled, "Why Evolution Is Stupid," featuring a presentation from a creationism superstar who bills himself as "Dr. Dino." Dr. Dino firmly believes that the world is only 6,000 years old and that, consequently, men and dinosaurs once lived together quite happily—the classic creationist view of the universe. This causes him to be in great demand in the thriving industry of "biblically correct science." He crisscrosses the nation for more than 200 paid speaking engagements a year, and Grove sees the "Evolution Is Stupid" presentation as a critical counter to the "devil's work" going on just beyond the big wooden doors of courtroom 2.

"Come to the meeting. You really need to hear the truth," Grove tells one of the legion of reporters headed into court. "You're sure not going to hear it in *there*." He turns to a young woman and hands her a flyer, his eyes locked on the courtroom doors, where some of the top lawyers and law firms in the nation have congregated to pit science against faith. "All you'll hear in there are the babblers," Grove promises. Then he launches into a complicated monologue on flood geology in Genesis and why the absence of feathered fish and finned birds proves that Darwin's notion about common ancestry is so much bunk. Soon a bubble of solitude has formed around Grove as almost everyone within hearing distance shies away. But the preacher, who has gained local notoriety for his militantly gory antiabortion signs and Halloween parade floats, is not easily deterred.

He has come to bear witness to a trial that began as an obscure dispute over science textbooks in the rural Pennsylvania township of Dover. Now pastoral Dover sits firmly astride the front lines of America's culture war, occupying the uneasy space between America's religious faith and its long-standing fondness for scientific progress, between an idealized past and an uncertain future, between education and indoctrination, between the natural and the supernatural. For the next several months, the ninth-floor courtroom in the Ronald Reagan Federal Building will belong to *Kitzmiller et al. v. Dover Area School District*, an unintentionally epic lawsuit filed by a group of parents against their evolution-doubting school board. The case does indeed have much in common with the 1925 Scopes Monkey Trial, a public spectacle in which Clarence Darrow and the American Civil Liberties Union unsuccessfully challenged a Tennessee law banning the teaching of evolution. But unlike its illustrious predecessor (which, popular imagination and classic films notwithstanding, had exactly no impact on the law or educational practice at the time), the Dover case is positioned to define (or redefine) for decades just what children are taught about where we come from.

Reverend Grove, on the other hand, seems more interested in where we're *going*, which is the true bottom line of this battle, at least for the side that most reviles evolution. He appears to be keeping a mental list of all those he meets in the courthouse hallway, as well as those who testify in the case—a list of who will be headed to heaven when Judgment Day arrives, and who will, in Grove's estimation, be taking a more southerly direction. His world is a comfortably (or terrifyingly) straightforward place. Either you believe in the literal truth of the Bible—that God created heavens, earth, and man in six literal days—or you are damned. "Those Darwinists won't know what hit them," Grove says quietly as the voluble head of the National Center for Science Education stands a few feet away. A crowd of reporters listens intently as she presents evolutionary theory as the most verified and mature theory in all of science, the very cornerstone of modern biology, as well as the key to all sorts of lifesaving medical and pharmaceutical research. Grove shakes his head, astonished that anyone could buy such talk. As he espouses a theology that predicts the eternal damnation of this spokeswoman for science, along with dozens of people around him, his tone is not nearly as sorrowful as he intends. He juts his bearded chin forward and brings his cowboy boots down on the hard courthouse floor and sounds more triumphant than sad; after all, he says, his God is a vengeful God.

He also knows that America is largely on his side, at least according to opinion polls. Nearly half the citizenry accepts the idea that God created man in his present form, just as the Bible holds. Only a third believes that there is valid scientific evidence to support the theory of evolution. And an overwhelming majority supports what the Dover school board purportedly sought to do: teach evolutionary theory while also taking pains to point out what critics call its "flaws" and "gaps." In Dover, this was to be accomplished by informing biology students of "scientific" alternatives—specifically, the new kid on the origins theory block, "intelligent design." Accepting ID, as it has come to be known, requires a belief that empirical evidence—real, hard science—shows that the complexity of life on Earth cannot be explained without the intervention of some sort of master designer. The designer is not identified as God—or identified at all—by IDers; they insist that all they can do as scientists is point to evidence that life is intelligently "designed," a distinction proponents believe should make all the difference when courts weigh issues of church and state. There is a bit of a nod and a wink to this, as everyone involved knows that they're talking about—or, more precisely, *not* talking about—God, which is why such a large segment of Americans approve when ID is explained to them. America, alone among Western nations, is overwhelmingly God's country. "Majority should rule," Grove says. "Isn't that the way America is supposed to work?" The parents who filed suit beg to differ.

A pale auburn-haired woman walks stiffly past the reverend, looking straight ahead as she plunges through the doors into court. She is a former member and past president of the Dover school board, but Grove does not offer her a green flyer or any sort of greeting; he just purses his thin lips and once again shakes his head slightly. It appears Casey Brown is on Grove's second list. That she has done good works, serving her community and her church for many years, doesn't much matter. She has, in the view of Grove and many other citizens of Dover, betrayed her own, taking the side of the godless ACLU, the evolutionists, the poisoners of young minds.

"Evolution is the road to atheism," Grove explains. "Our children's future is at stake."

Our children's future is at stake: This is probably the only statement Reverend Grove will make during the course of the trial with which Carol Honor "Casey" Brown can unequivocally agree.

As she comes to court to testify about her long tenure on and rancorous departure from the Dover school board, she can still scarcely believe it has come to this: a community divided, families divided, friends divided. Children have been ridiculed in the school yard for being open to the concept of evolution, taunted and mocked for being related to monkeys. A Dover High School student's senior project, a sixteen-foot mural depicting the ascent of man from lower forms, which had been donated to the school and displayed in a science classroom, was taken down and burned—not by vandals, but by a school district official, with the tacit, if not gleeful approval of school board members. Casey, meanwhile, having expressed her opposition to introducing religious ideas into public school science classes, has received hate mail and crank calls and angry stares in the street. And now she sees it all spinning out of control: Dover is no longer just a small town, a small school district, a tightly woven community. Dover has become a cause.

It is a cause for the Seattle-based Discovery Institute, progenitor of the intelligent design movement and the seductively reasonable argument that schools should of course teach evolution, but that they also should "teach the controversy." It is a cause for the Americans United for Separation of Church and State and the American Civil Liberties Union, which has fought every significant battle over evolution in the country since (and including) Scopes, and is facing dozens more across the country. It's a cause for the massive Pepper Hamilton law firm of Philadelphia, which has donated the manpower and monetary muscle to champion the view of parents who sued, and who see "teach the controversy" as a stealthy method of injecting religion into the classroom instead of keeping it at home and in church. And it's a cause for the Michigan-based Thomas More Law Center, a law firm launched by the billionaire founder of Domino's Pizza with a mission to return religion "to the public square," billing itself, rather immodestly, as "the sword and shield for people of faith." Together, these outside forces have lined up experts in science, philosophy, history, and religion prepared to deliver several months' worth of testimony. Outside legal teams have descended on the community; millions of dollars are being expended; the media glare is relentless. Even Casey's husband, Jeff Brown, a fifty-four-year-old electrician who also once served on the Dover school board, landed on television on Comedy Channel's *The Daily Show* in an episode titled "Evolution Schmevolution." The expenditure of energy and time seems staggering to Casey in a battle neither side feels it can afford to lose. Casey is one of the few in town who seem to realize that everyone in Dover has already lost a great deal.

This had been building for years. It began with the annual school board "retreat"—a fancy name for what amounted to a cafeteria-catered, help-yourself buffet at a teachers' lounge in one of the elementary schools. Board members were supposed to feel free to air their thoughts and concerns in an informal setting, without nosy reporters or members of the public listening in. That's where the local version of an ancient conflict took root, in January 2002, when a new board member, Alan Bonsell, an auto and radiator repair shop owner with whom Casey had campaigned, announced that he was very concerned about issues of morality. He wanted to bring prayer and faith back into the public schools. We need the Bible in the classroom again, he argued strenuously, and we need to teach creationism to achieve a "fair and balanced curriculum." More than budget cuts, more than textbooks, more than school construction or any of the other mundane but critical issues facing the district that they had all campaigned on, Bonsell seemed to care most about creationism. That, he said, was his number one issue. School prayer was second on his list.[1]

Casey shrugged off these comments as the concerns of someone new to running a school district and new to public office, someone with deep religious convictions who meant well, but somehow remained unaware of Supreme Court rulings and of the constitutional separation of government and religion that made his ideas inappropriate for public schools. He'd learn, she figured.

But at the following year's retreat, Bonsell, by then in line to become board president (a position traditionally rotated among board members), again voiced his concerns. And he now had a solid majority on the board joining him—one that left Casey out on the margins. It was wrong to teach evolution in biology classes without even mentioning the alternatives, Bonsell announced. They should be taught fifty-fifty.[2] Anything less would be unfair to families and schoolchildren.

As Casey haltingly testifies about her recollections of these times, her voice quivers. She sips water and sits almost painfully erect. In the gallery, sitting on one of the hard wooden benches that look almost indistinguishable from church pews, Alan Bonsell watches—a thickly handsome man with a bushy mustache and the hint of a grin, as if he knows something most others don't know. Occasionally, the back of his neck turns red at some bit of testimony, but he almost never loses that small grin. It turns out that he, too, kept a list, of the same sort Reverend Grove kept.

"He told me," Casey Brown testifies in her shaking voice, "that I would be going to hell."

She had long considered Bonsell a friend, and recalling this moment still has the power to crush her even after two years have passed. When she speaks of it later that evening, away from the intimidating ritual of the witness stand and ensconced at the kitchen table in the Browns' eclectic country home, she can recall her friend's prediction of eternal damnation only in a whisper. But though Casey Brown may be easily wounded, she is nobody's pushover. She has always been her own person, strong in her beliefs, a military brat, former educator, former journalist, mother, and volunteer. She even went through a hippie phase that, as luck would have it, has been preserved for posterity: Open an old copy of the Woodstock record album and there inside, that photo of a lithe young woman skinny-dipping is Casey of thirty-seven years past. She is not the least embarrassed, and neither is her husband, Jeff, who likes to pull out the venerable LP to show off like a trophy. Today her convictions about religion are just as strong and unyielding as her school board compatriots', but with a critical difference: Casey argues fervently that it is wrong—morally, legally, spiritually—to impose one's beliefs on others. Even as her fellow board members repeatedly quizzed her about whether she had been "born again"—making it clear that answering *no* would render her suspect, or worse—she felt compelled to demur, responding instead that their sworn duty as elected officials was to serve *every* member of the public equally, and to take pains to avoid marginalizing those of different faiths or those who have no religion at all. "However, it has become increasingly evident that it is the direction the board has now chosen to go," she scolded her colleagues in her last official act as a member of the Dover school board. "Holding a certain religious belief is of paramount importance."

For that, Casey heard herself pronounced an atheist, after which the school board proceeded to put previously obscure Dover, Pennsylvania, on the national map by voting six to three in favor of a remarkable new biology curriculum. By then the board members had received advice (along with videos, DVDs, and books critical of evolutionary theory) from the Discovery Institute and Thomas More, and had decided to take the legally prudent course of avoiding overtly creationist or religious teachings, much to the consternation of Bonsell's chief ally on the board, William Buckingham, who railed darkly about those "liberals in black robes" driving God from public institutions. Instead, the board adopted a subtler threefold strategy designed to pass constitutional muster.

First, they would alter the curriculum to inform students that there were "gaps" and "problems" with evolution, which was, after all, just a "theory, not a fact."

Then they would introduce (but not actually teach, as they saw it) the new "science" of intelligent design. Proponents of ID assert that the blind and random forces of nature could never have built such complex structures as a man or a bacterium or a strand of DNA without outside help—without some sort of intelligent designer pulling the strings. The theory, without actually identifying this designer, implies the existence of God.

And, finally, Dover's students would be told that, as a result of the first two points, there was a genuine and growing scientific controversy about the validity of evolutionary theory.

Thus evolution would be dissed in the classroom (much to the satisfaction of biblical literalists on the board and in the community), while students would be exposed to the notion that there could be *scientific*, as opposed to religious, evidence for a "designer" of all life on Earth. And this would be accomplished, the school board promised, without any of the religious or biblical classroom references that in the past had been ruled unconstitutional by the U.S. Supreme Court. As the board majority saw it, all this would do was *improve* science education in Dover. They represented it as an inclusion of new and exciting theories, a commitment to accuracy and fairness by referencing "both sides" of the evolution question, and a lesson in critical thinking added to all that tired, materialistic Darwinian dogma. Who, the board majority maintained, could argue with that?

They might have had a point, too, but for two big problems. Their own in-house experts—the entire science faculty—informed them that they believed intelligent design was neither credible nor a science, that it was a form of creationism in all but name, and that they adamantly opposed its introduction into the curriculum. And then there were the official discussions leading up to the new policy that seemed to belie the board's subsequent insistence that it had no religious agenda—discussion and debate that featured extensive references to God, creationism, the "myth" of church-state separation, and Buckingham's outraged attempt to shout down critics by saying, "Two thousand years ago, someone died on a cross. Can we have the courage to stand up for him?"[3]

The lawsuit from parents that followed wasn't just predictable; it was anticipated. The Thomas More Law Center had been searching the nation for just such a test case, which is why it offered to represent the cash-strapped school board for free should it decide to make Pennsylvania the first state in the country to introduce intelligent design into its curriculum. And the Dover school board was happy to oblige.

In the courtroom, Casey Brown finishes her testimony and Bonsell continues to smile. Later, during a break, David Napierski, who was appointed to the Dover school board after the evolution vote, stands, stretches, and expresses his support for intelligent design and a "teach the controversy" approach. Like other members of the school board, he says his support for "balance" is not based on any extensive knowledge of what intelligent design, or evolution for that matter, is all about. Indeed, he clearly fails to grasp that evolutionary theory in no way claims that man descended from apes, but only that today's men and today's apes share a common ancestor in the distant, prehistoric past. "I've discussed this with my daughter," Napierski says, "and one thing's for sure, we didn't come from any monkey. She knows that as well as I do." What Napierski and his colleagues still do not seem to understand is that Charles Darwin would agree entirely with that statement.

"This country wasn't founded on Muslim beliefs or evolution," Buckingham exhorted the citizens of Dover during the pivotal school board meeting. "This country was founded on Christianity and our students should be taught as such."

That was the moment when Casey began to be afraid. Not of Buckingham or of any one individual. She was afraid of what this controversy would do to a community that she loved.

"I was afraid Dover would never be the same," she says after court, finally relaxing in her own home, away from the glare and the attention.

"And I was right."

It is those who know little, and not those who know much, who so positively assert that this or that problem will never be solved by science.

—Charles Darwin

The intuitive mind is a sacred gift and the rational mind is a faithful servant. We have created a society that honors the servant and has forgotten the gift.

—Albert Einstein

PART I

ORIGINS

EVANGELIST: SOME PEOPLE WOULD HAVE YOU BELIEVE DINOSAURS EXISTED MILLIONS OF YEARS AGO. IT'S JUST NOT TRUE. GOD CREATED THE EARTH SIX THOUSAND YEARS AGO. I TELL MY KIDS, YOU HAVE TO REMEMBER, DINOSAURS AND HUMAN BEINGS LIVED ON THE EARTH AT THE SAME TIME.

TONY: WHAT, LIKE *THE FLINTSTONES*?

EVANGELIST: IT'S IN THE BIBLE.

CHRISTOPHER: T-REX IN THE GARDEN OF EDEN? NO WAY. ADAM AND EVE WOULD BE RUNNIN' ALL THE TIME, SCARED SHITLESS. THE BIBLE SAYS IT WAS PARADISE.

—The Sopranos

Chapter 1

BALANCING ACT

To the combatants, the conflict in Dover seemed new and dangerous, even epochal, but in truth it was but the latest iteration of a battle spanning five centuries and continuing still. It began when Copernicus launched *the* scientific revolution, removing humanity from the center of the solar system and revealing the Earth, despite all appearances and assumptions and faith to the contrary, to be a mere mote adrift in a vast cosmos, no longer the apple of God's eye. Then came the Age of Enlightenment and the learned deists who founded America, men like Jefferson and Franklin and Washington, who envisioned a creator setting the universe in motion but then letting matters unfold on their own—one reason, perhaps, why the Founding Fathers so adamantly fashioned a nation in which religion and government were never to interfere with each other. A century later, the paleontologists and geologists began to unearth a past no one ever had suspected, of long-extinct jungles, giant reptilian monsters, and an Earth that appears to be billions of years old instead of the 6,000 years carefully calculated from the Bible and assumed to be true for most of a millennium. That bedrock beliefs could crumble so quickly and easily in this new age of science was disturbing, to say the least, yet the western world took comfort in the one great truth that stood through it all, dating back to Plato and before: the grand design of life that laymen and scientists alike could observe everywhere around them. They witnessed the amazing delicacy and aerodynamic perfection of a bird's wing; the fish's sleek body so astutely fashioned to swim; the miracle of the human eye, a complex assemblage of innumerable parts that far outstripped anything man could ever hope to build—marvelous machines and breathtaking beauty in form and purpose, all of it evidence that a master engineer of infinite power had breathed life and purpose into creation. Science, it seemed, couldn't alter that fundamental truth; indeed, as the power of microscopes and telescopes and man's insight into nature increased, the purposeful design underlying creation seemed not less but more obvious. By the middle of the nineteenth century, scientific proof of the existence of God seemed achingly, gloriously within reach.

And then Charles Darwin took all that away, too, delivering in its place a world built in part by accident, in part by the brute, blind drive to survive—a purpose, to be sure, and a direction, but not a design. Chance, adaptability, and good fortune ruled this new world, where each species could not be seen, after all, as a master composer's symphony, but as a desperate mechanic's jury-rig of used parts. Dolphins (but not fish) have vestigial fingers inside their fins, and a bat's wing (but not a bird's) closely resembles the structure of the human hand not because such adaptations make anatomical sense from a design point of view, but because all three sets of limbs were derived from the same basic mammalian model: arms, wrists, phalanges, parts recycled and reshaped by variation and natural selection across vast stretches of time. Darwin and those who embraced and perfected his theory perceived an even greater grandeur in this view of life, of a nature so full of wonder that a simple, primitive life-form, no more than a germ, could evolve across the ages into a butterfly and a tiger and a man. To them, this suggested a God infinitely more subtle and magnificent than ever before imagined, having fashioned a creation that creates itself.

But the implications were also fairly horrifying when it came to man's place in this Darwinian world. Higher purpose was gone. Made in God's image—gone. And what of the soul? Only men had souls, it was said, but if humans shared a legacy with apes and sharks and garden slugs, did that even leave room for a soul? For an afterlife? For something greater than the flesh? The logic of Darwin, notwithstanding his own invocation of a creator in his writings, suggested that man's ascendance was nothing more than a happy accident, the flip side of which was this: If you could turn back the clock and do it all over again, humanity, which had the arrogance to fancy itself the pinnacle of creation, might not even come to exist the second time around. Life, intelligence, consciousness, and love were not gifts from God; it was all just a lucky break, a roll of the dice. And there it was: Darwin, alone among scientists in the new age, had finally provided the proverbial last straw for the faithful. It was one thing for science to destroy geocentrism, or to turn the Bible from literal history into lovely metaphor, but when it tried to dethrone man as God's masterpiece and render him no better (or worse) than marsupial or mollusk, then science simply had gone too far. The war that ensued has not really abated since.

"Will you honestly tell me (and I should be really much obliged) whether you believe that the shape of my nose (*eheu!*) was ordained and 'guided by an intelligent cause'?" an exasperated Darwin wrote to the pioneering geologist Charles Lyell amid the outcry that followed publication of *The Origin of Species*. It was as succinct a put-down of what has come to be known as intelligent design as has ever been delivered in science. But a century and a half later, Darwin's unfortunate proboscis (*eheu!* being Latin for "alas!") has still failed to stem the howls of protest from men and women who see evolutionary theory as both sacrilege and threat.

Tucked away in a mostly rural swath of York County, best known as the birthplace of the Articles of Confederation and as the temporary capital of the fledgling United States for nine months during the Revolutionary War, Dover began its life as a waypoint on the road to larger towns and markets. Even today, with a population just shy of 20,000, it remains a spacious place of hay fields, leafy roads, and redbrick houses, where the outside world can seem both pleasingly and frustratingly distant, and where the weekly high school football game captures far more interest and attendance than the pronouncements of the school board.

A half century before the Revolution, Old World immigrants settled the area, mostly Lutheran Germans from the Rhine River valley, farmers, carpenters, and brewers who fled religious persecution in their homeland, drawn by the tolerant Quaker philosophy that William Penn had established in his namesake state. Their family names still dominate phone books and maps in Pennsylvania. Rohler, Emig, Yingling, Kitzmiller. The economy they built here, once vibrant with corn and peaches, strawberries and wheat, along with a web of cigar factories, tanneries, quarries, and liveries to support the Shippensburg trade route once vital to the colonies, is patchy these days at best. Slowly but surely, Dover is becoming a bedroom community, with new housing developments cropping up year by year, an invasion of suburbia, though much of the old Dover remains, quaintly graceful in parts, and just plain faded and peeling in others.

Despite its roots as a refuge for the religiously persecuted, Dover is not new to conflict over its schools and religion. Precursors of the latest battle date back to the first good "inventory" of Dover in 1783, when the township reported the presence of 1,367 free men and women, four slaves, 219 horses, 146 barns, seven mills, and 23,811 acres of land. There was no mention of schools, though apparently there were a few, all of them religious and conducted in German, deliberately separate and apart from the rest of a young nation. In 1834, the Pennsylvania state legislature passed a public school law that instructed local governments to impose a tax to finance public education. The law—and the perceived interference from Harrisburg—proved deeply unpopular in Dover, which resisted the change for nearly two decades. Two "free schools," as they were then known, finally opened for six months a year beginning in 1850, where immigrants' children learned lessons in English only. The two schools competed furiously for students, but public support for non-religious education remained questionable. Opposition came to a head in 1881, when an arsonist torched both schools on the same night. No arrests were ever made. After that, public education in Dover was conducted in large part through a network of fifteen one-room schoolhouses. This setup engendered little controversy and continued through the mid-1950s, when the current school district absorbed the surrounding smaller jurisdictions and a larger, modern Dover High School was built.

It was around this time that the country, under President Dwight Eisenhower's leadership and in a sudden panic over the Soviet Union's successful launching of the satellite Sputnik in 1957, embarked on an effort to ramp up science and math education as a matter of national security. America needed to train a new generation of scientists, engineers, and researchers in order to meet the technical challenges of a modern society, along with the security challenges of the Cold War—the space race, the nuclear arms race, the missile race—and that meant shunting aside some old cultural and religious limits on science education. As new and up-to-date courses of study, textbooks, and curricula were created, a long-neglected subject that had been integrated for decades into classrooms elsewhere in the western world began to see the light of day in many American public schools. Evolutionary theory had been barely mentioned, or avoided entirely, in most schools and textbooks for the previous three decades, mostly owing to public and political pressure against textbook publishers, even though the teaching of evolution was formally banned in only four states and had been widely accepted by American scientists since the early 1900s. Finally, though, more than thirty years after the Scopes trial, evolution got its day in class, not through the work of atheists, liberal activists, or the ACLU, but as part of the country's Cold War against communism—an effort driven primarily by political conservatives. Still, the plotting by *religious* conservatives to drive evolution back into the shadows (or at least balance it with creationist objections and alternatives) began the moment a federally funded initiative started releasing updated biology textbooks and lesson plans in 1960 with major sections on evolutionary theory. A series of pro-creation educational statutes adopted in a few southern states sought to counter evolution's newfound influence; but these were soon followed by court decisions overturning every one of the statutes as unconstitutional attempts by the government to establish a state religion. The lasting effect of those legal rulings was to ratify the theory of evolution, not creation, as the standard scientific (and constitutionally acceptable) account of life's diversity and development for America's public school classrooms.

Holy wars, however, are notoriously unresponsive to legal niceties and judicial rulings. Conflicts between the forces of science and faith have continued unabated, mostly at low intensity, involving individual teachers or schools or districts rather than national policy, though sometimes there have been flare-ups that led to lawsuits and national media attention before they faded again. But the embers of discontent never entirely died out—not once in the fifty years it took for the evolution controversy to move from Sputnik to Dover.

Christy Rehm was among the first to notice that the lines of this battle were being drawn in Pennsylvania's "little Bible belt," quietly but inexorably. It was no small school board spat, destined to fade as tempers cooled, as many in Dover initially believed, if they even noticed at all. Christy sensed something bigger brewing, the same age-old battle that had demonized Charles Darwin for most of his life, and that, centuries earlier, had led the Inquisition to place Galileo, the father of modern astronomy, under house arrest until the end of his life for championing the preposterous and ungodly theory that the Earth orbited the sun. That could be where all this was headed, Christy Rehm feared—a local meltdown of historic proportions. She had grown up in Dover, knew its rhythms, knew the good heart that pulsed at its center—the ready unity and common cause that are a small town's strengths. This is what brought her back to her hometown after college, and led her to convince her husband they should raise their family here. But that strength could also be a weakness, as it could be twisted into a stubborn and closed-minded parochialism. Now Christy could feel something ugly coming.

Christy graduated from the same high school where the evolution war was later to erupt; but in her day, Darwin was never an issue (and not a particularly memorable subject for a future English teacher like Christy). The head of the science department at Dover High was herself a force of nature, having taught there long enough to instruct not only Christy, but Christy's father and aunts and uncles and cousins and virtually everyone Christy knew, and any one of them would tell you that the science teacher Bertha Spahr could not be bullied, bought, or badgered. "Bert" had the proverbial heart of gold, but early in her career, as a diminutive young woman in her first classroom of hulking football players and other unruly teenagers, she had learned the value of terrorizing her new students with an occasional blast from what she calls her "Sicilian temper." Instruction in evolution was the law of the land, it was in the Pennsylvania standards, it was in the official textbooks, and it would be taught. "End of story," as Bert Spahr likes to say, and woe to anyone who tried to do otherwise in her department. Evolutionary theory was taught without a blink throughout her four decades at Dover.

But in recent years, a new wave of younger, outspoken evangelicals came to settle in and around Dover, erecting a passel of new churches and helping make this region into a Keystone Bible belt. They expressed their discontent with the secular world with renewed vigor, very different from the low-key religious outlook of Dover's old-line residents. When evolution was taught to the newcomers' children, they most certainly *did* blink.

That something new was brewing became apparent to Christy Rehm when her husband, Bryan, a young science teacher at Dover High, began coming home from work uncharacteristically angry and upset. He's a tall man with a closely trimmed brown beard, and he has a tendency, even when content, to be on the somber side, particularly in contrast to the bubbly Christy's easy laughter. Lately, he seemed positively grim. Education, Christy says, is everything in their household, with three school-age children and another on the way back in 2003—and even though she taught in a district outside Dover, they still loved to talk shop. But shortly after classes began that year, Bryan's shop talk began to consist largely of "I can't believe they're doing this."

The school board member Alan Bonsell, having announced his interest in creationism and prayer in school at two consecutive annual board retreats, had by then achieved a position to do something about it. He headed the board's curriculum committee in 2003—giving him jurisdiction over textbooks and lesson content—and he would take over as board president in 2004. Soon the assistant district superintendent, Mike Baksa, cast in the classic role of nice guy caught in the middle, had begun visiting the science faculty to deliver a heads-up that Bonsell wanted "balance" in the curriculum. Baksa started at the top, with the nononsense Bert Spahr, informing her that the board appeared headed in a direction the teachers wouldn't much like. Baksa would later say he avoided using the term "creationism," but that was what Spahr and the other teachers understood.[1]

Bonsell wanted a very specific type of balance. Not Buddhist ideas on creation, or Navajo or ancient Sumerian or Islamic or any other creation story or myth, but Genesis-based, King James version, *"In the beginning"* Judeo-Christian creationism, taught fifty-fifty with evolution. Not in a philosophy or comparative religion class, but in science class.

There was no directive or new policy as yet, Baksa assured the head of the science faculty, but it was clear that this was more than just an idle wish that could be safely ignored until it simply went away.

At the end of March 2003, Spahr broke the news to the rest of the science faculty, putting them on notice that there would probably be some scrutiny from the school board on how they handled their evolution lessons, and that the board had creationism in mind. "They want us to teach *what*?" Rehm asked. He would later say he thought at first it was a joke. Court decisions dating back twenty years and more had outlawed creationism in public school classes. "They can't be serious, can they?"

Bert Spahr did not say anything, and Rehm had his answer. It was no joke.

What made this all a bit disconcerting was the fact that at Dover High School, students' encounters with Darwin were already highly abbreviated, with the topic covered in ninth grade at the very end of the school year, when spring breezes were in the air and getting students to pay attention to anything was a constant challenge. Even before Bonsell raised the issue, the teachers for the most part handled the radioactive subject of evolution in the most delicate way possible. They taught what the Pennsylvania state education standards required, but they also knew that Dover, with its sixteen churches packed into a very small space, was highly sensitive on this subject. They did not talk about origins of life. They referred to evolution primarily as "changes over time," and though the principle of common descent was taught, the teachers avoided getting into the evolution of man. That was a perennial land mine, a virtual invitation for a child to go home and say, "My teacher says we come from monkeys," and then the phone calls and tension and visits with the principal would begin. Like many teachers who know the limits of their students and parents, the Dover science teachers tended to soft-pedal evolution and then move on, giving only fleeting treatment to a theory that is considered one of the three most important in the history of science (along with Newtonian mechanics and the germ theory), and the central explanatory paradigm for the entire field of biology. "You do what you can do," Bryan Rehm says.

Rehm is a physics teacher by trade, but that did not spare him from the controversy. As the school board continued to focus on and question all things evolutionary, he found himself cutting back on certain class activities designed to get kids turned on to science. No one had ever criticized or questioned these lessons—quite the contrary, his students seemed to love them—but he started second-guessing himself. Such was the "chilling effect" of the board's stated concerns about evolution, he'd later say—a legal term for pressure-induced self-censorship that he had never used or thought about before. He didn't want to provoke the board into adopting policies he feared would be antiscience. And so he abandoned an essay assignment and class discussion called "What Is Science?" That's when he would have explained what a hypothesis and a theory and peer review were all about; how to tell real science from junk science; how careful observation and data collection can confirm a hypothesis sufficiently to make it a theory; and how a theory is not just a guess or speculation in the world of science, but a robust and tested explanation of natural phenomena. At that point in the lesson, the most familiar theory to the students (at least in name only), evolution, inevitably would come up, and some of the kids would mention various creationist objections to Darwin they had picked up at home or in church or on the Internet—such as the idea that evolution violates the second law of thermodynamics. Rehm would say no, that particular argument had been discredited by science for many years, and he'd explain why. Once the pressure began, however, Rehm just left this assignment out. He hated doing that, but felt he had no choice. If he didn't, things could get worse.

Other teachers quietly did the same. A colleague, the senior biology teacher Jen Miller, stopped her practice of having students create a long time line on rolls of paper in the hallway, aimed at demonstrating how species developed over millions of years—a time line that scientists overwhelmingly accept, but that creationists reject and consider offensive. Miller's decision was understandable: If certain board members believed in a "young Earth," why create an extra activity, however enjoyable and instructive it might be for the kids, that would be sure to provoke those board members if they found out? Another biology teacher, Robert Linker, similarly edited his lesson plans. He had always started the section on evolution by drawing a long line on the blackboard, then writing at one end "creationism" and at the other "evolution." These are two possible explanations for life on Earth, he'd say, and then explain why his class would study only one—because only one was science, while the other was religion, and it would not be legal or proper to teach religion in a public school. Linker dropped that approach like a hot potato.

All this was the result of a simple heads-up, without any actual directive to change a thing—preemptive actions by tenured teachers who could not easily be fired. Bryan Rehm just kept hoping that if they kept their heads down, it would all just go away, but Christy was getting angry. "It's not right," she told her husband. "This is not going to end well."

Sure enough, Baksa soon returned with more heads-up warnings. He was as smiling and pleasant as ever, but he continued to make it clear to the science teachers that Bonsell, and, later, the board member Bill Buckingham, who would be appointed by Bonsell to succeed him as chairman of the district curriculum committee, remained concerned about the way evolution was being handled. They still felt there needed to be balance. Rehm and some of the other teachers kept responding no, it's not appropriate for a science class to get into religious explanations of man and the world. Legally and professionally, they felt it would be wrong for them to do so, and the fact that Dover was a quite homogeneous, overwhelmingly Christian community did not make it okay. The teachers were Christians, too—the Rehms both taught Sunday school—but they believed strongly that religious teachings were for home and church, not public school science classes. The assistant superintendent would duly relay that information back to the board members, then he'd return to voice the same concerns. Soon he started bringing videos and literature on intelligent design for the teachers to review. They were told to view a Discovery Institute DVD attacking evolutionary theory called "Icons of Evolution," which some board members thought could be useful in classes. Its subtitle was: "Science or Myth: Why Much of What We Teach about Evolution Is Wrong." The science teachers found it worthless as a classroom resource and scientifically suspect. When this was reported back to the board, the teachers were told to view the DVD a second time.

Finally, a direct meeting was arranged between the teachers and

Bonsell in the fall of 2003 so that they could attempt to settle things without an intermediary filtering their words and sentiments. Christy and Bryan spent hours at home pulling science papers, court decisions, and legal opinions off the Internet—a huge packet of information showing that what the board was asking simply couldn't be done. It wasn't right, it wasn't professional, and it would never hold up once the lawsuits started flying.

The meeting, on the surface at least, seemed cordial. Bonsell explained his concerns in the context of the school district's "mission statement," which included a promise to "nurture the diverse needs of our students." That was a fundamental task of any educational system, Bonsell argued, but by teaching evolution as the be-all and end-all of human origins, they were failing that mission, confusing the students, and undermining their education. After the evolution lesson, Bonsell explained, kids will be going home and talking at the dinner table about what they learned that day in school, and their parents are going to have to say, "Sorry, your teachers are lying to you."

"Why would they say that?" Rehm wanted to know.

Because, Bonsell explained, the Earth is less than 10,000 years old, man did not evolve from monkeys, and evolution and the notion of common descent are against my religious beliefs and the beliefs of many other students and parents in Dover. This particular belief system, grounded in Genesis, is called "young-Earth creationism."

With that on the table, the teachers tried to do what they do best, what they do when one of their students doesn't understand something: They attempted to turn the meeting into a "teachable moment." They decided they would educate Bonsell about evolution and how it's taught at Dover High. They actually felt relieved at this point, certain that with a bit more information, everything would be fine. So the teachers went to great lengths to explain to the school board member how carefully they handled the subject in class, how they were always sensitive to the beliefs of their students, and how, while they could not get into religious doctrines in class, they always reminded the kids that if they had concerns they should talk to their parents or their pastors. And there was no "monkey to man" lesson plan, the teachers vowed—they always steered clear of human origins. In fact, the focus in class was far less on major evolutionary changes and more on smaller-scale "microevolution," because it was easy to grasp and there were many familiar examples: disease organisms evolving through natural selection to become resistant to antibiotics; or the evolution of pesticide-resistant insects; or the classic example of Darwin's finches, which evolved through isolation in the Galápagos archipelago to be different enough from other finches to be considered a separate species. This is the kind of knowledge that the annual state exams required, and it was the sort of information that was crucial for kids who wanted to pursue careers in medicine or the biological sciences, the teachers explained.

Bonsell also learned that at Dover High School, the teachers did not address the actual origins of life. This pleased him enormously and seemed to allay his concerns. The meeting ended quite amiably, with Bonsell seemingly satisfied and the teachers feeling that they had dodged a bullet. Things could get back to normal, Rehm told his wife that night. She did not celebrate, however, dubious that it could be that easy.

Sure enough, Mike Baksa returned soon after the New Year. There were board members who still wanted balance. And there was going to be a problem getting desperately needed new biology texts for the high school. Bonsell had taken over as board president at the start of the year, and had appointed a like-minded creationist, Bill Buckingham, to chair the curriculum committee of the school board, which had authority over textbook selection. Buckingham picked up where Bonsell had left off, complaining that the textbook favored by the science faculty was "laced with Darwinism." Instead, the board would go shopping for a textbook that featured creationism at least as prominently as evolution.

The understanding the science teachers thought had been reached with Bonsell turned out to be a monumental misunderstanding, one that would have long-lasting consequences. When the science teachers told Bonsell they did not teach the origins of life, they were referring to abiogenesis—the mysterious and as yet unexplained (scientifically, at least) process by which nonliving material in Earth's primordial environment led to the first living organism. This is what scientists consider to be the "origins of life," and the theory of evolution does not attempt to explain how it happened; Darwin's theory assumes that life is already present, and goes on from there. So there was never any reason to broach this topic in ninth-grade biology. But as Bonsell, Buckingham, and other board members used the term "origins of life," they were actually talking about what evolutionists call the "origins of species" or simply "speciation"—how different *forms* of life evolved from other forms. Biblical creationists believe that God created animals in their current forms all at once—dinosaurs, for instance, along with all other creatures and man—as in the biblical Genesis story. So the Darwinian theory of how species originated gradually, through evolution, posed a big problem for Bonsell, Buckingham, and like-minded board members, one that they felt needed to be balanced. Later, when it became clear there had been a misunderstanding and these sorts of origins were being taught after all, the original objections resurfaced. The board saw this as the "monkey-to-man" principle all over again—it didn't matter if human origins were omitted. If one species was evolving into another, this raised the same concerns, because the implication would always come back to humanity made not in God's image, but in an animal's. As Bonsell and Buckingham saw it, the kids were being lied to, and to correct that, they needed "the other side of the story."

When Bryan Rehm trudged in the door and told Christy that she had been right to remain skeptical and that nothing had been resolved, she all but exploded. "You're not getting through to these people," she complained. "You have people who know nothing about science trying to tell you how to teach science."

"They don't want to teach the kids," Bryan said morosely. "They want to indoctrinate them."

Funny thing was, the board members who favored creationism were saying the same thing about the teachers.

Chapter 2

WHAT LIES BENEATH

While conducting one of his popular "Time Walk" tours through the rolling landscape around Dover, Jeri Jones is apt to point out a rocky escarpment or a gentle valley in the distance, then tell his charges, "But for a lucky break, all this could be part of Morocco."

This gets everyone's attention, and Jones launches into an evocative description of the destruction of the supercontinent Pangaea, which was in the process of giving birth to the world's current stable of seven smaller continents about 180 million years ago. An enormous rift formed pretty much right where Dover and the rest of York County sits now, Jones says, where the landmass that was to become Africa was getting ready to peel off just outside the van window. If that rift had been successful, this little bit of cool, green Pennsylvania would have sailed off to become the gritty edge of the Sahara Desert. Instead, a larger rift developed to the east and the newborn continents split at what is today the eastern seaboard and the continental shelf, leaving York County in place for another age to play with.

"Don't you just *love* geology?" Jones chortles. Around the bend he'll wind the clock back a bit further, past the Jurassic and into the Triassic period, before Pangaea split, before the dinosaurs took over the world, back when the mid-Atlantic region of Pennsylvania occupied a subtropical latitude, more like the Miami of today. Jones anticipates the inevitable question: How can scientists possibly know this? Easy, he says, thanks to the study of paleomagnetism—the seemingly obvious but only recently understood characteristic of ancient, iron-bearing rocks, in which the iron molecules aligned themselves with earth's magnetic field at the time the rocks originally solidified. Such rocks acted like fossil compasses, their iron molecules "pointing" north, then freezing in place as the molten iron ore inside cooled, forever aligned with the earth's north pole as it existed eons ago, even as the rocks—and the continents on which they sat—shifted positions. By uncovering these rocks, scientists can calculate where they were located when they first cooled hundreds of millions of years ago, which in Dover's case was about 1,100 miles south of the township's current location. The climate was tropical.

"Imagine a huge, shallow lake over there, more a swamp really, like the Florida Everglades," Jones says. "It's hot, steamy—put on your sunblock. And then you'd see, hunting along the banks and the shallows, something that looks like a salamander. A seven-foot salamander. We've found its bones, right here. That was York County, 225 million years ago."

One of the greatest ironies in a conflict virtually filled with them can be found in the very landscape that Jeri Lee Jones loves and knows so well: Dover and environs are a virtual gold mine for geologists, paleontologists, and fossil hunters, home to some of the richest fossil finds of prehistoric life in the state of Pennsylvania. The very ground deep beneath Dover High School, shovelful after shovelful, contains what most scientists consider absolute proof of an Earth far older than a literal reading of the Bible will allow, and a mountainous source of evidence mainstream science sees as confirming Darwin's theory that all life shares common ancestry. Dinosaur bones, and even more primitive phytosaur and metoposaur fossils, have been found along the Little Conewego Creek, which borders Dover Township. Within a short drive, the fantastic sea life of the ancient Cambrian period is on display in a rare instance of ancient geology laid bare.

Not far from Dover Township, just across the York County line, is Swatara State Park, where a deep cut for an interstate highway revealed the rich fossil deposits of a Paleozoic era seabed from 440 million years ago, containing uncountable numbers of the ubiquitous, now extinct, tiny arthropod known as the trilobite, a humble organism yet a key figure in the evolutionary scheme of things. It was among the first arthropods—the line of creatures that include insects and lobsters and spiders, about 90 percent of all creatures that have ever lived on Earth. Trilobites are believed to have been among the first creatures with genuine complex eyes as we know them, which some scientists argue may have sparked an explosion of evolution and diversity. Long before the rise of the Chordata—the first backboned animals, which would eventually occupy the top of the food chain—tiny trilobites, many no bigger than a dime, essentially ruled the planet for millions of years. With their segmented bodies and stubby legs, the trilobites scuttled through the shallows, feeding, breeding, and spreading worldwide. The types of sediments and fossils exposed at Swatara, chock-full of the remains of trilobites, are believed to lie deep below the surface of most of the region, but the park is a special place, a rare access point to the past, where the action of Swatara Creek, a tributary of the Susquehanna River, carved a large gap in the fossil-bearing rocks and the remains of the past were brought unusually close to the surface, there for the taking. For paleontologists, geologists, and evolutionary biologists, this terrain is a primer, and it allows them to paint a portrait of a lost world. So plentiful and important is the trilobite that it was named Pennsylvania's state fossil.

A burly, jovial geologist and former university professor, Jones conducts field trips through local history—his "Time Walks" for the York County Parks Department. Crowds of school kids, hikers, birders, nature lovers, and the idly curious trudge along with him to old quarries and railroad ravines and retired iron mines—places where the Earth is laid bare and its treasures are exposed. He teases out for his listeners a vivid past of giant, toothy amphibians and saw-snouted protocrocodiles, then points and says, "Look over there. See that old boulder with the indentations in it? Those are dinosaur tracks." Most people would walk right past those markings without noticing, but once seen, the little tracks are unmistakable and evocative, a vivid message from long ago. For many, it will be the first, stunning exposure to the notion that the world has existed for billions of years, that the vast majority of creatures that have lived on Earth are long extinct, and that the landscape around them, which seems so permanent, is in fact young, having moved halfway around the globe, its living residents wiped out, restored, and wiped out again in mass extinctions and Earth-shattering disasters. They will be confronted, most for the first time, with the humbling idea that man's reign on Earth occupies only a tiny fraction of 1 percent of the time that the planet has existed, and that we are part of a great, repeating cycle.

Men like Jones, scientists and realists, churchgoing but not married to literal interpretations of the Bible, see the explanatory power of geology, paleomagnetism, and evolutionary theory as keys to unraveling how life has negotiated these cycles, to understanding the complexities of biology, biochemistry, medicine, and even the very tree of life that biologists use to chart genetic relationships. Evolution is not a religion or a philosophy to Jones, nor an outlook or a dogma. It is simply an explanation that makes sense of data—that's what a good theory does, and evolutionary theory explains a lot in Jones's book. In this case, the data in question are the entire living world around us, from why humans have a vestigial tail (it's called the coccyx) to the evolution of a trilobite's multifaceted, insectlike compound eye to the fact that the same complex of genes that regulate the segmentation of insect bodies (a trait that appears to have begun with trilobites) is also present in humans, adapted to regulate the development of the primitive hindbrain. A master designer, evolutionary biologists say, wouldn't need to build the human brain from ancient insect DNA, because the best engineering designs are fresh and clean, without all the compromises that come from adapting older systems (as any home owner who has added a second floor can attest to). Yet that sort of cobbling together is exactly what must have happened if evolutionary theory is true—creating a tangible line that binds life today with those tiny fossil sea creatures, and with those dinosaur footprints preserved for the incidental benefit of Jeri Jones's "time walkers."

Jones will pause a moment to rub his fingers lightly on those indentations in the smooth, cool rock, connecting with the long-toed little creature that stepped in the warm mud, hunting or fleeing or foraging before moving on. Night and sun and the tireless cycle of years turned mud into rock into fossil. Then wind, ice, rain, and flood buried it, washed it away, and finally threw it up again, revealed by the restless motion of the Earth, an inexorable tide hidden within rock and dirt. Jones knows this secret tide, the geology and early history of this corner of Pennsylvania, better than he knows his route home from work; he jokes that he remembers details of the Triassic period better than last week's Phillies' box scores. When he talks about local history, he's not thinking of German and Irish settlers in the eighteenth century; he's thinking about millions and hundreds of millions of years ago—MYA, to use the scientific style. Except, for certain audiences who join his geologic tours, the MYA goes MIA from his spiel. For them, he says simply, "That was York County, *a long time ago.*"

"You have to know your audience," he explains with a shrug. "Some people are Bible literalists; they don't want to hear millions of years. You want them to learn something, not just tune you out." As it happens, Jeri Jones knows quite a bit about the evolution-religion conflict that has whipped through Dover and that makes him circumspect with certain tour-goers: He's married to an evangelical minister. "I don't just understand the conflict," he laughs. "I live it. Every day."

The preacher gazes at the sea of faces turned up at him as he holds aloft a well-worn copy of the Bible, waving it at a packed church the size of a concert hall.

"I look forward to the day when every teacher is teaching out of *this* book," he shouts, and he is answered by a loud chorus of hallelujahs. "And there will be no separation of church and state . . . We will live in a theocracy. And what a glorious day that will be!"

His audience nods and murmurs approval, a scattering of amens. They have come to hear this Texan preacher, Dave Reagan, of Lamb and Lion Ministries, deliver a message both fearsome and hopeful: of the end of the world (coming very soon); of Satan's big lies (evolution and humanism top the list); and of the second coming of Jesus—"no namby-pamby savior" meekly walking to his crucifixion this time, but a sword-wielding warrior king astride a magnificent white horse descending from heaven, a literal figure of the apocalypse. This is not a metaphorical horse, Reagan assures his audience. "This is real, as real as you and me." As he describes this moment of scripture come alive, the theater-size movie screens behind him flash images of a cartoon Jesus riding a great white horse, clouds parting. Then Reagan asks his audience to imagine them all beamed up from Earth at that rapturous moment, the true believers, transformed from flawed and sinful humans into "glorified beings"—like angels, only better; perfected, eternal, all-knowing, at one with God—all of them riding their own white horses, swords at their waists, their dazzling white linen robes tinged with blood. And this time, Reagan says with satisfaction, it's not Jesus's blood being spilled. It's the blood of the unfaithful, the humanists, the damned (approximately 60 percent of the world population, according to the most up-to-date biblical-based calculations provided by one of Reagan's predecessors on-stage), who will be either destroyed or sent fleeing in terror at the sight of the Lord's cavalry of "saints." Then the world will be ruled for 1,000 years by Jesus and his army of all-knowing saints in an earthbound the-ocracy (which will include an unspecified number of the aforementioned teachers who will use the Bible as their only textbook). "I get really ex-cited about that," Reagan says with satisfaction. "This is talking about you and me."

"Hallelujah!"

There are just a few catches on this particular road to eternal saint-hood, Reagan tells his audience. You must embrace "God's word—all of it." That means the Old Testament as well as the New, every bit of it read literally. Genesis, Adam and Eve, the great flood, the Fall, all of it, word for word. And that means rejecting Satan's lies, the notion that the universe is billions of years old, that man evolved rather than being in-stantaneously created on the sixth day of Genesis. Rejecting those decep-tions of science, of "materialistic naturalism," is the only route to heaven, Reagan preaches. Being a good person won't do it. Doing good works won't help. Thinking that way amounts to embracing the lie of human-ism, which isn't listed among the seven deadly sins, though Reagan and the vast majority of those present seem to think it ought to be right at the top. Humanism in this gathering is a code word that needs no explain-ing, as it embodies virtually all the ills plaguing society today. In more neutral quarters, humanism is defined simply as a belief in human ratio-nality and a rejection of religious revelation as the path to truth, wisdom, and justice. But here, in a setting where questioning God or biblical lit-eralism or a 6,000-year-old Earth would be considered a sin (not to men-tion tasteless, unpatriotic, and downright rude), humanism is nothing more than, as Reagan calls it, "the fundamental religion of Satan." As he sees it, humanism—this notion that people, rather than God, can make their own rules for the common good—lies at the root of all that is bad in America. Such faith in human nature only serves to usurp faith in God, Reagan says. And so crime rates go up. Promiscuity, abortion, and atheism are also its consequences, because at its core is a belief that justifies immorality, selfishness, and even murder—a belief Reagan identifies as evolution, with its bedrock principle of survival of the fittest. If you believe in evolution, why not kill, or get an abortion, or reject the notion of God, salvation, and redemption? Why not just live for the moment, since we're no better than animals? That's why such thinking is so tempting and so evil, and why right-thinking Christians must reject it, Reagan avows. When the world ends, only the people who accept the word of God, who know the Bible is word-for-word true, will be saved. They'll get to ride the white horses and live forever: "There will be a whole generation that does not die," Reagan promises, "and we are very likely that terminal generation."

And the others, the outsiders, the people who believe in Satan's lies, the humanists who think the Bible is metaphorical and can coexist with Darwin and natural selection, they're going to be lost, left behind, burned away. Reagan sounds not the least bit mournful at this point. His round face, which bears a striking resemblance to the nation's forty-sixth vice president, Dick Cheney, is rapt as he says, once again, "What a glorious day that is going to be."

In other words, the devil with Charles Darwin. Literally.

Dave Reagan delivers these and other choice comments on the coming apocalypse to a warmly receptive crowd as a headliner at a gathering called "Steeling the Mind," a conference advertised "for the Believer who loves the Lord with a passion, loves their Bible as God's infallible Word, and wants to be taught by the best Bible teachers on the planet." The title refers to the core mission of the gathering: to "steel" evangelicals with the knowledge they need to "defend" the Bible against the sorts of things Jeri Jones and the teachers at Dover High School and Charles Darwin's followers have to say. There are two types of people who must be met with this "steel," according to the organizers of the conference. First, there are those who are "assaulting" the Bible because they are Satan's unwitting dupes, and the conference attendees are urged to try to help these misguided souls see the light before it's too late. But others knowingly embrace the lies of evolution and atheism, and they are Satan's willing servants, those who must be fought tirelessly and mercilessly. These are the anti-Christian conspirators, the ones who have launched "a war against Christians"—with various speakers identifying the malefactors as scientists, liberals, humanists, Al Gore, Jimmy Carter, the ACLU, the Democrats, and anyone who harbors the mistaken belief that the U.S. Constitution requires separation of church and state. The notion that Christians, more than 80 percent of the population, are an oppressed group in America not only has gained currency among the thousand faithful in attendance here; it is simply assumed to be beyond question.

Four or more similar "Steeling the Mind" conferences are held every year around the country, staged by the Idaho-based Compass International, which is in turn part of a vast network of like-minded gatherings and conferences with titles such as "The War against Christians" and "Creation MegaConference." The annual "Justice Sunday" is a weekend-long indictment of the judiciary for its rulings against prayer and creationism in school, against posting the Ten Commandments in government buildings, and, of course, for repeatedly upholding *Roe v. Wade*. Together such gatherings make up a well-attended, well-financed, and well-oiled machine that is delivering a much-sought-after and unabashedly militant view of the world far different from that offered by mainstream science—a parallel universe of knowledge and inquiry that begins with an unquestionable belief in the divine origins of both the Bible and the universe.

These conferences mix old-fashioned preaching, new-age salesmanship, multimedia PowerPoint presentations, and what is represented as cutting-edge alternative science. The presentations claim to refute notions of an ancient Earth, of common descent, and of evolution. And they purport to prove scientifically that the Earth is young; that Noah's worldwide flood is responsible for the creation of fossils and the geologic "illusion" of an ancient Earth; and that man was created as he is today, not through evolution or adaptation, but with a garden paradise and a tree both forbidden and tragically irresistible. This is the religious heart of anti-evolutionism in America, where the ideas are first broadcast to a most receptive audience, then percolate outward to school boards, to state legislatures, and to Washington. A frequent speaker at such conferences, the former House majority leader Tom

Delay, the disgraced ultraconservative who longed to impose a "biblical worldview" on the nation, once famously blamed evolution for a laundry list of social ills, including shootings at schools. "School systems teach the children that they are nothing but glorified apes who have evolutionized out of some primordial soup of mud," he railed.

This "Steeling the Mind" meeting, as far from an old-fashioned tent revival as can be imagined, is set in the rolling hills of San Juan Capistrano in coastal southern California, where million-dollar homes seem as plentiful as the swallows that flock each year to the town's namesake mission. More than 1,000 affluent, conservative, well-educated evangelicals have turned out this day, each attendee paying $59 and waiting in a line snaking a city block out the entryway. The massive Ocean Hills Community Church, which looks more like a modern convention center than a place of worship, may be 3,000 miles from Dover, yet it is here, and in the many other conference settings like it nationwide, that the terms of the conflict in that Pennsylvanian town are rooted. The roster of influential speakers, from Dr. Dino to Tim La-Haye, the best-selling author of the "Left Behind" series of Christian Rapture books, feeds a fervent and growing sentiment against teaching the theory of evolution to schoolchildren. And it explains why the Dover school board was so adamant in staying its course, despite the normally respected opinions of its teachers and the legal challenges that inevitably awaited its controversial decision. The stakes, at least as they are outlined by Dave Reagan, something of an evangelical superstar, are just too high: It's not about disagreements over science or philosophy, and it's certainly no gentleman's conflict in which the two sides can agree to disagree. This is a battle for the souls of children, nothing less. This is about being afraid that you're going to heaven, but your kids, infected by teachers preaching the false "religion" of evolution, are going to burn in hell, separated from you for eternity, suffering forever. What could be worse than that? Risking a lawsuit from the ACLU is nothing compared with the bone-chilling fear that your children's very souls might be in jeopardy, just from sending them to school. As Bill Buckingham put it, "You allow your children to believe in evolution, they are doomed."

"That's why I'm here," one young woman attending the conference with two small children in tow tells another woman sitting in the row behind her. The kids are utterly bored and unable to sit still, despite a large collection of crayons and coloring books. "We've got to do something about the lies they're teaching in public schools. If I could home-school my girls, I would, but I have to work." The other woman nods sympathetically and jots down the name of a private Bible-based school in the area that has a generous financial aid program "for working moms of faith."

"You can't send your children to the public schools," she says as she hands over the note. "They have made God illegal. They don't want our boys and girls to have faith."

The mother with the two bored children looks about to cry. She nods and mouths the words, "Thank you," then turns her attention back to Reagan just in time to hear him explain with gusto that there will be no trials or lawyers or Miranda warnings for sinners, blasphemers, and hu-manists after the second coming: "Jesus will rule with an iron rod. . . . Justice will be instant."

The two children look up for a moment from their coloring books, one of which features cartoons of a (recent) prehistoric landscape shared by man and dinosaur, and they watch silently as their mother murmurs, "Amen." Then they go back to work with their crayons.

Dover may have moved to the front lines of the evolution-creation battle, but it is far from alone. At the same time the Dover controversy erupted, in town after town and state after state, from California to Kansas to Ohio to New York, a new and concerted attack was under way on the teaching of evolution and other bedrock principles of modern science. An uneasy balance had more or less held since 1987, when the Supreme Court last ruled on the subject by banning "creation science" from public schools, but now all bets were off. From the age of the Earth to the big bang origin theory of the cosmos, from global warming to stem-cell research, Americans are rethinking, politicking, suing, and fighting as never before about what we teach our kids and spend our tax dollars on.

As events began to spin out of control in Dover, policy makers in no fewer than twenty-one other states were considering proposals that questioned the scientific validity of evolution or introduced alternatives such as intelligent design, and lawsuits over the teaching of evolution and creationism were pending in some twenty others. In California, students at parochial schools are suing the state university system for not giving academic credit to creationist biology classes that, if they mention evolution at all, portray it as a failed religion. The Lancaster, California, school district has mandated the teaching of "gaps" in Darwin's theory. President George W. Bush's Justice Department, meanwhile, has even gotten the U.S. government deeply involved in this battle through a number of evolution-related actions undertaken by its new "defense of religious rights" legal team, which seemed to equate the teaching of evolution with a civil rights violation. In one case in Texas, this legal team investigated a university biology professor for requiring students to affirm a belief in a scientific theory of human origins in order to get a recommendation letter for medical school.

Meanwhile, furious officials in Cobb County, Georgia, were appealing a federal court's decision to order the removal of stickers placed on science books saying, "Evolution is a theory, not a fact." Legislators in Alabama, Missouri, Wisconsin, and a dozen other states were considering laws to allow teachers to "debunk" evolution in class. Still other states continued to dodge controversy by making the term "evolution" disappear—Virginia's education standards, for instance, use only the phrase "changes over time" when addressing the origins of species. And even in states where evolution is a firm requirement for school lesson plans, many teachers avoid it like the plague, fearful of angry parents and First Amendment lawsuits—or because they personally disbelieve evolutionary theory. About 25 percent of science teachers nationwide, with higher percentages concentrated in the South and Midwest, reject evolution as a valid explanation of human origins. And even greater numbers report feeling pressure from parents, students, or administrators to avoid Darwin's theory as much as possible. In one high school in California, four of five biology teachers simply skip the chapters in their textbook dealing with evolution, despite state mandates to teach it.

The subject of human origins had become so sensitive that the IMAX theater chain decided during the Dover controversy to cancel showings in Texas and throughout the South of a documentary science film about undersea volcanoes, for fear that its brief references to evolution and well-documented similarities in DNA between ocean creatures and man would be considered blasphemous by a majority of moviegoers. Meanwhile, in the San Diego area, classes are booked solid at the Institute for Creation Research—"Creationist U," as it's widely known—where undergraduate and graduate degrees in Bible-based science are conferred and have never been more popular. The school provides a base for a cadre of experts who fan out across the country to lecture and testify on the fallacies of fossils and the evidence in support of man's having been created in his present form a few thousand years ago.

Opposition to the teaching of evolution seemed to be invigorated in 2004 by the reelection of a conservative Congress and a president who has publicly endorsed the teaching of creationism and intelligent design in public schools—"Both sides ought to be taught," he has said on more than one occasion. The movement has been propelled by the growing political muscle of the religious right, backed by millions of dollars in contributions from conservative philanthropists and eager evangelicals, fueled by Americans' notoriously poor grasp of both science and biblical scholarship, and encouraged by the solid lead of anti-evolution in public opinion.

The nearly unanimous opinion of the scientific community is that evolution is the bulwark of modern biology and medical research, from the development of new antibiotics to the fight against cancer, and that America's next generation of scientists will fall hopelessly behind the rest of the world if evolutionary theory is watered down or banished from our schools. But the general public does not seem to see it that way. Two-thirds of Americans, Gallup polls and other surveys have found, would prefer our public schools to present some form of Genesis in addition to evolution. A third want creationism *in place* of evolutionary theory. Yet polls also find that solid majorities of Americans are essentially ignorant of evolutionary theory and the scientific evidence that supports it, and are nearly as clueless when it comes to the details of the literal biblical stories of creation (with many unaware that there is more than one creation story in Genesis alone).

The "Steeling the Mind" conference is a prime example of this disconnect between science and religion—there are no negotiations, no compromises, no attempts to stand in the other side's shoes. It is war without end, without sympathy, without pity—a war fought against stick figures and cartoons, with each side ready to demonize its opponents without hesitation. It would not occur to the organizers of "Steeling the Mind" to invite an evolutionary biologist to the conference so that attendees could hear a different perspective, just as it would not occur to the organizers of the annual conference of the National Science Teachers Association, which met a few weeks later in the same vicinity, to invite a panel of creationists to join one of its sixteen separate sessions on teaching evolution and dealing with related controversies in public schools.

The scientific community sees the creationist critics of evolution as yahoos, religious zealots, and scientifically suspect charlatans. The creationists see the evolutionists as immoral and dishonest purveyors of a pseudoreligion called Darwinism that makes God superfluous. They vilify and abhor one another in speeches, at conferences, on websites, and in blogs. There are occasional civil debates and attempts to cross the lines with bipartisan conferences, but they are rare; for the most part, any reaching out between these warring factions is intended to gather information to refute. Scientists try to ignore creationists, though occasionally an anti-evolutionary movement will catch fire with the press and the public—intelligent design being the leading example—and the scientific community will rally to refute its claims. The creationists are less likely to gather information from the opposition, mostly owing to distaste and disinterest, but there are other reasons at work, reasons that animate the public school debate: the fear that evolutionists can "brainwash" young people with evolutionary "dogma." Creationist leaders have read the same surveys as evolution proponents: poll after poll that shows belief in evolution rises with education level, while areas in the United States with the strongest commitment to creationism tend, as a whole, to have the lowest levels of education. Decades of Gallup polls unwaveringly reveal a remarkable correlation between education and what Americans believe about origins: 65 percent of Americans who attended graduate school believe evolutionary theory is scientific and well supported by the evidence; 52 percent of those with bachelor's degrees accept evolution; but only 20 percent of Americans with high school educations or less believe evolutionary theory is well supported by the facts.[1] One speaker at "Steeling the Mind" put it this way: "Kids go off to college and give up on God. Start worrying less about where your kids are going to go to college, and send them to a Christian school now." A young man in the audience turned that concern on its head, admitting that he preferred to bank on ignorance: "I'm really afraid to learn too much about evolution, because it might make me doubt my religion. And then where would I be? What would I tell my family? My girlfriend?"

This was the polarized environment in which the Dover school board took up the question of teaching about origins; its members picked a side and never turned back.

A majority of Americans, regardless of their religious convictions or churchgoing habits, are in the same position as this young man. They don't understand what evolutionary theory is about, while at the same time they express disbelief in its principles—that is, they have opinions about evolution, but very little knowledge. Most would be surprised to learn that evolutionary theory does *not* mean that life and humanity arose through random chance alone, which is how creationists often characterize Darwin's thinking. In fact, evolution is the opposite of a random process. The power of natural selection to weed out unsuccessful traits and species—a process commonly mischaracterized as "survival of the fittest"—may be unconscious and unplanned, but it is among the most directed and nonrandom forces in nature. Even more Americans would be surprised to learn that the word "evolution" appears nowhere in Charles Darwin's seminal work, *The Origin of Species*.

Instead, before the "Steeling the Mind" conference ends, those in attendance will learn that Roman Catholics are destined to go to hell (because the Catholic church is said to place its doctrine above the literal word of the Bible); that widespread belief in UFOs and alien visitors is the work of the devil (it contradicts the Bible's assertion that God created life and sent his only son to Earth); that God loves Jews more than any other people and that they will rule at God's side once Jesus returns (one small bit of fine print: Jews have to convert in order to get the good seats); and the creation of the modern state of Israel signals the approach of the end-time—Armageddon. Speakers at "Steeling the Mind" were very much into predicting the imminent end of the world on the basis of biblical prophesy. It affected their outlook, making them supportive of aggressive U.S. military intervention in the Middle East (more signs of the approaching apocalypse); dismissive of classic Christian notions of charity and doing good works as useless gestures; and contemptuous of environmentalism, fears of global warming, and efforts to seek world peace. "These are the closing days of the history of the world," said Gary Frazier, a Texas-based minister and collaborator with Tim LaHaye in "Left Behind" seminars around the country. These seminars capitalize on the juggernaut best sellers about Christians swept up in the "Rapture," while the rest of humanity gets what's coming to it, left behind on a war-torn, anti-Christ-ridden earth. "War will come at the end," Frazier promises with a small smile. "That kind of blows away the world peace crowd. The truth is, we are never going to have world peace. God says there will be wars and desolation until the end."

Most of the eight speakers at "Steeling the Mind" made claims about the wrongheadedness of evolutionary science, although there was only one working scientist among them. Larry Vardiman, who holds a PhD in atmospheric science, explained how his work at the Institute for Creation Research near San Diego has uncovered evidence casting doubt on the accuracy of widely accepted radiometric dating techniques—the process of measuring radioactive decay in various rocks and fossils in order to assess their age (and by extension the age of the Earth). His highly technical presentation on the nuclear decay of uranium 238 inside zircons sailed over the heads of his lay audience, except for the part about how he started by assuming the Bible was literally true and proceeded from there in evaluating the scientific data. The scientific community has rejected his and his institute's findings for years as flawed in concept and execution; but at this conference, such a distinction serves to enhance, not damage, Vardiman's prestige and credibility. He delivered with aplomb the expected applause line at the end of his eye-glazing talk: "It's a young Earth after all."

Intelligent design, the latest trend in anti-evolutionism, and the movement that features the most scientific expertise, is given short shrift at this conference. This is because ID is in public presentations insufficiently biblical, and because its scientific star, a Pennsylvania-based biochemist named Michael Behe, is neither a biblical literalist nor a "young-Earth creationist"—he simply believes that life was designed somehow, rather than evolved, and he claims to have identified the molecular goods to prove it. Only Mike Riddle, the former Marine captain who spoke about "UFOs and the Bible," talks up the "good science" behind intelligent design as he stands at a table full of videos and books for sale from the creation ministry he works with, Answers in Genesis. "It's not biblical, but it's a good step in the right direction," as Riddle sees it. But he also fears it could lead the faithful astray, because the designer is not clearly identified. "Someone might think it's space aliens," Riddle says darkly.

Rather than intelligent design, old-line creationism is the order of the day at "Steeling the Mind," which is why the superstar on the agenda, the last speaker of the weekend, is Kent Hovind, the creationist with the biggest drawing power, the 200 annual speaking engagements, and a unique position as sideshow attraction for the legal showdown in Dover, Pennsylvania. He even has his own creationist theme park in Pensacola, Florida—Dinosaur Adventureland Park, "Where Dinosaurs and the Bible Meet!" The conference attendees were clearly waiting to hear Dr. Dino all day as they patiently joined guitar-accompanied Christian songs, then marched to the stage to marvel at a 500-year-old Torah on display. They listened to seemingly endless pitches to book $3,000 "Bibleland Cruises" to Turkey, Rome, Greece, and Israel with Hovind, Reagan, and other celebrity creationists on board; to buy discounted Ahava hand cream ("the same hand cream Cleopatra used," regularly $23, but only $10 a tube for "Steeling the Mind" attendees); and to purchase some of the thousands of creationist and apocalyptic books, DVDS, and magazines for sale outside the auditorium. Hovind's collection of DVD lectures and debates was easily the most extensive of all, with titles such as "Lies in the Textbooks," "The Dangers of Evolution," and "Dinosaurs in the Bible," available in English, Spanish, French, Russian, and Japanese. These were extremely popular at only $17.95 each, as were the package deals such as the set of twenty debates "with ardent evolutionists" on twenty videocassettes or DVDs for $169.

After the master of ceremonies makes one final pitch for biblical hand cream, Hovind at last bounds onto the stage. A former high school teacher with an aw-shucks midwestern accent and a boyish grin, Dr.

Dino launches into his stock speech, which starts off with plenty of red meat designed to please an audience of the faithful, or outrage an unfriendly crowd: "Evolutionary theory is the dumbest, most dangerous theory in the history of man. . . . There is no evidence to support that we came from a rock 4.6 billion years ago. . . . That's just stupid."

.And, of course, phrased that way, it does sound stupid, mainly because Hovind cheerfully confounds the origin of life, the big bang theory of the universe, and Darwinian evolution, and then distorts scientific theories about all three in order to make them sound as ridiculous as possible. Evolution predicts that humans descended from earlier, more primitive organisms, not "from a rock"—that would be the Bible, which asserts first that man and woman were created simultaneously as adults, and then later states that the first man was made from dust and the first woman, some time later, from his rib. But this is not a time for subtleties or for comparing inconsistencies in evolution and Genesis; his audience is eating up his stand-up-comedy style of presentation as he adds, "I've debated one hundred professors. They're a lot smarter than me, but I slaughter them because I'm right and they're wrong." This draws warm applause and laughter; if there is one unifying subtext to this conference, it is that those speaking and those in attendance feel that they have a special lock on the truth, a certainty that transcends expertise, education, or credentials.

For this crowd, Hovind omits most of his talking points on scientific objections to an old Earth and to evolution, instead referring the audience to his DVDs for sale if they want the lowdown on why evolution is, in his opinion, scientifically stupid. On those tapes, his arguments include the assertions that the second law of thermodynamics makes evolution impossible (systems decrease in complexity over time, he says, yet life has grown more complex, according to evolutionary theory); that there is insufficient cosmic dust on the moon for it to have existed for billions of years (the Apollo astronauts would have sunk out of sight in a sea of dust if the universe were really that old); and that there is no evidence of "transitional" life-forms in the fossil record that bridge the different groups of animals—fish, reptile, bird, mammal. He has many other arguments, too, but the vast majority not only have been refuted by mainstream scientists, they have been attacked by other leading creationists as poor arguments that discredit evolution critics. The cosmic dust accumulated on the moon, for example, is exactly as current measurements from space probes indicate it should be after billions of years; thousands of transitional fossils and life-forms have been discovered deep in the strata of ancient Earth (according to paleontologists and evolutionary biologists, though not creationists); and the second law of thermodynamics has to do with the behavior of nonliving matter in a closed system—whereas life, far from a closed system, continually gains energy from the sun and Earth, providing ample opportunities for growth and new, more complex life-forms. Physicists and engineers who actually work with the principles of thermodynamics consider Hovind's argument laughable and beneath even a first-year physics major, though to a receptive and scientifically unsophisticated audience of the faithful, such arguments are more than sufficiently convincing. They want to believe, and not one member of this well-dressed, well-spoken assemblage questions Hovind's pronouncements or asks for supporting evidence. The audience members already know evolution is stupid—they just like to hear Hovind say it. When he asks how many of the thousand attendees had previously seen one of his presentations or videos, half the audience members raise their hands.

Hovind chooses to focus this day's presentation on his moral and social objections to evolution—not so much why evolution is stupid as why it is dangerous. This hits much closer to home for the attendees at "Steeling the Mind." "Since the inclusion of evolution in textbooks in the 1950s," Hovind says, "there has been a moral collapse in our country. It's not the way it used to be."

According to Hovind, teaching evolution has led to increases in crime, premarital sex, adultery, and drug use. As he speaks, he shows statistics on his PowerPoint slides, reflecting jumps in crime in the 1990s that he attributes to the teaching of evolution. (His statistics end before the precipitous drops in violent and other crimes that have occurred in the past eight years, also while evolution was being taught.)

He also asserts that it is impossible to believe in both God and evolution, then reasons that this means it's impossible for an evolutionist to behave morally. "How do you tell right from wrong if you're an evolutionist? Where is the standard? The only standard then is . . . *I* decide what's right and wrong. . . . That's why evolution is so dangerous." From there, it's a short leap to link evolution to atheism, the shootings at the school in Columbine, Joseph Stalin's mass executions, the Japanese attack on Pearl Harbor, and the horrors of Nazi Germany. Slides of endless depredations flash across giant screens as Hovind speaks. "A straight line runs from Darwin to the extermination camps," Hovind says. "I don't think you can understand what happened to the Jews until you understand evolution. . . . If evolution is right, then Adolf Hitler is right."

Next he blames racism and ethnic cleansing on evolution, as if neither existed before Darwin, though history, stretching back to biblical times, is replete with endless examples of race war, enslavement, genocide, and ethnic hatred, much of it motivated by religious conflict. But Hovind places the blame squarely on Darwin: "What is the foundation of racism? They believe they have evolved farther. . . . Evolution isn't just wrong. It's dangerous. . . . The philosophy of evolution says there is no God—man is god. That's the cause of so many problems."

Up until this point, Hovind's presentation has been by rote, as the usually vociferous Dr. Dino seemed off his game, perhaps tired from his rigorous speaking schedule or worn down by his ongoing legal problems with the IRS (which raided his home, office, and Dinosaur Adventure Land theme park) and with Escambia County authorities in Florida (who closed the theme park's museum for being built without a permit). But he is always an accomplished showman, and as the amens and applause sound at regular intervals, he seems to draw energy from the crowd. "If there is no God, we're in trouble," he jokes. "We're hurtling through space at 66,000 miles an hour and nobody's in charge. . . . Seriously, though. . . . We are in the center of the battlefield of the greatest war in history, the battle between God and Satan."

He pauses then, and surveys the crowed. Then he says that Christians are fortunate to have a book that tells how it all turns out. "I read the last chapter. We win."

He winds up his talk amid wild applause with an appeal for good Christians to save souls, to vote Republican, and to join a local militia. "Evolution is the foundation philosophy for the New World Order," he says. "If America ever needed to be saved, it's now."

Hovind quickly exits, already thinking about his next speech. Soon he will take his message cross-country to another receptive audience—this time, in Dover.

Chapter 3

I'D RATHER TAKE A BEATING
THAN BACK DOWN

Bill Buckingham grew up in the tiny York County farm village of Strinestown, not far from Dover, his family of modest means and not particularly religious. His father set a strong and sometimes difficult example, a Pennsylvania Railroad track foreman and former army drill sergeant who would call Bill home from playing with his friends, his booming voice carrying upwards of a half mile. *Bill! BILL BUCK-INGHAM!* At age fifty-nine, Bill could still hear that drill sergeant's voice floating on the wind. He would race home and his father would be standing in their trim backyard, with a little piece of trash, maybe a cigar wrapper, at his feet. "Pick it up," the senior Buckingham would order, and Bill would do it. He loved his dad, but that was one tough bird, an honest, hardworking, taciturn man who got drunk every Christmas day, which always confused Bill, as his father was no drinker. Many years later, he learned that his dad had received serious injuries in a Christmas-day battle in the Philippines during the bloody island-by-island combat in World War II. The holiday onslaught in 1943 had left most of the men in his father's platoon dead. "I drink for them," he told his son. He was set in his ways and was not easily swayed from them—and it seems his son inherited those traits.

Bill's parents didn't bring him to church as a child, but some neighbor ladies took him to Catholic Mass on Sundays beginning when he was ten, and so Buckingham became a Catholic. Later his mother started to come, too. Buckingham stayed with the Catholic church for more than three decades, through his tour of duty in the Marine Corps in the mid-1960s (he still professes pangs of guilt at being assigned to guard a submarine base in Connecticut rather than going to Vietnam) and his return to York County, where he worked as a policeman. If he ever gave a thought to evolutionary theory, it was only in passing. The public schools he attended in his community still featured prayer, Bible reading, and Christmas pageants; he doesn't remember learning much, if anything, about Darwin or his theory. Nor was it an issue in his early religious life: Catholics generally have no problem with evolution, and church doctrine warns against taking the Bible too literally. Catholic theologians emphasize the metaphorical nature of the Old and New Testaments, so scripture and Darwin are not on a collision course as far as the Vatican is concerned. Buckingham didn't really think about it at the time.

As a young policeman in the 1970s, Buckingham worked at the four-officer police force of the borough of Red Lion, centered on a rural crossroads eighteen miles from Dover. He often worked alone, developing something of a swashbuckling approach—some might consider it rash—when forced to cross paths with the drug dealers and outlaw bikers who frequented his town at that time. When you were outnumbered and outgunned, Buckingham recalls with obvious relish, the trick was never to show fear or doubt—never hesitate, never back down. "I took on a whole motorcycle gang by myself once. They took over a bar and they wouldn't leave. I was new. I wasn't afraid of nothing. I went in and backed them down and they never came back."

Thirty years later, hobbled by innumerable health problems, three ugly knee replacements, and a very bad back, he can still sound as defiant, brave, and borderline reckless as he did as a young cop. "It must have been the look on my face when I walked in that bar. I said, 'You guys can take me down. But I have six bullets in my gun. Who wants to be the six?'"

No one, apparently. As Buckingham tells it, the rogue bikers shuffled out, revved their hogs, and sped off—anything to get away from that crazy Red Lion cop. "I had more luck than brains," he says. And then he adds a bit of insight into his character, an observation that explains not only that day in the bar, but his subsequent adventures as the curriculum chair of the Dover Area School District: "I'd rather take a beating than back down."

No one who tangled with Bill Buckingham during his tumultuous tenure on the school board would ever accuse of him of backing down.

He recalls his time as a police officer as marked by threats against him and his family, a failed ambush attempt, and the shotgunning of his house—all stemming from a narcotics ring he had attempted to bust. After ten years, with the welfare of a wife and three children to consider, Buckingham decided he had done enough time on the streets. He took what was supposed to be a safe job working as a supervisor at the York County prison, with regular hours, good pay, and predictable duties. But a routine tussle in the visiting room with an inmate hoping to gin up evidence for a brutality lawsuit got out of hand, and Buckingham ended up with a severe back injury. Then came multiple painful surgeries and a forced medical disability retirement in 1990.

It was around this time that Buckingham underwent a religious conversion. He had been primed for it during the waning years of his work as a policeman when he witnessed the deaths of two children. A drunk driver had plowed into a car carrying a mother and her little boy. The mother, her neck broken, died instantly, but the boy was thrown from the car, badly injured but still alive. Buckingham found him by the side of the road. The burly cop cradled the broken child with the mop of yellow hair and the pale blue eyes and told him to hang on, help was on the way. And that little boy just looked at him for a second, then his eyes rolled back and he was gone. There are few more terrible things on Earth, Buckingham says, than feeling the body of a child go limp in your arms. Not long after that, a 911 call summoned him to the scene of an eleven-year-old girl choking on a marshmallow. Her mother had tried to reach down the child's throat to pull out the blockage, but had instead pushed the marshmallow farther in. The marshmallow formed an airtight seal, and neither Buckingham nor the paramedics who followed could do anything in time as the eleven-year-old died in front of them. "Those were the only two times I cried on the job," he says. "I learned what it is to be human, and that I had to depend on God."

While laid up after his back surgery and in great pain, Buckingham asked to see his priest. Instead, a nun was dispatched to see the bedridden parishioner. This did not sit well with Buckingham, who thought a small-town pastor ought to be able to find the time to see a parishioner in such dire straits, especially one who had done considerable charitable work for the church. "The nun was a nice lady, but . . . when you're in that kind of shape, you want to see the guy you look at on Sunday mornings."

Solace came later from an unexpected quarter, however: The pastor of nearby Harmony Grove Community Church, a fundamentalist congregation of about 200, asked how he was doing. Touched, Buckingham accepted the pastor's invitation to attend services at Harmony Grove, and sitting there in a pew a few days later, he felt a warmth and sense of belonging that he had never experienced in his Catholic church. And so he and his wife, Charlotte, made the switch, and he accepted the fundamentalist belief that the Bible is the "literal inspired word of God."

Now he says, "To me, the Earth is six thousand years old. I believe in the book of Genesis as being literal. And two thousand years ago, more or less, someone died on the cross for us. It's either all truth or it's all lies. There's no in-between. . . . And I *know* it's truth."

Buckingham never fully recovered from his injuries or the seemingly endless series of surgeries, and they left him essentially unemployable, and at risk for reinjury with little provocation. Insurance companies did not want him in the workplaces they covered, he says. He went back to school and earned a paralegal degree, in the hope that this and his law enforcement experience would make him a valuable asset to the legal community, but work proved all but impossible to find. So he began to seek opportunities to volunteer—coaching for Little League, fundraising for the church, anything to keep from sitting in the house, vegetating, and popping painkillers. He was getting by on ever-increasing doses of the painkiller OxyContin, a powerful opioid time-release prescription drug known for both its addictive qualities and its ability to generate a heroin-like high. Oftentimes it left Buckingham unable to drive, and his wife would have to take the wheel. He began to suspect that he might be addicted, and he tried to wean himself from a drug—sometimes referred to as "hillbilly heroin," because its abuse has been particularly high in rural America. But he did not succeed; the withdrawal symptoms would become too much for him. Nevertheless, when a member of the Dover school board left before his term in office had expired, Buckingham applied for the spot, looking for something to throw himself into. He walked up to Jeff Brown at the close of a particularly contentious meeting, introduced himself, and said he supported what the Browns stood for. Jeff and Casey Brown had campaigned on a platform of fiscal responsibility and had proposed reining in what they deemed to be an excessively expensive high school renovation project. "I'd like to be a part of that," Buckingham said.

Jeff Brown didn't know much about Buckingham, but he instantly warmed to the earnest, down-to-earth ex-cop and promised to support him for the open position. "Bill was different then," Brown would recall wistfully a few years later. "He was thoughtful, levelheaded. There was nothing to suggest the Bill we'd all get to know later on a bit too well."

Jeff Brown was always something of a loose cannon—you could never be quite sure what he was going to say on any occasion—but he was smart, hilarious, popular, a member of a respected longtime local family. He was also a fine electrician, who didn't charge an arm and a leg like some of the younger guys in the business, and this counted for a great deal among the thrifty old-line residents of Dover. At that point, before the controversy over evolution began, if Jeff Brown spoke for a prospective board member, that was usually good enough for most. The rest of the members welcomed Buckingham with open arms; and when Alan Bonsell became president of the board in 2004, he named Buckingham chair of the curriculum committee.

It so happened that at this time, the school needed new biology textbooks. The existing texts were more than six years old and had quite a few outdated sections, and there were too few books to go around—the ninth-graders had to share them as classroom sets. The science faculty had reviewed several of the most widely used textbooks and had enthusiastically requested the Prentice Hall volume *Biology*, by Kenneth Miller and Joe Levine, better known among teachers as the "Dragonfly Book," for its vivid, luminous cover photo of a dragonfly in flight. The school already used an older edition of the same text, so the transition would be simple. The book is considered so engaging for students and so helpful to high school teachers that Miller, an affable and articulate professor at Brown University (and a vociferous champion of evolutionary theory), is treated like something of a rock star by biology teachers when he appears at conferences and panel discussions; many will approach him and ask him to sign their teacher's editions of the Dragonfly Book.

The practice in Dover, as in many small school districts, is that all textbook purchases must be voted on by the full school board, and the books themselves, after recommendations from the faculty, are to be reviewed by the curriculum committee before the board votes. The new chairman of the committee appointed himself to review *Biology*. Once he was through, he most decidedly did not want the authors' autographs. What he wanted was a different book.

He was forced to hold off, however, owing to his increasing addiction to OxyContin. It was ruining his life, causing wild mood swings between euphoria and depression. A few months earlier, national headlines had exposed the radio commentator Rush Limbaugh's addiction to painkillers, and Buckingham decided it was time to take action. He made a public announcement at the school board meeting on February 2, 2004; asked for support and forgiveness; then took a leave of absence and headed off to treatment and detoxification. "Pray for me," he asked. One of the local newspapers, the *York Daily Record*, lauded his courage in an editorial and wished him well. He returned a month later, outwardly cured. Only much later would he reveal that he was still addicted (though no longer using the drug) and suffering from symptoms of withdrawal that left him moody, short-tempered, and, he said, afflicted with periodic memory loss. His memory problems would loom large in the months and years to come. One thing he did not forget, however, was his distaste for the Dragonfly Book.

When the curriculum committee met with the science teachers after his return from detox, Buckingham presented them with a list of complaints about the book. All of the complaints revolved around the section on evolution toward the back of the book—page numbers and paragraph references that discussed Darwin, common descent, and the notion that species evolved and changed over time.

"This book is laced with Darwinism," he complained, his voice contemptuous when he uttered Darwin's name. He then delivered a litany of complaints, similar to those previously expressed by Alan Bonsell, that the Dragonfly Book's evolution chapter presented natural selection, common descent, and "monkey to man" as fact, not theory. And there was no way he was going to approve a book that was so one-sided, that failed to present scientific problems with evolution, and that remained silent on alternative explanations for creation.

When Bertha Spahr heard Buckingham say "monkey to man," the head of the science department snapped to attention and gave him a sharp look, the sort that usually makes ninth-graders shrivel into their desks. Buckingham just returned a cop's practiced flat-eyed stare. It was an odd complaint, and it made Spahr suspicious, because she knew that *Biology* minimized references to "monkeys to man," or, more accurately, common descent of modern humans and modern apes from a shared ancestor. This omission, in fact, was one of the science teachers' considerations in choosing the text. The authors had calculated that the book would sell better if it went easy on human evolution, and Spahr and the other teachers proved them right when they concluded that this book was the best to purchase because it would be "the least offensive" to members of the community with religious concerns about evolutionary theory.

"Mr. Buckingham, we never did, don't now, and never will teach that man came from monkeys," Spahr said tartly. "And if you say 'monkey to man' again, I'm going to scream."

Buckingham then pulled out two photos of the huge mural depicting stages of evolution from primate to human that a student had painted as his senior project and then donated to the school—the mural that a district maintenance official had carted off and burned as offensive and blasphemous. In the previous school year, Spahr had complained bitterly about the fate of the mural, which had sat for a year propped on the ledge of a classroom blackboard because the maintenance staff refused to hang it up. Spahr had asked the district superintendent if the arsonist would be disciplined for theft and destruction of school property, but she was told to mind her own business; the culprit, to her knowledge, was never disciplined. Now here was Buckingham—gleefully, in her opinion—showing off the photos like trophies. "How can you say we don't teach monkey to man?" he said triumphantly. "It's right there."

"Where did you get those?" she demanded, but Buckingham wouldn't say. It would turn out that the maintenance official who burned the mural had taken pictures of it and given them to Buckingham as evidence that Dover High School was teaching its students that man evolved from apes. (The maintenance official attended the same church as Buckingham, but the curriculum committee chair said he had nothing to do with the burning, that he felt it had been wrong, and that the proper action would have been to return the mural to its creator.)

Spahr and the other teachers told Buckingham that the mural was not part of any course or lesson, but a gift, and they reiterated that they did not teach about the evolution of humans, focusing instead on other species. This was a voluntary omission, intended to avoid controversy, rather than any sort of requirement by the state or an indication that they had doubts about the science involved. Indeed, the science teachers, like the vast majority of working biologists, saw the fossil record of human evolution, from *Australopithecus* to *Homo erectus* to Cro-Magnon man, as some of the most compelling evidence for common ancestry from any species. They did not, however, mention this to Buckingham, whose face was already sufficiently red.

In any case, Buckingham's concerns were not assuaged. At the first of two school board meetings in June 2004, he made his objections more strongly than ever when a parent and former school board member stood up during the public comment period and asked why, after nearly a year of delay, there still were no new biology books for the ninth-graders. Buckingham bristled—he and this former board member, Aralene "Barrie" Callahan, had crossed swords in the past.

He repeated his assertion that the Dragonfly Book was flawed, one-sided, and "laced with Darwinism," and then said, "It's inexcusable to teach from a book that says man descended from apes and monkeys. We want a book that gives balance to education."

What was said next in a series of intense, sometimes angry exchanges, and in a second board meeting a week later, would become vital in determining the fate of this school district and its policies about evolution, for Buckingham's comments would be pivotal in divining the true intent of the school board. Were those policies secular and educational, or were they religious—and, therefore, unconstitutional? This point has been hotly disputed by Buckingham and Bonsell—they have repeatedly denied, in interviews and in sworn testimony, harboring any religious intent or advocating religion in public schools. But what most people heard and understood at that meeting in June—including two reporters for competing newspapers with no ax to grind—was that several board members, the most outspoken of whom was Buckingham, wanted creationism in the classrooms of Dover.[1]

This was the first heads-up to the public that a new front in the culture wars had been opened in sleepy little Dover. There had been the complaints to the teachers about the textbook, the interest in teaching creationism expressed at the behind-the-scenes retreats, and the private comments Bonsell and Buckingham made to fellow board members—such as Jeff Brown's recollection of Bonsell calling evolution "fiction," while Buckingham referred to it as "atheist propaganda." But none of that had made it into any open sessions of the board or into the pages of the local newspapers. So the board meeting of June 7, 2004, created something of a local sensation.

During the public comment period, a graduate of Dover High School stood up to voice his concern that the board's notions about creationism would end up trampling over the constitutional separation of church and state. Max Pell had just finished his freshman year at nearby Penn State University. He was a biology major, and he tried to explain to the board members what he had learned so far about the power of evolutionary theory to explain the natural world. Then he said, "Creationism is a religious theory. Why does it have to be taught in biology class?"

"Have you ever heard of brainwashing?" Buckingham shot back, suggesting that Pell had been overly influenced by a university that thought of Darwinism as fact instead of theory—proof that Dover High needed "balanced" textbooks and lesson plans. He added, "Separation of church and state is a myth. There is no separation."

Bonsell responded as well, if a bit more gently than his acid-tongued colleague, suggesting that Pell's concern was unfounded. There were only two possible explanations for life on Earth, Bonsell said: evolution and creationism. It was nature, or a creator. As long as both were presented as theories rather than fact, there would be no constitutional problems. The district simply would be providing a complete, balanced picture for students, not presenting evolution as fact, or as the only explanation out there, but putting it into the proper context. This was about making science education better, the board president argued.[2]

Bonsell and Buckingham made it clear that they were not prepared to approve the Dragonfly Book, and that alternatives needed to be explored. After the meeting ended, observers heard Buckingham offer one other telling observation that soon appeared in the news reports, delighting his supporters and galvanizing his opponents (even though he denied saying it): "This country wasn't founded on Muslim beliefs or evolution. This country was founded on Christianity, and our students should be taught as such."[3]

Speaking off-the-cuff and from the heart, Buckingham had expressed an argument that many Americans instinctively accept, and many others adamantly oppose as distinctly un-American. This is the belief that the United States was founded as a nation of Christians, not a nation of scientists, which is why, the argument goes, the Declaration of Independence, as well as the Founding Fathers themselves, continually referred to a creator. The counterargument is no less powerful—that the nation's foundational document, the Constitution, contains no such reference, a pointed and deliberate omission by the Founders. The idea of a "Christian nation" has been embraced, debated, discarded, and resurrected a dozen times over in the course of American history. It lies at the heart of a series of landmark court cases that provide the context for *Kitzmiller v. Dover*, addressing how public schools must deal with questions of evolution and creation, science and faith. These are well-known cases with an eighty-year history and storied names: Epperson, Waters, McLean, Edwards, and, most of all, Scopes. The intellectual leaders of the Dover school board might not have known it, but through their words and deeds, they were barreling toward reliving—and, if they could, reversing—every one of those cases.

Bill Buckingham undoubtedly would have liked John Washington Butler. Both men were bullheaded and indefatigable, self-appointed instigators of the evolution-creation debate, separated by eighty years yet sharing similar ideals and backgrounds. Butler was also a tough kid who grew up in a small rural town, and like Buckingham he worried that the children of America were being fed dangerous lies in school. He, too, had a notion that a young man or woman could go off to a university and come home an atheist, simply because some teacher or text presented evolution as a fact, and it horrified him.

Butler, however, wasn't interested in the sort of balanced treatment Buckingham had advocated for Dover. Butler wanted nothing less than to crush evolution and wipe it from the blackboards, teaching nothing but Genesis, as had been done since the early days of the Republic, before that cursed Darwin had sailed off to the Galápagos to find his inspiration. Butler lived in a different cultural clime and place, back in the days Buckingham still speaks of fondly—a time when public school children prayed and read the Bible and sang hymns in every class.

Butler decided to make a run for the statehouse in his small district of tobacco and dairy farms in northeastern Tennessee, near the Kentucky border. His campaign for the legislature featured one issue: getting evolution out of the schools before that godless Darwin drove all the youngsters of the Volunteer State into atheism.

He won his election handily, and afterward he penned an anti-evolution law that was passed overwhelmingly by the Tennessee legislature in 1925. Butler's law was a no-holds-barred advocacy of Christianity by government fiat, breathtaking in its straightforward religious nature and its censorship of science. It barred teaching in any publicly supported grade school, high school, or university "any theory that denies the story of the Divine Creation of man as taught in the Bible, and to teach instead that man has descended from a lower order of animals." The Butler Act, as it came to be known, made teaching evolution a misdemeanor, punishable by a fine of $100 to $500. There were other anti-evolution statutes scattered around the country, but Tennessee's was the only one to make teaching Darwin a criminal offense. It is safe to say that, without John Washington Butler, no one would have ever heard of John T. Scopes.

The Butler Act, in turn, had been inspired by a fire-and-brimstone anti-evolution crusade led by William Jennings Bryan, the progressive politician, former presidential candidate, and former secretary of state under Woodrow Wilson. Renowned for championing the rights of the poor and powerless, Bryan was called the "Commoner," a Democrat who combined fiery populist rhetoric with fundamentalist Christian views. His last battle, in the twilight of his career, was a tireless campaign against evolution as society's "great evil," a view he espoused time and again as he toured the nation on what was known as the Chautau-qua circuit—a series of town meetings, tent revivals, political debates, plays, and lectures that were a kind of preelectronic mass culture of the era before the Great Depression. Although his political star had long since faded, Bryan remained one of the brightest stars of the Chautau-qua culture, second in demand as a speaker only to the duet of Helen Keller and her teacher, Anne Sullivan. He used his circuit appearances to deliver thundering denunciations of Darwin and the geologists who insisted that the Earth was far older than the Bible allowed, commend-ing crowds of thousands in town after town to "trust in the Rock of Ages, not the age of rocks." Over time, his movement inspired the intro-duction of anti-evolution legislation in more than two-thirds of the states (though only a few states actually enacted the laws). He personally ap-plauded Butler and the governor of Tennessee for signing the Butler Act, although before its passage, he had opposed criminalizing the teaching of evolution as potentially too controversial.

Bryan was a curious mix of liberalism and radical conservatism. A progressive on many issues, particularly those pertaining to social and eco-nomic justice, he was not normally opposed to scientific advances, for he saw in them great potential benefits for ordinary Americans—better medical care, better educational opportunities, better jobs. But evolution was another story. He saw it as offering nothing good and as holding the power to do great harm in the everyday lives of Americans, and so the validity of the science behind it was quite irrelevant to him. What raised Bryan's hackles was the *effect* of teaching and believing the "absurdity" that man could somehow be related to apes and other lower animals. He had read a book, written in the early 1900s, which suggested that univer-sity students were less likely to attend church than other Americans, and that scientists and professors—particularly biologists who accepted Dar-win's theory—rarely believed in God. Undoubtedly, they would pass on their atheistic tendencies to young people, Bryan concluded. In the early 1920s he published a popular pamphlet, "The Menace of Darwinism," in which he wrote: "Under the pretense of teaching science, instructors who draw their salaries from the public treasury are undermining the religious faith of students by substituting belief in Darwinism for belief in the Bi-ble." Then he opined, in remarks that would be echoed eighty years later in Dover, Pennsylvania, that he had no problem with the teaching of evolution as a theory, so long as its alternatives and its flaws were also taught.

His denunciations became more strident as the years passed. He blamed World War I and German militarism on "social Darwinism"—a flawed application of the theory of natural selection to societies that Darwin never intended, and that could be used to rationalize social injustice, racism, fascism, and intolerance. "The Darwinian theory," Bryan said, "represents man as reaching his present perfection by the operation of the law of hate—the merciless law by which the strong crowd out and kill off the weak." Providing inspiration for later evolution critics, such as Dr. Dino, who would regurgitate the Great Commoner's claims, Bryan advanced a grim vision of evolution that, in truth, stemmed more from his dislike of the dog-eat-dog laissez-faire economics of America during the Gilded Age than from the original biological theory Darwin had proposed. (Darwin's theory predicts, among other things, that a species often will be most successful if its members *cooperate* with, rather than attack, one another.)

On the surface, it might seem odd that anti-evolution sentiment struck America in the 1920s, just as the country had entered a truly modern age in science, education, invention, and culture, from jazz to films to the rise of radio and a dynamic new American literature. Scientific attacks on evolution had pretty much been exhausted in Europe decades earlier. In the United Kingdom, for example, the debate continued for years immediately following the publication of *The Origin of Species*, but Darwin's theories were widely accepted in his native land by the time he died. America, on the other hand, had initially been much more accepting of Darwin, and his theory had arrived on the New World's academic and intellectual shores with little initial controversy, soon filtering into public school lessons without arousing organized opposition or even much notice. Perhaps this was in part due to the fact that evolution was taught in high school and above, and at the end of the nineteenth century, relatively few Americans would have received those lessons—only 6 out of 100 seventeen-year-olds were even in high school at the time. That situation soon changed, however, as high school attendance rapidly increased in the twentieth century, bringing new and dangerous ideas like evolution to a far wider American audience. The timing of the anti-evolution crusade also coincided with the discovery of extensive fossil remains of creatures that appeared to be early ancestors of humans—more than apes but less than men, creatures who used primitive tools, had a social fabric, and looked not all that different from modern humans. Suddenly a long-standing claim of the anti-evolutionists—that there was no physical evidence of man's descent from lower forms—appeared in jeopardy. These discoveries in the fossil record suggested that belief in biblical creation just couldn't hold up in the modern world.

Inevitably, there came a strong response against the forces of modernism from traditionalists in American society, spearheaded by a new religious movement that caught fire in this same era: fundamentalism. The movement first led to a crisis of factionalism between conservative and liberal theologies within the mainline Protestant denominations of America, then spilled over into society at large in matters of science, education, and politics—a spillover that has never really abated. The "five fundamentals" were put forth by a group of religious conservatives in 1910 at a major religious conference, where they created something of a sensation, and then were championed by many traditionalist religious leaders, Bryan among them, as an antidote for the moral decay that modernism and evolution were said to have brought to America. This doctrine of fundamentals, which grew in popularity during the lean times following World War I, held that true Christians had to believe unquestioningly in five things or face damnation: the inerrancy of the Bible; the literal truth of the virgin birth and the divinity of Jesus; the doctrine of atonement through God's grace and human faith; the resurrection of Jesus after the crucifixion as a historical fact; and the authenticity of Jesus's miracles and his promise of a second coming. If you accepted the five fundamentals, then you knew evolution could not be true—man had to have been created, not evolved from lower forms. And you also knew that teachers who tried to convince children that evolution was real and true were trying to harvest souls for Satan. Bryan had found the perfect religious foil for Darwinism.

Science and fundamentalism, tradition and modernity, were once again on a collision course. The age-old drama experienced by Copernicus, Columbus, and Galileo came to a new climax in 1925. This time, the setting would be Dayton, Tennessee, and the impetus would be the Butler Act.

More than a bit of stagecraft was involved before there could be a proper confrontation. The American Civil Liberties Union had made a widely publicized offer to represent anyone who sought to challenge the Butler Act—the organization in effect "fished" for a client—and this offer became a lively topic of conversation among town leaders at Robinson's drugstore in Dayton. Soon it was suggested that the town's ailing economy could get a much-needed boost from a trial involving evolution and the inevitable media attention it would command. A civic leader whose coal and iron company was about to go bankrupt was the most enthusiastic proponent, seeing this as a last chance to save his business. Among those present for this conversation was the local school superintendent, who liked the idea well enough to send for John Scopes, a young substitute science teacher and football coach, who hotfooted it over to the drugstore. Questioned by the town leaders, Scopes told the drugstore crowd that the biology textbook he used in class, Hunter's *Civic Biology*, had a section on evolution, and he might have taught a lesson from it. (The book was approved for use in schools by the state of Tennessee, the ban on evolution notwithstanding.) Scopes quickly agreed to become the "test case" for the Butler Act—it sounded like great fun. Besides, the regular high school science teacher, who was first asked to "volunteer," had declined. Two friends of Scopes's who happened to be attorneys said they would prosecute the case, and a legal legend was born.

As predicted, the case created an international uproar, though it proved uniformly unflattering to Dayton, which was condemned in most media reports for persecuting one of its own teachers. "Two months ago the town was obscure and happy," said the famous writer for the *Baltimore Sun*, H. L. Mencken; "today it's a universal joke." As publicity grew, Bryan volunteered to assist the prosecution, anxious to find a fitting capstone for his long career. America's best-known lawyer, Clarence Darrow, signed on for the ACLU (the Dayton "conspirators" had actually wanted the science-fiction writer H. G. Wells, but he declined). Darrow, ironically, was the third most popular Chautauqua speaker in the country, right after Bryan—and was America's most popular doubter of God.

The trial was a circus in nearly every sense of the word. A thousand people showed up to watch the first session in the roasting Tennessee heat of July 1925; the judge eventually moved the proceedings out onto the town center lawn for fear that the courthouse would collapse under the weight of those in attendance. Darrow demanded the removal of a large outdoor banner visible to the jurors, exhorting them to read the Bible. The judge obliged him, though he refused Darrow's additional request to abstain from opening each session of court with a decidedly Christian prayer. Media from throughout the nation and the world covered what has since come to exemplify the cliché "trial of the century." It was the first trial to feature live broadcasts (on radio) of the proceedings. The carnival atmosphere was made complete by parading chimpanzees, street-corner proselytizers and prophets of doom, refreshment stands, and a large kiosk on Main Street set up by the Anti-Evolution League of America to sell Bryan's books and a tome by one of the league's leaders, T. T. Martin, entitled *Hell and the High Schools*. Calling Martin's book over-the-top would not even begin to describe its shrieking, doomsday tone; in one of the milder passages, the author argues that allegations of German military forces poisoning French wells and children's candy during World War I did not even approach the evil of teaching Darwinism. "The Germans . . . were angels compared to the teachers, paid by our taxes, who feed our children's minds with the deadly, soul-destroying poison of Evolution." The book advocated cutting off funds to any school that taught evolution, and requiring a kind of anti-evolution loyalty oath from all public school teachers and officials. Mencken, whose riveting and acerbic coverage of the trial and its surroundings has become a classic of journalism (and whose newspaper paid for Scopes's expert witnesses), wrote that Martin "addresses connoisseurs of scientific fallacy every night on the lot behind the courthouse" in order to promote *Hell and the High Schools*. It sold briskly throughout the trial.

The deck was stacked against Scopes from the outset, as everyone in Dayton knew it would be. The judge, a conservative Christian, had conspicuously attended church services conducted by Bryan shortly before the trial began. Almost every one of his rulings favored the prosecutors, and he destroyed the defense by refusing to let Darrow call scientific experts to attest to the validity of evolutionary theory. That would be irrelevant, the judge ruled: The state had made teaching evolution illegal, so what did it matter if evolution was legitimate science or a fairy tale? Darrow then sought to use the experts to show that evolution really didn't conflict with any reasonable reading of the scriptures, nor was it inconsistent with belief in God; but again the judge ruled against allowing the experts to take the stand, leaving, in essence, no defense at all. Without experts in the case, all that was left was Scopes's naked admission that he had taught evolution, and a parade of six of his students to attest to it. The judge, with a Bible at his side, made certain the only question that the trial would address was whether or not Scopes broke the law, not whether the Butler Act itself was proper or constitutional.

The refusal to allow the experts was not entirely unexpected; the ACLU's purpose in the case was not necessarily to win the trial—in fact, winning would have been inconsequential, because the influence of the case would then stop at the Tennessee border. The ultimate goal was to create a record for appeal to the U.S. Supreme Court, so that anti-evolution statutes could be struck down or prevented everywhere. The judge's denial may have slightly reduced the courtroom theater, but it had also offered additional ammunition for appeals.

A highlight of the trial was to have been a much-anticipated speech by Bryan, in which he argued against allowing the experts on evolution to testify, but the normally riveting speaker sorely disappointed spectators, whom he addressed directly rather than speaking to the judge. The oratory was disjointed and in parts silly, as when Bryan curiously insisted that man was not a mammal, and he relied on too many shopworn jokes from his Chautauqua days, such as his ridiculous complaint that evolutionists didn't even have the decency to assert that Americans descended from "American monkeys," but instead insisted that we originated from "Old World monkeys." He received only polite applause at the end (the judge allowed regular outbursts from the audience).

Darrow's co-counsel, Dudley Field Malone, who had been Bryan's undersecretary of state, then delivered what many, including a crestfallen Bryan, felt was the best oration of the trial, in which he argued that no one has anything to fear from truth, whether it be religious truth or the scientific truth of evolution: "The children of this generation are pretty wise. If we teach them the truth as best we understand it, they might make a better world of this than we have been able to make of it." He got the longest applause of the trial, though his audience was primarily on the anti-evolution side. Even old Butler, who was there to watch his law in action, congratulated Malone and opined that, in the name of fairness, the judge should have allowed the defense experts to speak.

This set the stage for the now legendary confrontation between Darrow and Bryan, who allowed the wily defense lawyer to bait him into testifying in the sweltering heat, no doubt in hopes of recapturing the crowd's and the media's adulation after his flop of a speech. Bryan's two hours on the stand were an unmitigated disaster for him. Darrow's unrelenting cross-examination—he questioned everything from Jonah's being swallowed by a whale to whether the six days of Genesis were literal days—led Bryan to concede that it was not reasonable to read everything in the Bible literally, eliciting gasps from the audience and bringing a grin of triumph to Darrow's face. Darrow had goaded Bryan into forsaking fundamentalism. If the Bible wasn't to be taken literally, how could evolution contradict it?

Bryan grew furious, shaking his fist at Darrow at the end of their overheated duel, accusing the defense lawyer of attempting "to cast ridicule on everybody who believes in the Bible."

"We have the purpose," Darrow shot back, "of preventing the bigots and ignoramuses from controlling the education of the United States."

At that point, the judge quickly adjourned for the day. He then ruled the next morning that he was throwing out all of Bryan's testimony as irrelevant to the simple question of whether Scopes had violated the statute. Darrow then surrendered, asking for the jury to deliver its verdict, and Bryan was thereby denied the chance to make his summation, which he had been preparing for days. Mencken, who detested Bryan, wrote that the man who had almost been president had failed in his effort to become a kind of "peasants' Pope," adding, "It is a tragedy, indeed, to begin life as a hero and to end it as a buffoon."

Crushed by the rigors and embarrassments of the trial, Bryan died in his sleep five days later.

Scopes was found guilty by the jury after eight days of trial and only nine minutes of deliberation. The judge fined him $100, which Bryan offered to pay. The conviction was later overturned on a technicality by the Tennessee Supreme Court, which found no separation of church and state issues to trouble its collective mind, instead focusing on how the judge had erroneously set the amount of the fine instead of letting the jury handle that task. The decision ended the ACLU's attempt to get the case before the U.S. Supreme Court on the broader constitutional question, undoubtedly as the Tennessee high court intended. Nothing else could have safeguarded the state court's absurd finding that there was nothing unconstitutional about making it a crime to contradict the Christian account of creation. In the end, although the Scopes Monkey Trial (Mencken's name for the case) captivated the nation, it changed nothing.

Across the decades, the Scopes trial has achieved mythic proportions in American society and culture, in part because of the larger-than-life characters it attracted and in part because most people's understanding of it is derived from the film version of the award-winning play *Inherit the Wind*. This was a fictionalized, idealized rendition of the Scopes case, written during the height of the McCarthy era in the 1950s and incorporating themes drawn from that era as much as from the 1920s.

The most common myth—one that has no truth to it—is that the evolutionists won, if not technically, then morally, with the anti-evolution crusade left deflated and bereft. The confusion is understandable, because the outcome and effect of the trial have become conflated with Bryan's unmitigated loss to Darrow in their famous duel. But contrary to conventional wisdom, the anti-evolution crusade did not fold up its tents after Scopes. In the years following the trial, states continued to contemplate and occasionally adopt new anti-evolution laws, and—more important—textbook publishers began to omit evolution from their biology books, not because of laws like the Butler Act, which remained few and far between, but to avoid controversy. They knew that school districts, if given a choice, would select the textbook with the least evolution in it to avoid offending students, parents, and politicians. The bottom-line purpose of textbook publishing is not to make a case for any particular scientific theory, but to sell books and make money, so the post-Scopes calculus was simple and ruthless: Darwin was the kiss of death when it came to sales. The next edition of the book Scopes had used, *Civic Biology*, had almost all references to Darwin removed by 1926, the year after the trial. The researchers Judith Grabiner and Peter Miller did a systematic study in 1974 of science textbooks in the Scopes era and found that references to evolution were drastically reduced in biology books published after the trial; the year of publication, before or after 1925, was obvious without looking at the copyright date. By 1930, seven out of ten high schools omitted any mention of evolution.[4]

The anti-evolution crusade did peter out around the time of the Great Depression, not because Clarence Darrow had made its crusader in chief look the fool, but because the crusade had won—even without new laws or prosecutions. Evolution had disappeared from the schools, a situation that continued for more than thirty years, until the space race and the Cold War initiative to make the United States a center for world-class science. It was such a persistent problem in the minds of working scientists that, in 1959, the Nobel Prize–winning geneticist Hermann J. Muller, during the centennial celebration of the publication of *The Origin of Species,* angrily told his colleagues, "One hundred years without Darwin is enough!"

Evolution did finally return to biology books in the 1960s, when the National Science Foundation gave money to a nonprofit task force, the Biological Sciences Curriculum Study (BSCS), which assembled a team of educators and scientists to create a series of model, state-of-the-art biology textbooks. The resulting texts covered a full range of topics in the life sciences and included evolution as a major theme, reflecting its importance to working scientists and researchers. Within two decades, 50 percent of the biology texts used in American schools were BSCS books, and the rest were modeled on them.

The reappearance of evolution in America's schoolbooks led to the resurgence of the long-dormant anti-evolution crusade, and the conflict soon landed in the courts again. There were still anti-evolution laws on the books in three states. Tennessee had voluntarily repealed its Butler Act. Arkansas, however, had a statutory prohibition against teaching evolution dating back to 1928. In forty years, not one person had been prosecuted for breaking that law. But in 1967, Susan Epperson, a Little Rock science teacher and believer in theistic evolution (which accepts Darwin's theory, but as a tool through which God acts), challenged the law's constitutionality. She taught tenth-grade biology at Central High School, which ten years earlier had posed the first major test of Washington's willingness to enforce the Supreme Court's landmark desegregation decision in *Brown v. Board of Education.* Now Central High would be the setting for a stunning reversal for the anti-evolution forces in America.

A trial court in Little Rock found in favor of Epperson and over-turned the ban against evolution. That led to a statewide outcry by anti-evolutionists and an appeal by state prosecutors, who persuaded the Arkansas Supreme Court to reinstate the law. But the U.S. Supreme Court had the final word, and it declared such bans an unconstitutional establishment of religion, throwing out every such anti-evolution law in the nation. As the ACLU had urged in its pleadings, the Supreme Court had finished the job that the organization had begun in the Scopes case more than forty years earlier. This was yet another landmark opinion of the Court presided over by Chief Justice Earl Warren, the former Republican governor of California whose positions on civil rights, racism, and separation of church and state had seemed radical at the time, but eventually became mainstream values. It was the Warren court that out-lawed compulsory prayer in public schools in 1962. Five years later, in the Epperson case, in an opinion written by Justice Abe Fortas, the Court found that states could not single out evolution for special treat-ment simply because it offended fundamentalist Christians. Fortas, who knew something about religious discrimination, having grown up a Jew-ish schoolboy in the overwhelmingly conservative Christian city of Mem-phis, reasoned that the Arkansas legislature, by banning evolution for clearly religious reasons, had in essence created a government religion—something Fortas declared an affront to everything America stands for. Fortas wrote:

> Arkansas' law selects from the body of knowledge a partic-ular segment which it proscribes for the sole reason that it is deemed to conflict with a particular religious doctrine; that is, with a particular interpretation of the Book of Genesis by a particular religious group.... The antecedents of today's decision are many and unmistakable. They are rooted in the foundation soil of our Nation. They are fundamental to freedom. Government in our democracy, state and national, must be neutral in matters of religious theory, doctrine, and practice.

With bans on evolution overturned, anti-evolutionists responded first by mandating "balanced treatment." Such laws required schools to teach creationism and evolution equally. The goal Bill Buckingham and Alan Bonsell seemed to be setting for Dover in 2004 had been dealt with in Tennessee in 1973, when it adopted a balanced-treatment law requiring all biology textbooks that discussed evolution to give equal time and space to alternatives, which the law defined as "including, but not limited to," the biblical Genesis account. The statute charged the state textbook commission with the task of choosing properly balanced books (no evolution and no creationism at all would also constitute balance under this law), exempted the occult and satanic origins from having to be included in the texts, and also defined the Holy Bible as a reference work, not a textbook, so that it would not have to be banned for failing to give equal time to evolution.

In *Daniel v. Waters*, a group of science teachers sued the Tennessee textbook commission charged with carrying out this law, asking that it be overturned for the same reasons cited in *Epperson*. The case went as high as the U.S. Sixth Circuit Court of Appeals, one rung below the Supreme Court, and once again the anti-evolutionists lost. The court ruled in 1975 that such balanced-treatment laws violated the constitutional ban on governmental establishment of religion, adding drily that the Tennessee statute was absurd on its face:

> Throughout human history the God of some men has frequently been regarded as the Devil incarnate by men of other religious persuasions. It would be utterly impossible for the Tennessee Textbook Commission to determine which religious theories were "occult" or "satanical" without seeking to resolve the theological arguments which have embroiled and frustrated theologians through the ages.

This decision, while removing overtly religious teaching from public school science classes, left the door open for more scientific alternatives to evolution, and so "creation science" was born in the early 1980s as an attempt to provide Bible-friendly scientific explanations of carbon dating, the fossil record, geology, and the apparent similarities between various animals and humans. "Flood geology," for example, attempted to explain the various geologic strata and fossils as products of the worldwide flood described in Genesis and survived by Noah and his ark full of animal pairs. In the *McLean* case in Arkansas and the *Edwards* case in Louisiana, which both involved laws that mandated the teaching of creation science, the anti-evolution forces once again failed. Creation science was found to be an unconstitutional introduction of Christian religious beliefs into public schools—a dodge intended to get around the Supreme Court's proscription against straight creationism. In seeking to establish a scientific basis for creation theory, lawmakers in Louisiana—their law being the only one of the two to make it all the way to the U.S. Supreme Court—relied on an erroneous definition of evolution, claiming falsely that it attempts to explain how life arose from inanimate chemicals. Next this law created a false dualism, making the claim that there are only two possible explanations for life on earth: evolution and creation science. The Louisiana legislature then asserted that there was far more scientific evidence to support a theory of supernatural creation than there was to support evolutionary theory. Finally, the law claimed that evolution was being taught as a "fact," not a theory; that it had become a kind of "religion" to a cabal of powerful scientists; that teachers had been "brainwashed" to believe in this religion and to suppress information about creation science in order to avoid exposing the weaknesses of evolution; and that teaching evolution as fact while "censoring" the alternative of creation science trampled on the religious beliefs of students. The legislators who crafted this Balanced Treatment Act relied solely on creation scientists to provide the support for these claims, ignoring a veritable legion of scientists prepared to debunk the creationists' assertions.

From this one-sided approach, the Supreme Court deduced in a 7–2 ruling in 1987 that the Louisiana legislature had a religious motive in passing the Balanced Treatment Act. The Court reached this conclusion notwithstanding the bill sponsor's claim that his only intention was to end "censorship" of creation science, which he described as a legitimate theory accepted by "hundreds and hundreds" of (unidentified) scientists. The scientific language of creation science changed nothing, the Supreme Court majority ruled, because in the end, the law "endorses religion by advancing the religious belief that a supernatural being created humankind." As in *Daniel v. Waters*, the Supreme Court made it clear that truly scientific alternatives to evolution could always be taught, but primarily religious alternatives could not.

One justice wrote a stinging dissent to this straightforward holding, deriding the Supreme Court jurisprudence on separation of church and state as "embarrassing," and mocking attempts to identify legislators' motives as a wasted effort. Justice Antonin Scalia, in only his eighth month on the court, wrote that his fellow justices had erred with shallow, antireligious reasoning and should have taken the Louisiana legislators at their word when they said their law had legitimate secular purposes. Then Scalia took issue with the very basis of the ruling, arguing, in essence, that even if the Louisiana lawmakers had religious motives, so what?

> We do not presume that the sole purpose of a law is to advance religion merely because it was supported strongly by organized religions or by adherents of particular faiths. . . . To do so would deprive religious men and women of their right to participate in the political process. Today's religious activism may give us the Balanced Treatment Act, but yesterday's resulted in the abolition of slavery, and tomorrow's may bring relief for famine victims.

Then Scalia gave a hint of where he would like to see the country headed if he were calling the shots: Such laws as Louisiana's ought to be considered constitutional even if they advance a religious cause, so long as they are doing so to protect an individual's religious rights or to "accommodate" religion. "If the Louisiana Legislature sincerely believed that the State's science teachers were being hostile to religion, our cases indicate that it could act to eliminate that hostility."

In other words, Scalia was sending a signal. Someday, if the stars realigned and a more conservative court membership took the reins, he would revisit the logic behind *Edwards*, which he derided as a "Scopes-in-reverse." The Balanced Treatment Act could be viewed not as an establishment of religion, but as a *protection* for it—from what, Scalia didn't say, but he could only have been referring to teachers with a propensity for excluding the supernatural from their biology classrooms.

As for the majority's decision, Scalia had nothing but insults:

I am astonished by the Court's unprecedented readiness
to reach such a conclusion, which I can only attribute to
an intellectual predisposition created by the facts and the
legend of *Scopes v. State* . . . an instinctive reaction that any
governmentally imposed requirements bearing upon the
teaching of evolution must be a manifestation of Christian
fundamentalist repression. In this case, however, it seems to
me the Court's position is the repressive one. The people of
Louisiana, including those who are Christian fundamentalists,
are quite entitled, as a secular matter, to have whatever
scientific evidence there may be against evolution presented in
their schools, just as Mr. Scopes was entitled to present whatever scientific evidence there was for it.

And there the law has remained since 1987, with the creationists
foiled at every turn, unable to ban evolution, unable to teach creationism,
unable to turn creationism into a science that the Supreme Court would
accept—exactly the opposite of the post-Scopes landscape. Little changed
in the next seventeen years, although a new type of creation-friendly science, intelligent design, slowly matured, its proponents strategizing, raising money, and readying their arguments for the public stage and for a
new legal challenge. They waited patiently for an opportunity to drive a
wedge in the wall of case law giving evolution sole domain over the nation's public classrooms—waiting for the right moment, when the country and the courts would be most receptive.

As the controversy began to unfold in Dover, the U.S. Supreme
Court was being remade through the death of one justice and the retirement of another. After the confirmation process concluded, only two of
the justices who had heard the *Edwards* case remained on the bench.
And one of them was Scalia.

The second meeting of the Dover Area School Board in June 2004
took place on a warm Monday evening, and it drew an unusually large
crowd, thanks to the fireworks at the previous meeting. Nearly 100 men,
women, and children packed the cafeteria at the elementary school
where the board convened. The gathering appeared nearly equally di-

vided between opposition to and approbation of the school board, with all factions sensing that a seemingly simple dispute over a textbook had become something much more fundamental and potentially much bigger. Everyone, it seemed, wanted to have a say—particularly Bill Buckingham.

He started off with an apology to anyone he might have offended with his comments at the previous week's meetings. His fellow board member Casey Brown, sitting next to him, bristled. She thought his words were the right ones—he needed to apologize for his rudeness to members of the public, in her opinion, particularly the student he had labeled as brainwashed—but his tone sounded anything but apologetic to her ears. As a result, she expected more fireworks that night, and sure enough, it wasn't long before they began, following her pointed observation that all board members had taken an oath to uphold the Constitution and the school code, and that included a strict separation of church and state.

Buckingham would have none of it: "Nowhere in the Constitution does it call for a separation of church and state." Later, he recalled how his generation prayed and read the Bible in public schools, and it never hurt him. But now "liberals in black robes" were taking away the rights of Christians. When other board members and members of the public expressed concern about injecting religion into the classroom, Buckingham made the statement, "Two thousand years ago, someone died on a cross. Can't someone take a stand now?"[5]

The parallels with the old Supreme Court creationism cases and the Scopes trial seemed eerie at times. There was Buckingham in the role of William Jennings Bryan, complaining that Christians were being oppressed, then playing the role of the Tennessee legislator with his claims about brainwashed evolutionists. And then there was Bonsell, backed by others on the board, calling for balance, so that students could be exposed to all views rather than just one theory—a theory many perceived as hostile to their constitutionally protected religious beliefs. It was as if the board members were exactly following Scalia's reasoning in his dissent in *Edwards,* when he argued that legislators (or school boards) should be able to protect students from religious hostility with just this sort of balancing act with evolutionary theory. There is no indication that any of the board members—other than, perhaps, Casey Brown, who opposed the push for creationism—had ever read Scalia's dissent, or any majority opinions in any of the creationism cases. If anything, this board was totally oblivious of the history and jurisprudence that had addressed these same concerns decades earlier.

The meeting was something of a free-for-all once the public comments began. When the question of the textbook purchase came up again, Buckingham said that the text, *Biology*, which the science department had thought was so great, had received a grade of F from a think tank in Seattle. He didn't name it at the time, but he was referring to the center of the intelligent design movement, the Discovery Institute, which rates textbooks according to its own anti-evolutionary standards. (Discovery had actually given *Biology* a grade of D.)

Buckingham also engaged in an angry exchange with Jeff Brown after Jeff's wife, Casey, suggested that the board consider an elective course in world religions rather than injecting creationism into science classes. Such a course could examine religions around the world, their differences and similarities, including their beliefs about creation, and thereby avoid constitutional pitfalls and potential lawsuits. Public schools could teach about religions, she said, they just couldn't preach one of them. Jeff Brown said that anything more than such an elective course would be illegal and would lead to costly lawsuits, prompting Buckingham to call him a coward, unwilling to take a chance for his principles. "I'm glad you weren't fighting during the American Revolution," Buckingham said to the man who sponsored him for the board. "Because we'd still have a queen."

Then Buckingham's wife, Charlotte, rose and gave a lengthy statement calling for creationism, not evolution, in class. The board president, Bonsell, allowed her to exceed the normal time limit of five minutes for members of the public; some observers said that she took two to three times that amount. She quoted scripture, called evolution nothing but lies, demanded that Genesis be taught in class, and asked, "How can we allow anything else to be taught in our schools?" Then she quoted from the gospels on how those present could be "born again." Casey Brown would later recall hearing board members on either side of her—Buckingham and Heather Geesey—mutter, "Amen," several times during Charlotte Buckingham's speech. Casey could hardly believe her ears. She thought she had been transported back to the period of the Scopes trial and was listening to an old-time Chautauqua. "It was like being at a tent revival," she said later. "I just didn't know what to say."

Later, Buckingham upbraided a representative in the audience from the organization Americans United for Separation of Church and State, who had come to caution the board about the legal ramifications of introducing religion into science classes. "Americans United?" Buckingham snapped. "You wouldn't know what a good American is."

Buckingham also scolded members of the audience who believed in the evolutionary principle of common descent: "I challenge you to trace your roots to the monkey you came from," he mocked.

The meeting ended, once again, with no action taken on the new biology textbook—it hadn't even been on the agenda, although the subject had hijacked the meeting. After that night, news of the Dover school board's actions and plans began to spread beyond York County, with Buckingham's comment about standing up for the man who died on a cross featured prominently. And then Bill Buckingham got a phone call from the Discovery Institute, and things began to change.

Chapter 4

DARWIN'S NEMESIS

I t is fitting, perhaps, that the meteoric transformation of an obscure
conservative think tank in Washington state into the most politically
potent, media-savvy, and pugnacious challenger of evolutionary theory
in a generation would be initiated by a lawyer, not a scientist. In this age
of universal litigation of all matters great and small, who better than an
attorney to attempt to upend the legacy of Scopes, Epperson, and Edwards for the twenty-first century? Seattle's Discovery Institute may
have fashioned itself into the center for intelligent design, a would-be
giant killer pitted against the scientific establishment's embrace of
Charles Darwin, but the attorney Phillip E. Johnson, a law professor
from Berkeley, is this movement's godfather. And it is a *movement*, a
word not often used in conjunction with science but one that routinely
figures in religious discourse.

The Discovery Institute had been positioning itself for more than a
decade to deal with controversies such as Dover's, flooding the airwaves
with its own experts in biology, geology, and theology to criticize evolution as "dead," to support school boards that wanted to "teach the controversy," and to attempt to redefine science itself to include recognition
of a supernatural designer. Discovery's directors, fellows, and literature
all assert vehemently that the institute is secular, not religious, but this
assertion seems to be contradicted by its own twenty-year plan to transform science and society. This plan, which has been dubbed the "wedge
strategy," appeared on the Internet several years ago without the insti-

tute's permission. The stated goal of the wedge strategy is to rescue America from rampant, destructive "materialism," which is supposedly fostered by belief in "godless" evolution and the "purposeless" world it describes. The wedge strategy seeks to replace materialism with a world-view and a science that are more "consonant with Christian and theistic convictions," allowing the Discovery Institute and its allies to lead America into a spiritual and cultural "renewal."

Phillip Johnson is the architect of this supremely ambitious plan, and he describes his 1991 book, *Darwin on Trial*, as "the thin edge of the Wedge." Building a movement that focuses initially on criticizing evolutionary theory—putting scientists on the defensive and reinforcing the existing public distaste for Darwin—is the first step, the "wedge." Hammer that wedge into the tree of science hard enough, Johnson says, then add the alternative of intelligent design, and the tree will fall. More explicitly Christian theories can follow later, according to Johnson, but early on, a "big tent" approach must be taken to draw in as many adherents as possible.

Darwin on Trial has been so pivotal to twenty-first-century anti-evolutionism that most if not all of the main proponents of intelligent design credit Johnson with leading their movement. The creationist governor of Alabama in the mid-1990s, Forrest Hood "Fob" James, who famously imitated an ape during a presentation before the state board of education, considered Johnson's message so imperative that he used his discretionary funds to buy *Darwin on Trial* for every public school science teacher in the state. The beginning of a glowing profile in the *San Jose Mercury News* a few years later summed up Johnson's stature: "Phillip E. Johnson is a dangerous man. At least that's what his critics think."

For an affable, balding law professor in his sixties with a conspicuous gray comb-over (since turned snowy white), this was a pretty hot description, and a welcome one for his movement. If intelligent design was ever to have a chance to supplant evolution, it needed a gunslinger, a strong advocate, a *dangerous man*.

Though he clearly relishes his David-versus-Goliath role in taking on the scientific establishment's near-unanimous embrace of evolutionary theory, Johnson seems a most unlikely figure to have earned the moniker "Darwin's Nemesis." He grew up in Aurora, Illinois, born the year before the attack on Pearl Harbor, the smartest kid he knew and unafraid to let others recognize it. He recalls growing up a rationalist, barely tolerant of religion and attending church only because it was the "nice" thing to do, like joining the Boy Scouts. Every Sunday morning his mother drove his father to the local golf course and young Phillip to the local Methodist church, "an experience," he told one interviewer, "that gave me an abiding love for the game of golf."

At sixteen, tired of small towns, small churches, and high school life, he applied for early admission to Harvard University in lieu of the twelfth grade. He was accepted and entered the university in 1957, the year of Sputnik, the space race, and the return of evolution to American classrooms. Johnson endured an unremarkable career at Harvard as an English literature major, spent the year after his graduation teaching English in Kenya, then returned to his home state to attend the University of Chicago law school. He graduated first in his law class, earning the opportunity to clerk for the most famous—and liberal—chief justice of the century, Earl Warren. Passing up an opportunity to teach at Yale, he instead went to the University of California, Berkeley, in 1967 as a professor of law at Warren's alma mater, the Boalt Hall School of Law. As a professor, Johnson had virtually nothing to do with the law as it applied to religious rights, the separation of church and state, or the teaching of creationism versus evolution. His specialties were torts and criminal law, and he excelled at them, though he felt bored and stifled, plodding along through tenure, writing articles and teaching seminars.

If his résumé is an unlikely one for a hero of the religious right, his new home—Berkeley—fit the plot even less, a polyglot center for leftist politics, alternative lifestyles, political correctness, and, at the time of his arrival, student protests against the war in Vietnam. Johnson joined a peaceful protest against the war, but was soon turned off by the young leftists at Berkeley, though he found himself envying their passionate commitment to their beliefs, however misguided he thought them to be. At least they believed in something—a quality he missed sorely in his own life.

The breakup of his first marriage in 1977, coincidentally timed with his eleven-year-old daughter's invitation to a Bible school banquet, brought Johnson back into contact with church and his long-abandoned religious roots. The lifelong agnostic, who had relegated Christianity to a set of pleasing myths with little relevance to the modern world, began to reconsider his religious convictions, and he started attending church regularly. There he met his second wife, Kathie, another recent convert. Over time he became increasingly devout and realized that he felt a greater intellectual fulfillment among his fellow churchgoers than in the "materialistic" world of academia. All that was missing, he would later say, was a way to combine the two worlds.

Johnson found that way during the school year 1987–1988, while on sabbatical in London. He had an enviable arrangement: He worked three days a week as a visiting professor at University College in London, then saw the sights with his wife on his days off. Every day as he walked to work, he passed a bookstore. A pair of volumes in the window kept catching his eye: *The Blind Watchmaker*, by Richard Dawkins, one of the best-known (and famously atheistic) evolutionary biologists in the world, whose book seeks to explain why natural selection gives the appearance of design in nature where none actually exists; and *Evolution: A Theory in Crisis*, by the Australian molecular geneticist Michael Denton, who asserts that there is no evidence that one species of creature has ever evolved into another, and that, in fact, there is no known natural process to account for Darwin's principle of "common descent."

Johnson bought both books, and they were a revelation to him. First he began to understand and embrace the complaints long made by creationists, dating back to William Jennings Bryan and earlier, that Darwin had introduced a materialistic naturalism to scientific and academic thinking, that this philosophy had pervaded all of modern culture, and that it sought to replace purpose and God with the blind forces of nature. Dawkins himself had stated that Darwin made it possible, for the first time in history, to be an "intellectually fulfilled atheist." Then Johnson wondered how it was that Denton could pose such fundamental, and seemingly unanswered, questions about a theory that had been around for nearly a century and a half, and that most of the scientific community treated as fact, unassailable and proved. Finally, he saw Dawkins's writing as beautiful, powerful, persuasive rhetoric—but he also recognized in it the same tool great lawyers use to defend bad cases: compelling arguments that, in Johnson's estimation, seemed very slim on supporting evidence. Yet the renowned evolutionist never doubted his theory or considered the consequences of what Johnson took to be missing, crucial evidence in support of Darwin's vision.

Johnson's central insight and inspiration—and the heart of his argument in *Darwin on Trial*—was his complaint that Dawkins and other scientists had stacked the deck and rigged the rules of the game by excluding evidence of the divine or supernatural, and that they had done so by defining science as exclusively the study of *natural* phenomena. In effect, evolutionists had blinded themselves to better explanations and to the holes in their own theory, and had then sought to impose those blinders on all of society. This, in his view, is what happened in the Edwards case—the views of those who saw design in nature were not even considered, since their opinions fell outside the rules of science set by the evolutionists. "The Academy thus defined 'science' in such a way that advocates of supernatural creation may neither argue for their own positions nor dispute the claims of the scientific establishment," he would write. "That may be one way to win an argument, but its not satisfying to anyone who thinks it possible that God really did have something to do with creating mankind, or that some of the claims that scientists make under the heading of 'evolution' may be false."

If science was truly the pursuit of truth, Johnson decided, it should follow wherever the truth led. Yet the scientific community did all it could to prevent this from occurring, Johnson came to believe, because leveling the playing field would fatally expose the weaknesses of evolutionary theory. In a fair fight, Johnson felt sure, Darwin and materialism would be overthrown. So Johnson saw not only a stacked deck, but a vast scientific conspiracy to prop up evolution by hiding the truth—that truth being, in his opinion, that the Emperor Darwin had no clothes.

After debating with his wife how crazy it would be for a law professor with no scientific training to write a book challenging evolution, both decided he should give it a shot, and while still on sabbatical in London he began writing a draft of a "brief" against the late Charles Darwin. Around the same time, he met a young philosopher of science studying at Cambridge University, Stephen C. Meyer, who had become interested in the new study of intelligent design, which arose after the Supreme Court's decision in the Edwards case to ban creation science from public schools. Meyer and a collection of like-minded philosophers, mathematicians, scientists, and students believed that intelligent design, with its focus on evidence of purpose and design in nature without (publicly, at least) referencing religious beliefs or the identity of the "designer," would not be at odds with *Edwards* if taught in public school. Now here was Johnson with a legal brief indicting the entire scientific community for excluding evidence of design. Meyer was so excited by Johnson's insights and slashing attacks on evolutionary theory (using a lawyer's logic rather than a scientist's expertise) that he brought back a draft of the work in progress to his fellow proponents of ID and introduced Johnson to other up-and-coming stars of the nascent ID movement.

When Johnson returned to Berkeley, he circulated his draft more widely—even to prominent evolutionists—and wangled an invitation to a private meeting at a Jesuit retreat in Massachusetts. About two dozen prominent scholars, scientists, and theologians representing all points of view had agreed to come together in 1989 to discuss evolution and creationism in public schools in the aftermath of the Supreme Court decision in *Edwards*. Johnson got a colleague at Berkeley to vouch for his "sanity" with one of the meeting organizers (who happened to be the colleague's former college roommate), and Johnson was in. On his arrival, moreover, he decided that he was not as far out of his element as he initially feared. The respected paleontologist David Raup of the University of Chicago, who first proposed mass extinctions as important forces in evolutionary history, had been using Johnson's manuscript in his college seminars, much to the law professor's delight. Raup said he had come to the retreat specifically to meet Johnson.

Johnson received a far less cordial greeting from one of the foremost American evolutionary theorists and writers at the time, the late paleontologist Stephen Jay Gould of Harvard University, who also had received one of Johnson's drafts and was not amused. The group cajoled Gould and Johnson into an impromptu debate on the merits of intelligent design and evolutionary theory. As they were about to begin, Johnson remembers Gould saying, "I'm going to destroy you. But it's not personal."

As Johnson recalls it, Gould was actually the more nervous of the two. Johnson attributes this to the fact that only one of them had anything to lose. "I was nobody. I was just some law professor. He was the most prestigious Darwinist in the country at the time. He had plenty to lose."

The two attacked and parried for two hours, with each drawing blood as Johnson faulted Darwinists for drawing sweeping conclusions by extrapolating from very few data, then rigging the rules to keep out evidence of design. Gould accused Johnson of resurrecting old creationist arguments and mischaracterizing or misunderstanding a century's worth of accumulated evidence and observation that supported evolution at every turn. According to Johnson, when the heated match ended Gould was shaking with anger. The novice evolutionary critic judged the debate a draw. Gould would have the final word, however. When *Darwin on Trial* was published two years later, Gould wrote an early, savage review for the influential magazine *Scientific American*, panning everything from Johnson's scientific falsehoods to his failed logic to his "abysmal" writing style. "No wonder lawyer jokes are so popular in our culture," Gould quipped. Apparently, it was personal after all.[1]

To Johnson and the rest of the growing coterie of intelligent design advocates, however, the debate had put them on the map—the dangerous man had done his job. Johnson's book went on to become a best seller, and the Discovery Institute in Seattle (then newly founded) and its Center for the Renewal of Science and Culture—the brainchild of Stephen Meyer; a former official in the Reagan administration, Bruce Chapman; and the technology writer George Gilder—took up intelligent design as its principal cause. In 1993 Discovery's scholars, philosophers, and scientists gathered in the beach town of Pajaro Dunes, California, just north of Monterey, to set the intelligent design movement on its current course. Johnson and the others perfected their wedge strategy at this meeting.

Michael Behe, then an obscure biochemistry professor at Lehigh University in Bethlehem, Pennsylvania, made his debut as a new star of the ID movement as well. He provided the gathering at Pajaro Dunes with its first scientific bona fides, sharing a new idea he called "irreducible complexity." This is his notion that certain complex biological structures at the molecular level could not have evolved, because there was no way for natural selection to assemble the various chemical "parts." The process of blood clotting, for instance, is a "cascade" of protein combinations and reactions. Natural selection can't create such a complex "machine" all in one step. Nor, according to Behe, could it gradually assemble it in the conventional evolutionary model, one bit at a time, across many generations, because the bits don't work on their own. Natural selection can't select for traits that do nothing—it can select only for traits that aid survival. And a piece of the blood clotting cascade does nothing, as Behe explained it. Darwin knew nothing of biochemistry—no one in his era did—and Behe told the gathering that neither Darwin's original theory nor any of the more modern amendments and patches applied to it over the decades could explain these irreducibly complex biological functions. Johnson, Meyer, and the others saw in Behe's idea a scientific silver bullet with which to bring down evolution, because if these complex structures could not evolve, then it would seem they had to be designed. The nineteenth-century dream of proving the existence of God through science was back on the table.

Soon, $750,000 in funding over three years for the wedge strategy, with the potential for more, came from Howard Ahmanson, who was the heir of California's Home Savings banking fortune and a prominent supporter of conservative religious causes and candidates (and whose child had been tutored by Meyer). Ahmanson had previously been allied with "Christian Reconstructionism," a controversial movement to make government and civic institutions function according to Christian principles, and the effort to supplant evolutionary theory with a more God-friendly designer theory seemed like a natural fit. Ahmanson would take a seat on the board of directors at Discovery.

Johnson's idea was to gradually insinuate intelligent design theory into the national discussion, building public opinion in its favor by constructing the philosophical big tent that would welcome traditional creationists, biblical literalists, scientists who doubted Darwinian orthodoxy, and liberal Christians (or Jews or others) who disdained literal interpretations of Genesis but who nevertheless believed in a God who took an active interest and role in the world and in humanity's prayers and fate. These factions rarely cooperated, but in the interest of ousting Darwin from his position of primacy, Johnson argued that it was time to put aside differences and pull together, concentrating on driving the wedge into scientific orthodoxy. Everything else could come later if they would just all focus on targeting the evil of evolution. In public, the discussion would always be focused on problems with evolution and on evidence for an unnamed designer. In private, and among true believers, however, the "wedge warriors" admitted that the designer virtually all of them were referring to was the Christian God. In 1999, at the "Reclaiming America for Christ Conference" in Florida, Johnson said:

> Darwinian theory of evolution contradicts not just the Book of Genesis, but every word in the Bible from beginning to end. It contradicts the idea that we are here because a creator brought out our existence for a purpose. . . . When I am preaching from the Bible, as I often do at churches and on Sundays, I don't start with Genesis, I start with John 1:12. In the beginning was the word. In the beginning was intelligence, purpose, and wisdom. The Bible had that right. And the materialist scientists are deluding themselves.

It was brilliant. Everyone in the club knew that intelligent design could be viewed as a restatement of the gospel of John from the New Testament, using the tools of scientific discourse and molecular biology to bring it into the twenty-first century. But to the general public—and the school boards it wished to persuade—the wedge took the essence of creationism, the idea that a biblical deity had designed the universe and all its inhabitants, and removed from it any explicit reference to God, leaving only the "evidence" of design, which in turn could be described as "flaws and gaps" in the theory of evolution. A distinctly religious idea thereby became a secular one, at least in the view of the proponents of the wedge, who had the Supreme Court's decision in *Edwards* firmly in mind as they crafted their strategy. Later, to drive this distinction home, the wedge was refined with the Discovery Institute's catchy slogan, "Teach the controversy." Having manufactured an evolutionary scientific controversy that previously did not exist, Discovery's two dozen fellows—scientists, scholars, theologians, philosophers, and engineers who wrote and spoke on behalf of the institute—fanned out across the country and demanded that schools teach this controversy in biology classes. This demand was couched as an educational innovation and improvement, an exercise in critical analysis; Discovery's position was that it would be incorrect to teach children that evolution was an uncontroversial and thoroughly supported theory.

In 1998, the wedge strategy was distilled into a formal strategic plan with five-year and twenty-year goals, which included obtaining major news coverage and favorable op-ed pieces; running active campaigns to get ID into schools in at least ten states; and conducting an active research program to construct a working, unified theory of intelligent design to supplant the systematic explanation of life on earth offered by Darwin. Discovery began compiling a list of scientists who doubted evolution and passing it out to the media. These were ambitious, even audacious plans, and the cocky young IDers confidently predicted that they would "win" in far less than twenty years. And in those heady early times, there were enough victories to give them hope that they might just pull it off.

Right on schedule, Discovery persuaded more than a dozen school boards and state education authorities to consider adding intelligent design or their "teach the controversy" curriculum to public school science classes. Ohio and Kansas moved most aggressively to limit or "balance" evolutionary theory in the public school curriculum. By then, nearly twenty books on intelligent design for a popular readership had been published. Every major news organization had covered the movement in mostly positive terms—a headline in the *Los Angeles Times*, "Enlisting Science to Find the Fingerprints of a Creator," was typical. And ID had been the subject of a PBS broadcast, just as the wedge strategy called for, when the conservative icon William F. Buckley personally championed it on his show, *Firing Line*.

The Discovery Institute took a more aggressive stance in 1999, moving beyond championing intelligent design academically. It took on a role more akin to a reverse-ACLU in a case that nearly erupted into the second coming of Scopes well before the trial in Dover. Roger DeHart, a science teacher in Washington state, had been quietly skipping standard evolution lessons for nearly a dozen years in order to teach from an intelligent design textbook, *Of Pandas and People*. This book harshly criticizes evolutionary theory and makes a case for intelligent design.

Several parents were shocked to learn what was being taught—without official permission—in DeHart's classes. When the ACLU complained on their behalf, Discovery had its first martyr. The institute rushed to defend DeHart on grounds of academic freedom when the school district curtailed his creationist lessons. DeHart was cast in the role of John Scopes, although this time it was the ACLU, not William Jennings Bryan, gunning for a teacher who was attempting to fight officialdom. Senior officials from Discovery flocked to the Burlington-Edison school district, where DeHart worked, speaking publicly on his behalf, and at one point attempting to recruit anti-evolutionist candidates to run for the local school board. Materials from the Discovery Institute that were circulated among the students at DeHart's school included bookmarks headed "Ten Questions to Ask Your Biology Teacher about Evolution." There was also a template for homemade warning stickers for biology books; the template listed Discovery's answers to the ten questions, among them: "The subject of human origins is very controversial, and most claims rest on little evidence." Kids were supposed to paste these inside their textbooks.

After two years of back-and-forth between opposing factions in the community, DeHart finally left his high school, where a lawsuit would have been inevitable had he stayed. A legal showdown was avoided; De-Hart joined the faculty of a Christian school in California, and ended up being featured in a documentary by the Discovery Institute, *Icons of Evolution*—the same DVD that Dover teachers would years later be instructed to watch twice.

Throughout all the "wedge strategy" campaigns across the country, evolutionary scientists fought back, always arguing that Discovery's arguments about flaws and gaps were specious and just a recycling of old creationist ideas that had been debunked by mainstream science and rejected by the courts as unsuitable for public schools. The critics pointed out that creationists and ID proponents shared almost everything except the identification of the designer: They made the same objections to "materialistic" theories as both scientifically and morally insufficient. They made the same arguments about the "evil" influence of evolutionary theory, linking it to fascism, racism, and crime. They both claimed that life-forms appeared abruptly on Earth, with no evidence of gradual transitions between species, and that this implied a creator (or designer). Both wanted to teach evolution and its alternatives in school.

But as with DeHart, this tit for tat, even when Discovery ultimately lost the battle, elevated the public standing of the intelligent design movement, making its ideas seem not only important but equal in stature to evolutionary theory. Every time mainstream scientists mobilized to counter the institute or some other ID drama, the credibility of intelligent design rose, the ID people looked reasonable, with their "teach the controversy" mantra, and the scientific establishment came across as inflexible, bullying, and worst of all antireligious. This was the most brilliant aspect of the wedge strategy, because the general public had no trouble deciding whom it would prefer to trust: a survey reported in *Nature* in 1998 of members of the National Academy of Sciences, the most esteemed (and evolution-friendly) body of scientists in the United States, found that only 10 percent believed in God. At the same time, 90 percent of the general public not only believed in God but felt certain that God had a role in creating life on Earth, either in conjunction with evolution or without it. The seductive message of intelligent design was meant for those members of the public, not for the National Academy of Sciences.

But science is not a democracy. Science is a brutal arena where ideas are picked apart, attacked, and tested to see if they hold up. Those that do hold up live to fight another day. Those that don't are dragged off and discarded. To survive, a theory must be supported by vibrant, meaningful, replicable research. And here was the Achilles' heel of ID, the one area where the wedge strategy badly lagged: There was almost no actual, hard laboratory research in support of intelligent design. Despite all the politicking, public relations, strategizing, and fund-raising, by 1999 there were no peer-reviewed articles and no startling discoveries attributed to ID theory. No one had yet figured out how the idea of intelligent design could become a broad, unifying theory with practical benefits for medical and scientific research, as Darwin's theory had served for more than a century. So far, there was only Behe's work on irreducible complexity, a slim portfolio indeed for an entire new science—and for the enormous publicity and public expectations that were being generated by other, more successful parts of the wedge effort.

Then in late 1999, in the midst of the DeHart controversy, a copy of the "wedge strategy" document found its way onto the Internet and into the hands of ID's critics. After initially ignoring the document, then claiming it was a stolen fund-raising proposal of little import, Discovery officials eventually acknowledged its authenticity in an article on the institute's own website entitled, "The Wedge Document: So What?" This appeared only after critics had seized on the contents of the document and it had been featured in a book, *Creationism's Trojan Horse: The Wedge of Intelligent Design.*[2]

To evolutionary scientists and teachers who were trying to put out the Discovery Institute's fires all over the country, however, this was a very big deal indeed, for the document revealed what appeared to be the starkly religious core of the intelligent design movement. And this revelation came not in the form of accusations from evolutionists, but in the words of the movement's leaders. The wedge document states:

> Debunking the traditional conceptions of both God and man, thinkers such as Charles Darwin, Karl Marx, and Sigmund Freud portrayed humans not as moral and spiritual beings, but as animals or machines who inhabited a universe ruled by purely impersonal forces and whose behavior and very thoughts were dictated by the unbending forces of biology, chemistry, and environment. This materialistic conception of reality eventually infected virtually every area of our culture, from politics and economics to literature and art.... The cultural consequences of this triumph of materialism were devastating....
>
> Discovery Institute's Center for the Renewal of Science and Culture seeks nothing less than the overthrow of materialism and its cultural legacies.... Design theory promises to reverse the stifling dominance of the materialist worldview, and to replace it with a science consonant with Christian and theistic convictions.

The two "governing goals" of the wedge strategy: "To defeat scientific materialism and its destructive moral, cultural and political legacies" and "to replace materialistic explanations with the theistic understanding that nature and human beings are created by God."[3]

In the years since the wedge strategy became public knowledge, representatives of Discovery have taken ever greater pains to project a secular image, even changing the name of the Center for the Renewal of Science and Cultures to the less provocative Center for Science and Culture ("renewal" being a concept frequently associated with evangelical Christianity). The article posted on the institute's website, though acknowledging the wedge document, dismissed as irrelevant the religious convictions of intelligent design's proponents. So what if the men and women behind the movement were people of faith? That didn't mean the theory they advocated couldn't be secular. The argument is a plausible one—it mirrors almost exactly the logic Antonin Scalia applied in his dissent in the Edwards case—but just the fact that the Discovery Institute had to make it showed that the intelligent design movement was on thin ice. The movement's own words, and its founding document, were being used against it to paint its proponents as slicked-up creationists willing to place faith above science as soon they thought no one was looking. The wedge had worked wonders, but if it was going to continue doing so, the Discovery Institute had to pick its battles—and its bedfellows—with great care. If the institute allied itself with the wrong spokesperson or the wrong case, the whole movement could be irreparably damaged.

Against this backdrop, in June 2004, a young lawyer and analyst at the Discovery Institute named Seth Cooper received an e-mail that contained news stories on a controversy over creationism involving the school board in Dover, Pennsylvania, and a plea for information and assistance from one William Buckingham. Buckingham had heard something, he wasn't sure where, about intelligent design. An Internet search led him to the Discovery Institute website, which enabled him to send an e-mail asking for help. It fell to Cooper to respond.

Like so many of the complex events that began to unfold in Dover in the summer and fall of 2004, the telephone conversations between Cooper and Buckingham have become matters of dispute. Putative allies can no longer agree on what was said or when, or even whether the phone conversations were strictly about legal advice, thus rendering them privileged and confidential, or if they were less formal conversations, making them fair game for the press and the public.

Bill Buckingham walked away from those initial chats thinking that the Discovery Institute would back the school board if it decided to add intelligent design to its curriculum. Cooper sent him the DVD *Icons of Evolution,* which the science teachers of Dover were told to watch; an anti-evolution book; and several fact sheets on intelligent design, which included Discovery's adamant position that ID was definitely not a form of creationism. Buckingham, who has maintained that he never spoke of creationism anyway, saw in intelligent design just the sort of wedge its boosters in Seattle intended him to see: It would be a way to expose kids to a viable alternative to evolution that, if not exactly biblical, still left plenty of room for the God Bill Buckingham knew. If that's what it took to pass constitutional muster while still saving Dover's youth from the godlessness of Darwin, then he'd start pitching intelligent design.

Cooper, however, has a different recollection of the conversations. He warned Buckingham that the school board would be inviting a lawsuit and a losing battle in court if its members persisted in talking about creationism. And he says he recommended against introducing intelligent design into classes at Dover, instead suggesting the more nuanced "Teach the Controversy" approach—pointing out the flaws and gaps in evolutionary theory for the ninth-graders, and leaving it at that. The same Discovery Institute that confidently predicted it would win the evolution wars, and that had passionately fought four years earlier for the right of Roger DeHart to give lessons on ID, lacked the confidence in 2004 to advocate including ID in a school curriculum. "I sought to provide him and the board with a way of handling the topic of evolution without mandating the teaching of the theory of intelligent design or reading aloud any disclaimer mentioning it," Cooper said. "Buckingham . . . later told me that the materials I sent him were the solution for their situation. Our correspondence thus ended, as I was led to believe that the Dover Board would not be requiring instruction in creationism or in the theory of intelligent design."[4]

Buckingham remembers things quite differently. He says that the Discovery Institute was gung ho at first, then seemed to get cold feet as the inevitable controversies mounted: "Seth Cooper knows damn well that he sent me information to be used in the case supporting . . . including intelligent design into the curriculum. Mr. Cooper was like a rat jumping from a sinking ship."

A few weeks after this conversation with Cooper, Buckingham had a very different sort of exchange with Richard Thompson, head of the Thomas More Law Center, which Buckingham knew specialized in litigating Christian causes. Thompson informed Buckingham that there was a textbook available that taught intelligent design—*Of Pandas and People*, the book that Roger DeHart had been using in Washington state and that the Discovery Institute had defended. And Thompson told Buckingham something else: His law center had been looking for a test case on teaching intelligent design, and the time might be right to litigate this issue, perhaps right up to the Supreme Court.

More important, Thompson said, in the likely event that the Dover school board was sued should it decide to adopt an intelligent design program, Thomas More would help. The Christian law center—which positions itself as a kind of anti-ACLU, opposed to most everything the civil liberties union supports and vice versa—would represent the school district, free of charge.

Buckingham, bolstered by Discovery's materials and Thompson's promise, decided to draw a line in the sand. He wanted *Of Pandas and People* as a supplemental textbook in the biology classrooms, a companion to the textbook "laced with Darwinism" that the teachers wanted, the Dragonfly Book. He had purchased a copy of *Pandas* and shared it with several of his allies on the board, and they liked what they saw, including the references to "abrupt appearance" of species in the fossil record, with no evidence of evolution: fish with scales, birds with feathers, appearing in nature as if they had suddenly popped into existence. Evolution was presented as riddled with problems and unresolved questions. Here was the science Bill Buckingham knew had to be out there if they just looked hard enough—science that found evidence to support, not attack, the Bible. In his judgment, *Pandas* didn't get into religion per se, but the textbook made clear that man was created as he is, and that Darwin's "from goo to you" idea of common descent was a fairy tale. If the teachers wanted to receive the biology textbook they had selected, Buckingham figured, then the school district would have to buy "my book" too.

Chapter 5

CLASS ACTS

As the weather turned sultry and the classrooms emptied for the summer break, the residents of Dover finally seemed to notice en masse that their school board's odd little flap had evolved into an entirely new species. It took on a circuslike life of its own, replete with shouting matches in the boardroom and fire-and-brimstone rhetoric unlike anything heard in these parts in many years, if ever. Without much forethought by the school leadership, the teachers, the civic leaders, or anyone else in Dover, all the ingredients had fallen into place for a reenactment of *Inherit the Wind* in their peaceable little community.

Suddenly the news media in Harrisburg, Pittsburgh, and even Philadelphia, two hours away, along with the wire services that fed news to the nation, had started paying attention to Dover. The local newspapers, based in the much larger city of York, often barely noticed the township. Now the part-time correspondents who had covered the Dover school board were being augmented by full-timers from the main offices. Reporters were calling board members, talking to parents, even camping outside churches on Sunday to gauge local feeling on the controversy. Suddenly, everyone had an opinion. At the Dover United Church of Christ, across the street from Dover High, the minister assured one newspaper reporter that evolution was no threat to faith.[1] "What I like about evolution is the upward movement," Reverend Michael Loser, said helpfully, ". . . that God is calling the whole Creation to be like what God wants it to be." His eighteen-year-old parishioner

Matthew Rupp, who had just graduated from the school, emerged from services to tell the reporter he didn't really know what all the fuss was about—the word "evolution" never even came up in class that he could remember. (He allowed that he vaguely recalled mention of Charles Darwin and natural selection.) Matthew's younger sister, who had just completed ninth-grade biology, said she wouldn't object to Bill Buckingham's idea for balanced treatment, though the school year had ended before her biology class even reached the section on evolution. Her only exposure to the concept was in history, when her class covered the Scopes Monkey Trial. What seemed clear is that, even before the evolution war began in Dover, the anti-evolutionists were far ahead and arguably had won without lifting a finger. Darwin was already an educational outcast, barely dragged out of the attic at the very end of the year, the bearded old man dusted off and paraded about it in the drabbest of clothes for a moment, then quickly forgotten by the students, if they noticed him at all.

Across town, however, the same reporter discovered diametrically opposed sentiments at the Dover Assembly of God Church, where the associate pastor said he supported the school board's creationist leanings and didn't see how the scriptures left any room for Darwin to be on the side of the angels: "Whatever the Bible says is what I stick to," he announced. The mere mention of evolution was too much for some people. The pastor's wife, Yvette Sproull, said that she home-schooled her five children to avoid, whenever possible, exposing them to "ideology" such as evolution. Unfortunately, some exposure to the concept is inevitable, she said, such as when reading to her children from their dinosaur books. On those occasions, she discusses the topic with her kids, then together they tear out any book page that mentions evolution. That strategy will go only so far in shielding her brood from evil, however. She added: "Eventually, we are going to have to tell them about homosexuality and premarital sex."

With such polarization of opinions in Dover, residents turned to a tried-and-true method of expressing themselves while pounding their opponents: letters to the editor flooded the two local newspapers. Many of the writers castigated school board members for being zealots, being "an embarrassment," and, in the words of one writer, driving the town "down the Taliban Road"—an uncomfortable and, to many citizens, outrageous parallel between anti-evolutionists and the former tyrannical fundamentalist rulers of Afghanistan. Other writers praised the board and Bill Buckingham in particular as visionary, willing to do something about an America "that strays from its values," as one woman put it. Another writer, a member of the school board, complained about all the wrongheaded criticism. She wrote passionately to her fellow citizens: "Our country was founded on Christian beliefs and principles. We are not looking for a book that is teaching students that this is a wrong thing or a right thing. It is just a fact." Then this board member, Heather Geesey, added a fateful point: "You can teach creationism without it being Christianity. It can be presented as a higher power." Her letter provoked even more responses pro and con, as well as intense interest from civil rights organizations (this clipping promptly landed in a thick and growing notebook of evidence on the school board's allegedly religious intent). The controversy gained steam with each issue of the newspaper.

During the summer and fall of 2004, the terms of the debate never really shifted from creationism versus evolution, though the school board members tried mightily to shift it, having received a crash course in constitutional imperatives—and the ingredients of a winning case—from the Discovery Institute and the Thomas More Law Center. Their own terminology did change, for the most part, from biblical to New Age—from Genesis to intelligent design—but in the minds of most people in Dover, the two had become inseparable and synonymous. Eventually, the board members would deny even using the term "creationism" in public meetings. Members of the public and the school faculty, who remembered hearing board members use precisely that term, were astonished, then appalled to learn that audiotapes routinely made of each school board meeting had been erased, and that no board member ever intervened to rescue the evidence. Relations with the press degenerated rapidly after that—reporters who had quoted the discussions about creationism were blamed for the "confusion" and were mocked openly by board members during public meetings (though neither of the major newspapers serving York County received any formal requests for clarification or correction). Meanwhile, the sardonic news columnist of the *York Daily Herald*, Mike Argento, began channeling the spirit of H. L.

Mencken, at one point writing, "It would be tempting to say that the Dover Area School Board has been taken over by a bunch of clowns, but I won't, mostly because I don't want to receive angry phone calls and e-mails from clowns offended by the comparison." Early on, he pointed out that the Dover school board seemed to be forgetting that the host city of the first "monkey trial"—Dayton, Tennessee—had endured a difficult and humiliating lesson. City leaders there had deliberately manufactured the Scopes case as a vehicle for community improvement, but in spite of these motives, Dayton ended up a national laughingstock.

Christy Rehm, for one, wasn't laughing. Driving home from a particularly vituperative board meeting—she was then nine months' pregnant—she nearly shook with anger. She could not get Bill Buckingham's comments about America being a Christian nation out of her head. She taught at a school in York with a very diverse student population of immigrants from Latin America and the Middle East—a sharp contrast to the homogeneous and overwhelmingly white Dover school district—and she could imagine how hurt and confused some of her students with different faiths and traditions would have felt had they heard Buckingham's pronouncements. America, to them, was a place where their beliefs were their own and the government had nothing to say about it, beyond the simple promise that they could worship as they pleased, or not worship at all. She had students who were first- and second-generation Americans, from families who knew firsthand that there are places in the world with no such freedom, places that have national religions or religious bans or oppression and violence—and so these young people loved America in a very special way. Christy could see it in their eyes, hear it in their words. And there was Bill Buckingham, without realizing he was doing so, in Christy's view renouncing one of America's great virtues. The outspoken board member had said that he had to be true to himself, that he wouldn't be forced into adopting "political correctness." But to Christy's thinking, he was engaging not in plain talk but in un-American talk. She turned to her husband as he drove them home: "How could you tell that to a student, 'Oh, we can't express your belief, but we can express *that* person's belief in the classroom.' This is so wrong!"

Christy believes that everything a teacher does in the classroom is teaching—not just the specific lessons, but everything. What she wears.

What she says. What she doesn't say. How she greets and treats people. And, certainly, the decisions she makes, both in class and out. What was the Dover School board telling its young people with these sorts of religious statements about science and faith? That was teaching, too. Buckingham's classic quip was already filtering down to the school yards, perfect for recess torture sessions: *Which side of your family has the monkey in it?*

At the school district offices, while the citizens of Dover waited to see what sort of fireworks would erupt at the board meetings in July, Assistant Superintendent Mike Baksa just tried to keep up with the demands of various board members, who had told him to investigate biology textbooks and curricula in use at local parochial schools. In the quest for alternatives to the Dragonfly Book, he also obtained materials from Bob Jones University, the fundamentalist Christian college in South Carolina best known for its championship debating team, its anti-Catholic teachings, and its controversial ban on interracial dating (hastily abandoned in 2000 after George W. Bush, then a presidential candidate, visited and praised the school, sparking a national uproar).

The list of textbooks Baksa developed for the board was enlightening. The biology program at Bob Jones has its own texts and manuals produced by the university's own publishing wing, all of them based on young Earth biblical creationism. According to the university's academic catalog: "While most secular biologists are committed to evolution as the basic principle of biology, Bob Jones University trains Christian biologists who see the living world indelibly marked with the fingerprints of a God of limitless wisdom and power." The product profile for the textbook stated, "Those who do not believe that the Bible is the inerrant, inspired Word of God might find many points in this book puzzling." These "points" included the observation that any conflict between a literal interpretation of the Bible and any scientific findings could be due only to mistakes or biases by scientists. As appealing as this outlook was to Buckingham and some other members of the Dover board, the Bob Jones texts were rejected as too overtly religious to stand a chance of being held constitutional for public school classrooms.

In contrast, the three parochial schools in York that Baksa surveyed all used mainstream biology texts with substantial sections devoted to evolutionary theory, as is typical of private schools nationwide: Delone Catholic School of Eastern York used the same Dragonfly Book the teachers at Dover wanted; York Catholic School used McGraw Hill's *Biology: The Dynamics of Life*; and the evangelical Christian School of York used Holt, Rinehart, and Winston's *Modern Biology*. Ironically, the evangelical school had chosen the book with the largest and most detailed section on evolution of any of the texts, including a chapter on human evolution; the publisher appeared so interested in emphasizing the book's embrace of Darwin that its promotional materials billed the text as "The Natural Selection." Board members expressed little interest in either of these alternatives to the Dragonfly Book.

Baksa also was dispatched to a conference on evolution and creationism in Harrisburg, which featured Professor Edward Larson of the University of Georgia, author of *Summer for the Gods*, a history of the Scopes trial. At the conference, Baksa obtained a chart outlining a variety of scientific and religious theories regarding the origins of species, which he gave to all the members of the Dover school board. The chart classified intelligent design as a form of creationism. This chart was not made public in Dover, however; instead, a newsletter was prepared for eventual public dissemination that assured the citizens of Dover that intelligent design was *not* creationism.

Meanwhile, Buckingham spent July attempting to line up board votes in favor of purchasing the intelligent design textbook that the Thomas More Law Center's chief counsel, Richard Thompson, had told him about: *Of Pandas and People*, with its extensive criticism of evolutionary theory and claims that life appeared created rather than evolved. Evolutionary biologists have long dismissed *Pandas* as presenting discredited pseudoscience, and they point out that the two principal authors were proponents of creation science who later embraced ID— but only after the Supreme Court's ruling in *Edwards* banned creation science from public schools. One of these authors and several other contributors to the text were also research fellows at the Discovery Institute.

The school board decided to hold only one meeting in July (customarily, it held two meetings a month), and to the surprise of most observers, nothing happened beyond routine matters. There were no showdowns, no lengthy tirades, no diatribes on church versus state. The agenda published before the meeting sounded so noncontroversial that only twenty members of the public showed up, rather than the 100 who had swarmed the last meeting. The only major news discussed by the board was the release of a new edition of the Dragonfly Book. The Dover science faculty had quickly reviewed it, with Buckingham's list of complaints in hand, and the teachers saw that the section on evolution had been shortened and refocused to dwell more on Darwin as an historical figure than on expounding his theory. All the scientific bases were covered, but just enough to meet the state standards—the publisher seemed to have gone out of its way to make the text as uncontroversial as possible without actually purging evolution. Bert Spahr and Mike Baksa were ecstatic. Bert quipped that one coauthor, Kenneth Miller, "must have been reading Mr. Buckingham's mind." It seemed all the concerns of the school board had been addressed in one way or another.

However, the board did not vote at the July meeting to approve purchasing the book. It did not even get on the agenda, to the teachers' disappointment; officially, their biology text remained under review. But the main reason for the continued delay had less to do with the Dragonfly Book and more to do with Bill Buckingham's plan to get *Pandas* into the classrooms. He simply wasn't ready yet to set his plan in motion. When a reporter asked him after the meeting ended if he would vote for the new edition of the Dragonfly Book, he refused to say. "You're trying to get me in trouble by asking that," he groused to the reporter. "If you want to know how I'll vote, come to the next meeting."

Not long afterward, Mike Baksa let Casey Brown know that others on the board had been studying *Pandas*, and that Buckingham planned to recommend it as a companion to the Dragonfly Book—part of his quest for balance. "We've got to see that book," Casey told her husband, Jeff.

Another member, Sheila Harkins—who had been Casey's friend and ally on the board for years—had been reading the review copy of *Pandas* that Buckingham had shared with his closest allies. Harkins's service on the school board had been distinguished in years past by a relentless drive to cut spending, and so Jeff wondered why she wanted to spend thousands to buy optional textbooks that the teachers didn't even want. "Sheila, you don't even want to buy the books that we're supposed to buy," he pointed out.

"This book was such an eye-opener," she answered, "about what's wrong with evolution."

Jeff Brown warned her that with all the publicity and the statements about creationism that had been made so far, the board was courting trouble: "If we even touch this subject we're going to end up in court," he predicted. But Harkins, whose daughter had taught in the Dover district for fifteen years, remained adamant—she would support Buckingham's quest.[2]

Buckingham hit another sort of roadblock, however. In preparation for the next school board meeting, he sent a memo to the district superintendent asking him to add to the August agenda a proposal to purchase 220 copies of *Pandas*, and to place it ahead of the Dragonfly Book, which once again would be coming up for a vote. Superintendent Richard Nilsen then asked him to stop by the office, and a tense meeting took place between the two men. Nilsen informed Buckingham that he definitely would not approve the purchase of *Pandas* as a companion textbook and that, therefore, he would not put it on the agenda. The teachers did not want the book, and Nilsen shared their concerns.

This complicated matters immensely for Buckingham. When the superintendent approves a textbook, the board can adopt and purchase it on a simple majority vote—that's all that was needed, for instance, to buy the Dragonfly Book. But when the superintendent rejects a book, the board must have a minimum of six votes to three in order to override the administration and make the purchase. And Buckingham felt there was no way he could get the six votes he needed. To his dismay, he also found that he could not persuade or bully Nilsen into seeing things his way. Instead, Nilsen proposed that they could work out a compromise: If Buckingham would support purchasing the new edition of the Dragonfly Book at the next meeting, Nilsen would back the purchase of a smaller number of copies of *Pandas* as reference books and make them available in the classrooms. Buckingham didn't like this one bit—he argued that they needed the two books in equal numbers, side by side—but in the end, he said he would go along with the compromise and would support, finally, the purchase of the standard biology text, delayed by then for nearly a year. Time was running out, the superintendent knew: The district had to order the new biology books in the next few days, or they'd never arrive in time for the start of the school year in September. With Buckingham's agreement, Nilsen could finally relax.

The deal was dismantled a few days later at the first school board meeting of the month, held on a stormy August evening, peals of thunder sounding appropriately in the distance as Bill Buckingham made his next move. The agenda called for a vote on the Dragonfly Book that the science teachers wanted and that the superintendent had recommended— the book which Buckingham considered "laced with Darwinism" but which he had told Nilsen he would reluctantly accept. Now Buckingham had another deal in mind. There were only eight board members present, and he had four votes against the book lined up, a deadlock. In the case of a tie, the negative votes would prevail under board rules, and the Dragonfly Book would not be purchased. That's when Buckingham revealed that he "controlled" the fifth and deciding vote of the absent board member, which he would "release" if the board would approve *Of Pandas and People* as a companion text, purchasing the 220 copies he wanted. "If we don't get our book, you don't get yours," he told the teachers and the opposing board members.

A loud debate ensued. Casey Brown said she had stayed up all night to do something no other board member had done—she had read *Of Pandas and People* "cover to cover," and found it to be riddled with bad science. Her husband, Jeff, was livid for other reasons. "I don't like blackmail," he said. "I don't like it that if we don't approve this other book, then that means no book." He accused Buckingham of "holding the students hostage," his words a hoarse shout before he was through.

Even Buckingham was taken aback at Brown's anger. He later said he was worried that Brown was so upset he would make himself ill. But Buckingham held firm. Even if he had to hold the Dragonfly Book hostage, he'd get the board members to agree to *Pandas* as a companion text, and the deal would be done. His tactics were well thought out. When several board members, including Alan Bonsell, complained that they needed to review *Pandas* before they could vote on it, Buckingham said, fine, let's put off the vote on both biology texts in order to get everyone up to speed. He wore the smile of someone who had outsmarted everyone in the room.

Until the meeting began, Superintendent Nilsen thought he had fi-
nessed matters and had Buckingham cornered. The compromise, he assumed, would get the textbooks the school needed while safely relegating
Pandas to a dark corner in the back of the science classrooms. He didn't
really know Bill Buckingham.

But Nilsen has his own strengths. A punctilious man—neat, reserved, professional—he doesn't go for angry displays. As he later put it,
he simply "voiced his displeasure" with Buckingham's surprising move.
He informed the board that their failure to adopt the book at this meeting would mean the school year would begin with insufficient biology
texts on hand, and those they did have would be dated and would no
longer meet state standards for science education. He also pointedly demanded to know just what the board intended by rejecting a standard
biology text that matched state requirements. Did they intend to change
the curriculum to teach intelligent design? Buckingham didn't answer
directly. He just said that an additional book—*Pandas*—would be an
asset. "We have an opportunity to level the playing field. What is everyone so afraid of?"

The two student representatives on the school board felt compelled
to speak up at that point. They have no vote—their presence is mostly
ceremonial—but both of them urged the board not to make evolution
into such a huge issue. One of the two couldn't even remember talking
about evolution in science class, and the other, Joshua Rowand, recalled
spending no more than an hour or so of class time on the topic. "We
need a biology book," he said. "Is all this really over a few pages out of a
one-thousand-page book?"

This would turn out to be a common sentiment among the students—
the supposed beneficiaries of the evolution debate in Dover. As one
ninth-grader said a few days after the meeting, "The only ones who
seemed to care about this are some board members and some parents.
The kids couldn't care less. They're spending more time talking about
this Darwin stuff in meetings than we're ever going to spend in class."

As the debate continued, Angie Yingling, a board member who had
voted with Buckingham against the Dragonfly Book, hesitantly raised
her hand and asked for a second vote. Yingling, a local real estate agent,
had been elected to the board in 2001, after running on a slate that in-

cluded Casey Brown and Bonsell. Buckingham considered her a reliable (and easily swayed) ally. Now, however, he watched, stunned, as she reversed herself—not for the last time—and the new biology textbooks were approved. "I can't believe you did that," Buckingham complained afterward. "Do you know what you've done?"

"I feel you were blackmailing them," she said. "I just want the kids to have their books."

Buckingham accused Yingling of ruining everything. The faculty and administration would never approve *Pandas*, and they would never get six votes to make the purchase. "That's why we needed to hold our ground tonight."

Yingling said she supported Buckingham philosophically, but she wanted to move the process along, indicating that he could still count on her in the future should *Pandas* come to a vote without the element of "blackmail." And Bonsell, who had voted for the Dragonfly Book despite his previously expressed desire for creationism in the curriculum, told Buckingham not to worry. Intelligent design would get its day, he promised. The board president seemed quite serene. "Six votes are not out of the question," he predicted.

Publicly, the issue seemed to fade. Another lull ensued for teachers who wanted to believe that the board would shift its focus, and for members of the public who had their own lives to worry about, and who would just as soon not trudge out on weeknights to an elementary school cafeteria to heckle or cheer the unpaid volunteers serving on the school board. The banter at the Dover Diner and in the dark, beery confines of the local taprooms was enlivened by the debate for a while, but pretty soon everyone got tired of monkeys and Darwin and Bill Buckingham's latest quote, and the conversation returned to more timeless subjects, such as the Phillies, where most of the patrons felt it belonged.

But behind the scenes, the controversy picked up steam. Late in August, the board's solicitor—a local attorney whose firm was on retainer to the board—wrote a memo distributed to board members that summarized a conversation he had with Richard Thompson of the Thomas More Law Center. The solicitor, Stephen Russell, expressed concern that Dover would open itself to a nasty lawsuit, just as Jeff Brown had warned. Russell wrote that Thompson's firm had been working with school boards in West Virginia and Michigan before coming to Dover, and those boards all had declined to put *Pandas* in their classrooms. Dover, in other words, would be out a legal limb, all by itself.

Nevertheless, and notwithstanding his loss at the last board vote, Buckingham convened the next curriculum committee meeting four days later with a renewed demand that *Pandas* be used in the classroom as a "side-by-side" text with the Dragonfly Book. The teachers argued strongly against it, with Bert Spahr complaining that *Pandas* taught creationism. When Buckingham would not relent, however, the wearied science teachers reiterated the superintendent's original compromise: They would agree to keep *Pandas* in the classrooms as a reference, which students could consult if they wished. But they would not teach from the book. Afterward, they could not tell if Buckingham had been mollified or not. "Maybe," Spahr told her teachers, once again, "this will just go away now."

The next board meeting, shortly after the start of the school year, was uneventful—there was nothing on the agenda about creationism, intelligent design, or *Of Pandas and People.* But word had leaked out about board members' continuing interest in the subject, and during the public comment period, a handful of residents once again stood up and spoke out on the issue, mostly inveighing against adding such ideas to the curriculum, as board members listened inscrutably. A woman who had taught biology at neighboring district high schools and at colleges in the area for thirty-five years urged the board to set aside their fears about teaching evolutionary theory. Not once in all her years at the front of the classroom had she ever heard that any of her students lost faith in God simply because they had learned about Darwin. And Barrie Callahan, the former board member whose daughter was taking biology that year, continued to probe the board with questions, trying to ascertain what was coming next and getting little more than exasperated comments in return. She had been hearing things of late—as a former board member, she still had her sources—and she had begun to fear the worst. Would the board, she wanted to know, give the public adequate notice before voting on *Pandas* or altering the biology curriculum? You should not try to railroad decisions about evolution through without public input, Callahan chided, or try to make any decisions "behind closed doors."

"Oh, Lord, it's a little too late for that," Bonsell retorted.

But Barrie Callahan's instincts and inside information were correct. The next board meeting, in early October, featured the announcement of a mysterious donation of sixty copies of *Of Pandas and People*. Richard Nilsen formally accepted the donation from an unnamed source and, because the books were to be labeled "reference" rather than used as classroom texts, no approval from the board was required to get them into students' hands. Nilsen's explanation for accepting books that he had previously rejected if purchased by the district amounted to this: Times are tough, and we never say no to free materials. The board members professed ignorance about the source of the textbooks, but whoever had provided the donation had come to Buckingham's and Bonsell's rescue just when it seemed they lacked the votes to purchase *Pandas* outright. When a former member of the school board stood up and demanded to know the names of the donor or donors who provided the books, the board members said they didn't know. Assistant Superintendent Baksa, meanwhile, had already received new marching orders. He was to prepare an addendum to the biology curriculum that would incorporate into the curriculum *Pandas* and the anti-evolutionist ideas it espouses. At the time, no one paid much heed to the timing of this effort to amend the curriculum, but it would take on added importance later, when every act of the board would be placed under a judicial microscope: The order to Baksa requiring him to develop the science curriculum addendum came a month *before* the anonymous donation of *Pandas* was made, but after Buckingham's play to buy copies of the book had failed.

As for amending the biology curriculum as it pertained to evolution, there wasn't much to work with. The Dover district did just what the state required in its teaching of evolution and nothing more, with ninety minutes of lectures, labs, and student activities that covered natural selection, the mechanism of evolution, and the origins of biodiversity. The official curriculum as it stood set five modest goals related to evolution.

1. Students will be able to discuss Darwin's observations of the living world. Students will be able to discuss the variability found in nature.
2. Students will be able to describe biomes and list the adaptations that organisms have to survive in this environment.

3. Students will be able to determine how limiting factors work to limit population sizes.
4. Students will be able to define types of competition and how they relate to population size.
5. Students will be able to list evidences used to support Darwin's theory of the origins of species.

And that was it: one historical section on Darwin's observations, which mostly concerned his voyage aboard the *Beagle*; three sections dealing with the related issues of biodiversity and the interaction of environment and organism; and one section—the last one, building on the previous four—on evolution directly. There was no mention of speciation, of origins of life, of human evolution, or even of the age of the Earth. From a creationist perspective, there just was not much to complain about. Nowhere did the goals state or even imply that students must accept evolution as the true explanation for the diversity of life on Earth, or even that they must believe Darwin was correct. The curriculum, in the artless language of the bureaucrat, required only that students be able to "list evidences" in favor of Darwin and evolution. This was not an arbitrary choice of wording. It is exactly what the Pennsylvania state legislature has ordered its public schools to teach, and what the state-mandated exams will test.

Under the direction of the school board, Mike Baksa came up with language for a sixth unit of evolution instruction: "Students will be made aware of gaps in Darwin's theory and of other theories of evolution."

This was the bare minimum Baksa figured he could do to meet the board's mandate while avoiding an open revolt by the teachers. And it also illustrated the scientific illiteracy of the Dover School Board, at least those members who were demanding this change. Indeed, they would later admit as much.

To most of the school board members, a gap in a theory meant that the theory had some big problem, like the famous "gap" on Nixon's Watergate tapes in which incriminating conversation was erased, or a gap in a criminal defendant's alibi that left time for him to commit the crime. They failed to realize that in science, gaps are different from those in murder cases. Every important and well-accepted theory has its share of gaps in the supporting data—and this is particularly true of historical scientific theories that rely on evidence such as ancient fossils (in the case of evolution) or other indirect observations, as contrasted with watching chemicals react in a test tube in real time. The theory of gravity, the big bang theory, the theory of relativity, quantum theory, atomic theory, plate tectonics theory—their histories all consist not simply of eureka moments in the lab, but also of a gradual filling in of gaps, a process that continues to this day. That is the nature of science, which continually tests its theories with new information. With large, explanatory theories such as evolution, the fact that there are gaps in the data is expected—problems arises only when gaps are filled and new information doesn't fit the theory. Then scientists say that a theory has been "falsified." This is why ancient Greek mathematicians and naturalists stopped believing the Earth was flat long before cameras were launched into space to photograph the globe—they knew the Earth couldn't be flat, because the available data did not fit the theory anymore. Ships sailed off in one direction but did not find or fall off an edge. On the other hand, the theory that the Earth is a globe was accepted centuries before it was actually "proved." That didn't mean there weren't gaps—such as why objects on the "bottom" of the globe didn't fall off into space, as the principles of gravity were not well understood until much later (and gaps in that understanding remain to this day). So gaps in theories are not only real but expected in science—and they do not in themselves disprove or discredit a theory.

The board members didn't grasp that distinction, however, and so they enthusiastically endorsed mentioning "gaps" in the belief that this statement represented a valid criticism of Darwin's theory. Still, this didn't mean that Buckingham's curriculum committee liked Baksa's language. On the contrary, the members rejected his proposal as not going far enough. During a meeting of the curriculum committee in early October, while the sole dissenter, Casey Brown, was absent, Bonsell, Buckingham, and Sheila Harkins came up with new language to replace Baksa's in a matter of minutes: "Students will be made aware of gaps/problems in Darwin's Theory of Evolution and of other theories of evolution, including but not limited to intelligent design. Note: Origins of Life is not taught."

Not only did this wording get intelligent design into the classrooms; it also was intended, with that final line, to eviscerate the specific lessons on evolution that Buckingham and Bonsell found most objectionable. Inadvertently, however, it left teachers an out—if they had the nerve to take it (as some did, to varying degrees). For the statement was muddled, and the primary effect of the final sentence added by Bonsell at the last minute was to enshrine the board members' confusion of the origins of *life* (the theoretical development of the first primitive organism from the nonliving primordial "soup" that was Earth's ancient environment) with the origins of *species*. They believed this provision would bar teachers from discussing macroevolution, speciation, common descent, and the theory that humans have common ancestors with other living creatures—which would have essentially put an end to teaching 90 percent of evolutionary theory. Teachers could have talked about small changes, like the evolution of pesticide-resistant bugs, but nothing more dramatic—no descendants of fish becoming amphibians becoming reptiles becoming mammals allowed. But to accomplish that, the curriculum change would have had to state, "The origins of *species* are not taught." No one corrected the board.

Indeed, this entire addendum to the curriculum was made without meaningful input from the science teachers. Usually, teachers in each department generate curriculum statements and changes for approval by the board—social studies teachers update history curricula, the English department updates the literature curriculum, and science teachers write their curriculum updates. This doesn't usurp elected officials' authority; it is a recognition that the laypeople elected to the board lack sufficient expertise in science or Shakespeare or the Treaty of Tripoli to write curriculum details, and are there instead to set policy and provide oversight. This time, however, the board members took the unusual, even unprecedented, step of reversing the time-honored process—they came up with the curriculum, then rammed it through without consulting their own in-house experts, the science teachers, or any other scientific experts in the community, state, or nation. Also unusual were the timing and the rush—every other curriculum change had been made one year in advance of putting it into effect, a common educational practice that minimizes disruptions and allows teachers to prepare before each school year begins. But this change was to take effect immediately. The only consultation the board members took seriously in crafting the addendum was advice from the Thomas More Law Center, which wanted a curriculum that pushed the legal envelope while remaining defensible in court—if not at the trial level, then in the Supreme Court, where the center's attorneys hoped Antonin Scalia and others with similar views on the separation of church and state might be receptive to allowing public schools to present intelligent design in science classes. The language of the addendum even had a bit of telltale lawyerly lingo in it—the phrase "including but not limited to," a bit of excess verbiage favored by many attorneys, though the "not limited to" is unnecessary and meaningless. Simply saying "including" would carry the same meaning.

Shortly before the next full school board meeting, Baksa shared the curriculum committee's proposal with the science teachers. They were stunned into silence. Then they desperately tried to come up with an alternative that the board might accept without committing what Bryan Rehm considered "an atrocity." Exhausted by the entire process and fearful that things might get even worse if they didn't capitulate to a degree, the teachers agreed to almost everything the board wanted: They would make students aware of "flaws and problems" in Darwinian evolution; they agreed that they would not teach origins of life (which they interpreted correctly to mean abiogenesis, something they had never taught anyway); and they would make *Of Pandas and People* available as a reference book, if reluctantly. They basically gave in on everything Buckingham wanted except for the specific subject of intelligent design.

But when the meeting convened on October 18, 2004, the champions of intelligent design were in no mood for compromise. In yet another wild, angry session, the board majority, led by Buckingham and Bonsell, presented its proposal just as Barrie Callahan had feared—without the usual notice to the public. The board's practice had always been to introduce major proposals at the first meeting of the month—the "planning meeting"—and then the board would vote on them at the following week's session, the "action meeting." This curriculum change, however, was being steamrollered through in one meeting, with no previous discussion. The agenda item had been as vague as possible, making no mention of intelligent design, so few had any notice that this was coming. Casey and Jeff Brown had been forewarned, however, and they adamantly opposed the proposed curriculum change, reiterating their belief that the board was inviting a costly lawsuit. They hoped raising the specter of harm to the district's pocketbook might have more effect than accusing their colleagues of injecting religion into the public schools. Not only wasn't the latter tactic working, but such accusations seemed to be taken as a compliment. After all, outside the meetings, when the Browns talked privately with Alan Bonsell, he would make pronouncements such as, "Evolution can't be true. God never changes." This would be uttered with such certainty and such conviction that it would bring Casey and Jeff up short. They had long since given up arguing with Buckingham, who was often difficult to deal with since returning from drug rehab, his temper hair-trigger. Their hope was that the others were just being pulled along by the force of Buckingham's and Bonsell's personalities, and that a financial argument could pull them back from the brink.

But the allure of a free ride with America's top Christian law firm was too powerful an inducement. Buckingham reminded his colleagues that he had negotiated free legal representation for Dover. Casey Brown suggested that this wouldn't necessarily insulate the district from legal costs. She pointed out that teachers as well as the district might need legal representation down the line and that, given their opposition to the curriculum change, their interests might not coincide with the school board's. In that case, the teachers could reasonably ask to be represented by the district's regular solicitor, the York County firm of Stock and Leader, which had prepared the memo warning the board of possible lawsuits—and whose representative was conspicuously absent from this meeting.

"If they requested Stock and Leader, they should be fired," board member Heather Geesey responded. "They agreed to the book and the changes in curriculum."[3]

There were mutters and gasps from several of the teachers and members of the audience at that: fire them because they might wish to assert their legal rights? Had they heard Geesey correctly? Bert Spahr then reminded the board—and informed the small audience watching the exchange like fans at a tennis match—that Geesey was dead wrong, that the faculty had done no such thing. They had tried to compromise, and had been rejected, Spahr sputtered. Not only did the science department disapprove of the new curriculum wording, the teachers had been deliberately excluded from the process. "We didn't know you were going to do this."

Barrie Callahan stood next and focused her wrath on *Pandas*. She said this textbook reflected science that was at least twenty years out of date. It was full of falsehoods and pseudoscience, she said, and it was both "pathetically simplistic" and "incomprehensible to a ninth-grader." Finally, she said, the board wanted a curriculum that barred teaching of the origins of life, yet *Pandas* was filled with discussions of that very topic. "The whole first chapter," Callahan pointed out, "is nothing but origins of life." She then read aloud excerpts of the book to illustrate each of her complaints.

"You know how to read," Buckingham jibed. "What's your point?"

Eleven people made public comments that night, ten of them in opposition to the board's proposed curriculum change, the meeting loud and angry, more so than any of the earlier clashes. The exchanges became so raucous that one board member, Sheila Harkins, rose repeatedly and shouted at people for being "out of order," though this wasn't her job—she wasn't then president of the board.

The one member of the community who spoke in favor of the curriculum change, Eric Riddle, assured the majority members of the board that they were doing a great public service, even if it did lead to a lawsuit. "It may cost us a little money to do what's right," he said, adding, "Maybe someday, I can feel good about putting my kids in this district."

Riddle's children, it turned out, were home-schooled.

Bryan and Christy Rehm attended the meeting that night. They had a sitter for their older three children, but Christy had given birth to a baby boy since the fireworks at the board meetings in June, and they had the infant with them that night. After several hours the meeting was still going strong. The Rehms weren't quite sure what would happen at the end. The superintendent had confided beforehand that he expected the whole thing to fizzle out, but this no longer looked very likely. "Let's get the baby home," Bryan told Christy, and she reluctantly agreed.

"Whatever's going to happen is not going to be good," Christy predicted.

The Rehms left as a recess was called and joined the crowd milling in the school hallway outside the meeting room. Bryan Rehm saw a clear path to where the board president was standing, so he edged through the crowd and confronted Bonsell. Bryan reminded him of his meeting the year before with the science teachers and how they had explained what was taught and the principles behind it, apparently to Bonsell's satisfaction at the time. "Why are you doing this, Alan?" he asked.

"It's the gaps and problems. We just want to cover the gaps and problems."

"What gaps and problems?" Rehm asked, exasperated.

"They're so big I can drive a truck through them," he recalls Bonsell answering, though the board president did not give any examples.

Not once during the meeting did members of the board explain to the public what intelligent design was about, or why they felt it would be a proper and useful addition to the curriculum. Much later, most of the board members would claim that they hadn't really known much at all about intelligent design when they mandated that every child in the district would learn about it.

When the meeting resumed, and Buckingham and Bonsell attempted to bring the matter to a vote, they had a surprising defection from their camp. Noel Wenrich, who did not accept evolutionary theory as valid and who was in favor of balancing the curriculum, decided he could not support the proposal. The thirty-six-year-old computer programmer and Army veteran offered motion after motion to defeat or delay or amend the curriculum proposal, using every parliamentary maneuver he could think of as the resident expert on *Robert's Rules of Order*. This infuriated Buckingham, who first expressed disbelief and then began insulting Wenrich, questioning his patriotism and his religious convictions. "The fact that teachers were cut out of the decision, the fact that they were made irrelevant bothers me to no end," Wenrich explained later. "After they compromised with us to allow a book they were uncomfortable with into the classroom only to be treated the way they were, well, I don't know why they would ever work with the board again like that in the future."

When Wenrich had exhausted all his ploys, and his fellow board members refused to budge, the vote finally came: six to three in favor of introducing intelligent design to the schoolchildren of Dover. Dover would be the first school district in the country to do so, catapulting the community to the forefront of the culture wars. As Bonsell had predicted months earlier, six votes were not out of the question after all. Voting for the change were Bill Buckingham, Alan Bonsell, Sheila Harkins, Heather Geesey, Angie Yingling (who had returned to Buckingham's camp, but then immediately began having second thoughts), and Jane Cleaver, who was leaving the board to retire to Florida. The dissenters, Wenrich and the Browns, resigned from the board. Wenrich's departure had been expected, as he had taken a job out of the area, but the Browns' resignation, particularly Casey's—she had been on the board for a decade—took everyone by surprise. Casey had suspected it would come to this, however, and she had a prepared speech typed up and ready, just in case. She read it tearfully but forcefully:

> Sometimes in order to fulfill the requirements of our office, we must put aside our personal feelings and beliefs. It is not always an easy thing to do—but it is what we must do in order to properly perform the duties and responsibilities of our office.
>
> In the past year, regretfully, there seems to have been a shift in the attitudes and direction of this board. There has been a slow but steady marginalization of some board members. Our opinions are no longer valued or listened to. Our contributions have been minimized or not acknowledged at all. A measure of that is the fact that I myself have been twice asked within the past year if I was "born again." No one has, nor should have the right, to ask that of a fellow board member. An individual's religious beliefs should have no impact on his or her ability to serve as a school board director, nor should a person's beliefs be used as a yardstick to measure the value of that service. However, it has become increasingly evident that is the direction the board has now chosen to go, holding a certain religious belief is of paramount importance.
>
> Because of this, it is quite clear that I can no longer effectively function as a member of this board . . . I shall pray for you all—pray that you will find the wisdom to separate your personal beliefs and desires from the proper fulfillment, within the law, of the duties and responsibilities of your office. I

shall pray that you will learn to represent all of the members of
our community, and all of their viewpoints—with impartiality
and with grace.

After the Browns tendered their resignations, Wenrich turned to
Buckingham and said, "We lost two good people because of you."

And as Casey Brown remembers it, Buckingham's response, laced
with profanity, was, "Good riddance to bad rubbish."

The next morning, with front-page stories on the board's action in
both local newspapers, the national media converging on the township,
and the ACLU among others already talking about lawsuits, a social
studies teacher at Dover High School, Brad Neal, sent an e-mail to Assistant Superintendent Mike Baksa. The note reflected the faculty's outrage at the curriculum change, and suggested that the board had changed
Dover "from a 'standards-driven' school district to 'the living-word
driven' school district." Neal sarcastically asked if there were a supplemental text that could be used to correct the mistakes of the U.S. Supreme Court and "to set our students straight as to the 'real' law of the
land."

Baksa's reply was chilling: "All kidding aside, be careful what you
ask for. I have been given a copy of *The Myth of Separation* by David
Barton to review from board members. Social studies curriculum is next
year. Feel free to borrow my copy to get an idea of where the board is
coming from."

The Myth of Separation, as the book's title implies, argues that the
Founding Fathers never intended to create a legal "wall" between religion and government, despite what the Supreme Court has ruled. Many
key elements of the book, including alleged quotations from the Founders, have been discredited. But Bonsell and Buckingham raved about the
book, and they had made it clear that once Darwin was put in his place,
the church-state separation principle being taught in social studies was
next on the list.

After the big vote, the winners exulted, realizing that the departure
of three members would allow them to appoint new, like-minded citi-

zens to the board, making compromise and circumspection completely unnecessary. But amid the celebration, a few disturbing realities became clear. For one, nobody had really thought out what would come next. When questioned by reporters, the superintendent said he was not sure how the new directive would be carried out. He would have to meet with teachers and figure out how a faculty that had never been trained to teach intelligent design and that thought it was unscientific was going to introduce this new subject in the classroom.

The Discovery Institute also continued to distance itself from the school board, not at all certain it wanted to make its big ID stand in Dover, though whether it was going to have much of a choice was not entirely clear. "That train," as Buckingham would later say, "had already pulled out of the station." After the donation of *Pandas* was announced, even before the curriculum change, the Discovery Institute issued an ambiguous and tepid press release praising the textbook and the idea of using it to explain to students the "weaknesses" of evolutionary theory.

As for legal repercussions, the Dover school board publicly dismissed concerns that the ACLU and Americans United for the Separation of Church and State were going to sue, despite indications to the contrary in follow-up articles in the two York newspapers. Board members insisted that intelligent design was not just "warmed over creationism" and that their actions were legal and beneficial to the district. Heather Geesey in particular seemed perplexed that anyone would be worried. "We're not going to be sued," she predicted airily. "I have confidence in the district's lawyers."

But in the event of a lawsuit, one persistent reporter asked, wouldn't Dover have to pay its opponents' legal bills if intelligent design went down in flames?

"My response to that is, 'What price freedom?' " Buckingham said. "Sometimes you have to take a stand."

"There seems to be a determination among some board members to have our district serve as an example, to flout the legal rulings of the Supreme Court, to flout the law of the land," Casey Brown said the next morning, sipping coffee and chatting with another reporter. "They don't seem to care. I think they need to ask the taxpayers if they want to be guinea pigs."

Sure enough, people began stopping Alan Bonsell on the street, asking him why he had put religion into the classroom and why he wanted to teach creationism. He expressed his frustration to the district superintendent and wondered why people were so confused, and why they did not understand that intelligent design was not religion but an alternative scientific explanation. Bonsell asked that a district press release be issued to set matters straight.

Somewhere along the line, the board also decided that it never intended teachers to "teach" intelligent design; the curriculum addendum called only for making students "aware" of ID, and this would be done by reading a prepared statement to all the biology students at the start of the evolution section of the class. This shift seemed less a matter of what the board members truly desired and more a matter of necessity. Although the board would later present the new statement as trivial and brief, as if this had been all they ever intended, they really were forced into this approach because there was no way the science teachers at Dover High could ever be persuaded to devise lesson plans on intelligent design. Mike Baksa, once again, was told to develop the ID statement for approval by the board. It would last only a minute of classroom time and be done—but it would get the subject in front of students, it would undermine the primacy of evolution, and it would direct students to *Of Pandas and People*, from which they could learn much more. "It's a start," Buckingham told himself. "Just a start."

Both the press release and the statement would prove to be public-relations and legal disasters for the school board.

Baksa asked the senior biology teacher, Jen Miller, to help him come up with the statement, and she agreed to make suggestions. The original statement Baksa drafted referred to evolution as "the dominant scientific theory." This is undeniably true, but it was removed by the school board, which had no members with any scientific training. Baksa also wrote that "there are gaps in Darwin's theory for which there is yet no evidence." The board removed the word "yet," dramatically altering the import of the sentence. Jen Miller suggested pointing out that there is a "significant amount of evidence" in support of the theory of evolution, but Baksa did not even bother including that statement—he knew it was true, but he also knew the board would never allow it.

After a variety of revisions, the statement the board approved to be read to every ninth-grader ending up carrying a connotation about Darwin and evolution very different from the language Baksa originally drafted:

> The Pennsylvania Academic Standards require students to learn about Darwin's Theory of Evolution and eventually to take a standardized test of which evolution is a part.
>
> Because Darwin's Theory is a theory, it is still being tested as new evidence is discovered. The Theory is not a fact. Gaps in the Theory exist for which there is no evidence. A theory is defined as a well-tested explanation that unifies a broad range of observations.
>
> Intelligent Design is an explanation of the origin of life that differs from Darwin's view. The reference book, *Of Pandas and People*, is available for students who might be interested in gaining an understanding of what Intelligent Design actually involves.
>
> With respect to any theory, students are encouraged to keep an open mind. The school leaves the discussion of the Origins of Life to individual students and their families. As a standards-driven district, class instruction focuses upon preparing students to achieve proficiency on Standards-based assessments.

After the statement had been approved, the press release went out. Among other things, it criticized the media and members of the community for making "confusing, conflicting and inaccurate statements" about the board's actions. The press release detailed the statement that would be read to students, explaining that its purpose was to "provide a balanced view, and not to teach or present religious beliefs." The press release also stated that the new curriculum change and the statement to be read to students had been created "in coordination with the science department teachers."

The science teachers immediately demanded a retraction, pointing out that the science department had provided input so as not to be deemed insubordinate, but had never supported or agreed to the statement approved by the board. Board members, in turn, were outraged by what they saw as defiance by the faculty. "What are they so afraid of?" Buckingham demanded. "Is evolution so weak that it can't stand up to a one-minute statement?"

But the board members were not the only ones who were outraged.

Steven Stough, a Dover resident for twenty years and a middle school science teacher working in another district, had followed the evolution controversy in the newspapers since it began. He avidly reads both local newspapers every day, and he took a special interest in the issue, not only because of his work but because his younger daughter would be starting ninth-grade biology the following year at Dover High. On November 19, 2004, the day the district made its press release, Stough read it on the district's website. Stough was fifty years old at the time, a Republican, not given to knee-jerk reactions, but as he would later recall, that press release "put me over the edge." He had thought all along the whole controversy would fade in time, that cooler heads would prevail. But when he read that statement, and saw what his daughter and all the children of Dover would be facing after the winter break, he simply could not bear standing by and doing nothing. H saw it as an assault on science and an attempt to foist a religious idea onto impressionable young people.

To Stough, it didn't matter if the board had adopted a one-minute statement or a weeklong class: it was wrong. So he called the ACLU. His daughter's school district might be violating her rights, he said. I'm looking for somewhere to turn. Then he left a number and waited for the ACLU to call back. He didn't have to wait long.

Half a mile away, Tammy Kitzmiller had also been stewing about the school board's actions. A young single mother, Kitzmiller was a slight blond woman whose eldest daughter, Megan, had just started ninth grade and would be affected by this new policy in just a few months. Kitzmiller was a naturally shy person, but that didn't mean she would tolerate anyone's overstepping his bounds when it affected her family. She felt it was her job, and hers alone, to discuss with her children matters of religion and spirituality, about whether there was an intelligence behind the creation of life and the universe. She felt quite certain that such discussions did not belong in public school classrooms. "How dare they!" she would rail to her daughters, always assuming that the board members would back down sooner or later. But they never did.

She asked her neighbor if he, too, was upset that the school board was trying to slip what she saw as some sort of New Age creationism into the schools. The conversation did not go well. It would be the first time—but certainly not the last—that someone in Dover called her an atheist.

Soon she, too, called the ACLU. A short time later, Kitzmiller found herself slipping quietly into a law office to discuss her options with an ACLU attorney. On the way in, a couple she knew emerged from the office. Tammy blushed and stammered. The husband said, "Look, you didn't see us, and we didn't see you." She nodded, realizing that this was going to be a hard thing to do.

Bryan and Christy Rehm called, too. Their opposition to the school board's policy was already common knowledge—Bryan had spoken out too many times at school board meetings for them to remain anonymous. Talk of a lawsuit was in the air and in the press from the moment the board voted on its policy. Everyone recognized Bryan—that "atheist teacher"—and assumed he'd be one of those suing. He had switched jobs and now worked at Christy's school in another district in the city of York, so there was no conflict, no reason to hold back. But the decision was a tough one for the Rehms. They would have preferred almost any other course of action. They did not want celebrity. They did not want to be part of legal history. They were both Sunday school teachers and regular churchgoers, yet they knew people would question their faith if they took legal action.

Already there were repercussions. They had pulled out of their driveway, kids loaded in the van, and there on the sidewalk was a crazy old man hopping up and down and scratching under his armpits. He was doing the monkey dance. Bryan and Christy didn't know whether to laugh or ignore him or be afraid.

They talked to almost everyone they loved and cared about: Christy's parents, who lived in Dover, their pastor, their friends, their employer. Their friends and family said: Go for it. And so they became two of eleven plaintiffs prepared to go to court to take on the Dover school board and, by extension, the proponents of intelligent design. Whether they wanted to or not, they would be making history.

And the next time they were treated to the monkey dance, Christy Rehm knew what to do: She laughed, loud and long.

PART II

SURVIVAL
OF THE FITTEST

I BELIEVE THAT GOD IS THE AUTHOR OF ALL
THINGS SEEN AND UNSEEN. . . . GOD IS THE
AUTHOR OF NATURE . . . AND EVOLUTION IS A
NATURAL PROCESS.

—Kenneth Miller, professor of biology,
Brown University, and coauthor of
Biology *(the Dragonfly Book)*

THE CURRENT DEPLORABLE CONDITION OF OUR SCHOOLS
RESULTS IN LARGE PART FROM DENYING THE DIGNITY
OF MAN CREATED IN GOD'S IMAGE. EVEN JUNIOR HIGH
STUDENTS RECOGNIZE THAT, IF THERE IS NO CREATOR, AS
TEXTBOOKS TEACH, THEN THERE IS NO LAWGIVER TO WHOM
THEY MUST ANSWER, AND THEREFORE NO NEED OF A MORAL
LIFESTYLE, MUCH LESS A RESPECT FOR THE LIFE OF THEIR
FELLOWMAN. . . . THIS IS SIMPLY UNACCEPTABLE.

—Foundation of Ethics and Thought,
publisher of Of Pandas and People

IT AIN'T WHAT YOU DON'T KNOW THAT GETS YOU
INTO TROUBLE. IT'S WHAT YOU KNOW FOR SURE
THAT JUST AIN'T SO.

—Mark Twain

Chapter 6

BROKEN WATCHES

Most Americans could not say who Gordon Gould was, or John Logie Baird, or Nikola Tesla, or Edward Jenner, or Jack Kilby, although the work of these scientists is ever present in our daily lives, part of our culture, our entertainment, our communications, our defense, our medical care—our entire modern technology. The world in general and America in particular would be unrecognizable without them, our lives immeasurably poorer and quite probably shorter. Yet most of us couldn't pick them out of a lineup, much less an encyclopedia.

Who are they? Gould invented the first laser, the ubiquitous device that makes possible CD and DVD players, fiber optics, military targeting systems, and bladeless scalpels that perform lifesaving surgeries unimaginable a generation ago. Baird invented television and changed the world, while Tesla powered it, having invented a way to deliver alternating current to home users and our modern electrical infrastructure—the one in which, day after day, without consumers giving it a thought, the lights go on and the toasters toast and the garage doors open and the respirators pump air into the lungs of premature babies. Kilby invented the microchip, another omnipresent device, the silicon brain inside computers and cars and nuclear reactors and iPods and GameBoys—the transitional techno-fossil bridging the mechanical and digital ages. And the Englishman Jenner saved lives and ended terrible scourges by inventing the first vaccination—defeating smallpox and pointing the way for an army of other disease vanquishers.

These great men, these scientists for the ages, dominate our households, but they are not household names. America loves to consume the fruits of science, but we are mostly oblivious of how the stuff works, and of the men and women who discover it for us. Even the giants among them are surprisingly anonymous.

Only three scientists of significance have achieved lasting name recognition among Americans in the twenty-first century, and none of them is directly linked with any of the technological wonders the nation so prizes. There is Einstein, who became a true celebrity notwithstanding the fact that few Americans even remotely understand his science; his fame endures mostly because his name has become a noun, a synonym for "egghead," the ultimate colloquial immortality, and because his crazy-haired photograph is a perennial favorite on posters. There is Newton, remembered not so much for his invention of calculus and physics as for the apocryphal story of an apple falling on his head to inspire the theory of gravity. Or perhaps it is for the venerable cookie that shares his name.

Finally, there is Darwin, the only scientist who has achieved both lasting fame and lasting infamy in America. He has inspired generations of scientists and evangelicals to do battle, and his thinking remains as relevant and provocative today as it was a century and a half ago. His writings, data, and reasoning are still plumbed and studied in the twenty-first century by biologists, paleontologists, and botanists. One of the nation's leading paleontologists and evolutionary biologists, Kevin Padian, a pioneer in studying the evolution of dinosaurs and birds, says that he and his colleagues still find new ideas and insights in Darwin's work. Padian teaches advanced graduate seminars on Darwin at the University of California at Berkeley. He finds that even his top students, already well on their way to becoming accomplished biologists and paleontologists, are "blown away" by what Darwin achieved and how little they really understood of his life, his scientific accomplishments, and the obstacles he faced.

Darwin's influence, of course, extends far beyond the world of science. Alone among scientists, Darwin has inspired continuing political and cultural movements in America—some in support of his views, but for the most part against him and against what he is thought to repre-

sent. Over time he has become an archetype, a mythic figure at once re-
vered and demonized—routinely ranked by scientists as one of the three
or four most important thinkers in history, and just as routinely ranked
with Hitler and Marx in the religious right's roster of evil.

As with most myths, the true Darwin and the nature of his research
have become clouded in the popular imagination, his philosophy mischaracterized, his quotations misused, his shy and gentle personality
vilified, and his science—the seemingly familiar theory of evolution—
mangled beyond recognition. This, too, is part of the story of Dover.

The simple truth is that Charles Darwin did not "invent" evolution;
he didn't even initially use the term, preferring instead the phrase "descent with modification" or the word "transmutation." His seminal
book, *The Origin of Species,* which laid out the basis of his lifework and
which he continued to build on for decades, uses the word "evolved"
exactly once. It is the very last word in the book.

Nor did Darwin think up the notion that life progresses, develops, or
evolves over time. That idea had been around long before he was born
and, in its most general sense of organisms adapting to survive and
thrive, was never particularly controversial. It was the *mechanism* of
evolution—the mysterious process causing it to occur, and the extent to
which a species could or could not be transformed—that got people
riled, challenged conventional wisdom and belief, and became the focus
of Darwin's research.

As for evolution itself, scholars and philosophers had conceived, discarded, and rediscovered the general notion long before the rise of Christianity. Like so many important advances in Western thought and
science, the idea seems originally to have emerged among the ancient
Greeks, particularly the group of thinkers led by Democritus and Epicurus known as the "atomists," who theorized that all living things and
all matter were made of invisibly small particles they called atoms. In
their view, the universe was born through the random and purposeless
combination, interaction, and crashing together of these atoms, a cyclic
and eternal process that did not require divine intervention. By conceiv-
ing the universe as without beginning or end, but instead undergoing

continual cycles of change, the atomists dispensed with the need for a creator. The gods were dismissed as the products of superstition and the all-too-human desire to blame others for misfortune; the atomists preached that personal responsibility, not appeasing false gods, was of ultimate importance. The Roman poet Lucretius advanced these ideas in *On the Nature of Things*, an epic in which he depicts the landscape, plants, and animals transforming and progressing from primitive to more advanced forms across the ages, including humanity's rise from savagery to civilization. The word "evolution" is derived from the Latin *evolutio*—to unroll like a scroll—an apt metaphor for the ancient concept of a nature unfolding.

The atomist point of view would turn out to be an eerily prescient anticipation of modern chemistry, particle physics, and the big bang theory, complete with a suggestion that stars and planets condensed from swirling clouds of cosmic dust. But it fell from favor over time because it postulated an underlying purposelessness to the universe, which other thinkers found repugnant, illogical, and contrary to evidence discernible within nature. Plato and Aristotle, whose ideas arguably have had more lasting impact on Western thought than any other two men in history, rejected the atomist view, arguing that there is direction, order, and purpose to the world and everything in it, and that this order inherent in nature could not possibly arise by chance from chaos. The random mechanism of crashing atoms proposed by the atomists simply could not construct a tree or a diamond or a human heart, these masters of logic and empirical thought contended, and their views prevailed and still carry weight to this day. Aristotle's view that a divine "prime mover" directed the development of the universe and life, coupled with his systematic approach to observing nature and the relationships between different living things, would dominate Western thought, religion, and scientific inquiry for the next twenty centuries (with time out for the Dark Ages and the nearly permanent loss of the ancient Greeks' body of work). It was Aristotle who first described the animal kingdom as a progressive array of life-forms that he categorized from the most primitive (plants and sponges) to the most advanced (humans), with each species possessed of a distinct and unchangeable essence—a description that would eventually be expanded and called in medieval times the "great chain of being."

In the thirteenth century, the most influential Catholic theologian, Thomas Aquinas, adapted Aristotle's empirical way of looking at the universe, along with his arguments for design and purpose in nature, to serve Christian philosophy. Aquinas accepted the notion that life and nature change over time (something previously considered to contradict scripture), but always with the goal of achieving the best possible result, rather than through mere chance. If nature behaved with purpose and made positive progress, it provided evidence of an intelligence that designed and guided it, Aquinas concluded, just as, to use his example, the archer guides the arrow. That intelligence is God, and this argument became the fifth of Aquinas's five famous proofs of God's existence, also called the teleological argument. Variants of Aquinas's Aristotelian views of science and nature were still being taught as state-of-the-art thinking about nature and the world five centuries later when Charles Darwin enrolled at the University of Cambridge.

But the state of the art was in flux at that time. A new age of reason and science had brought about an explosion of knowledge that had begun to challenge old orthodoxies by the time Darwin was born in 1809. In the study of the natural world, those new discoveries centered on geology—which in the nineteenth century included the study not just of rocks but of fossils and the biology of extinct life-forms, what today would be called paleontology.

The revolution began with the work of William Smith, a surveyor for a British canal company, in the late 1700s. Smith noticed at excavation sites that the various layers, or strata, making up the earth's crust were laid out in predictable patterns that could be recognized by observing the distinct marine fossils found in them. The same patterns occurred across the landscape of Britain—the deeper strata contained different-looking fossils, many of them extinct forms of sea life. From this consistent pattern, Smith developed a principle he called "faunal succession." He began obsessively collecting samples to prove his theory, earning the somewhat derisive nickname "Strata Smith." The findings were of immediate practical value—recognizing the patterns helped him plan and construct better canals. But Smith saw something of more lasting value, a clue to Earth's history. In 1815 he used his findings to draw the first modern geological maps of any kind, detailing much of the British terrain and its subterranean layers. His breakthrough ideas would eventually provide the basis for estimating the age of the Earth and would remain useful for the better part of a century, until more definitive radiometric methods were developed. But Smith was a man of humble family and education, and his scholarly "betters" plagiarized his work and claimed credit for his discoveries; he eventually went bankrupt and landed in debtors' prison for several years. More than a decade later, in 1831, he finally received the recognition he deserved from the Geological Society of London, whose president called Smith the "father of English geology."

Long before that overdue acknowledgment, Smith's ideas had opened up new lines of inquiry and thought, as geologists began to recognize that the Earth was not thousands of years old, as long-standing literal interpretations of the Bible suggested, but many millions of years old. There was no other way to explain Earth's many different strata, each with its own distinct (and now extinct) forms of life, and each taking what once seemed inconceivably long periods of time to form. And if the deep strata contained extinct forms of life that had come and gone millions of years earlier, then the biblical story of creation had to be metaphorical, not literal. There was no way around it. Partly in response to these apparent scientific verities, many Christian denominations developed doctrines of biblical inerrancy that did not require literal readings of the text of the Bible.

It helped that almost all these early geologists continued to frame their discoveries within the accepted notion of divine creation and design as driving the development of the Earth and life—even though it took place over a much greater time than had once been supposed. This is how they had been raised and trained to make sense of the world around them, and the appearance of design in life and the operations of nature seemed unquestionable. Charles Lyell, the most influential geologist of the era, produced convincing evidence of the great age of the Earth, but he explained the fossil record by asserting that each species had its "center of creation," and that its qualities were designed for its particular environment. When the environment changed too much, the species became extinct. One of the most powerful scientists of the era, the anatomist Richard Owen—who tutored Queen Victoria's children, gave dinosaurs their name, and acquired the famed fossil of archaeopteryx (now considered the bridge between dinosaur and bird, though

Owen didn't see it that way)—believed that life had an "organizing energy," divinely designed, that controlled its development and evolution. He asserted that each species—horses, humans, spiders—belonged to an immutable archetype. Evolution within the archetype was possible through the organizing energy, leading, for example, to the fossil evidence of horselike creatures evolving over millions of years from dog-sized forest animals to the plains-grazing *equus*. But there could be no crossing over between archetypes. Apes, then, were fundamentally different from men in Owen's view and not related—a "fact" he claimed to have proved with what was later revealed to be false evidence of differences in human and gorilla brains. (A scandal, contributing to Owen's eventual downfall, ensued when the anatomical evidence actually showed remarkable similarities in exactly the structures Owen claimed were different—the gorilla and human brain structure known as the hippocampus.)

An enormously important thinker, who preceded these new scientific discoveries about nature but influenced their interpretation, was philosopher and justice of the peace William Paley. He saw living creatures' abundant adaptations to their environment not as a challenge to faith, but as evidence of God's handiwork. Paley is the true intellectual progenitor of the modern intelligent design movement, as his interpretations of natural structures and processes are virtually the same as those of the twenty-first-century crop of design proponents, albeit without the knowledge of molecular biology. Paley's famous "watchmaker analogy" still provides the heart of the "argument from design" for the existence of God, an extension of Thomas Aquinas's fifth proof. Paley wrote in his 1802 book, *Natural Theology*, that no one who came across a pocket watch lying in a field would ever assume that it was created by chance or random forces—it is clearly a product of an intelligent designer. In considering the exquisitely engineered adaptations of nature, such as a sparrow's wing or an earwig's antennae, Paley argued, it is clear that they, too, had to have been designed to exactly meet the needs of the organism and the demands of its environment. "Every indication of contrivance, every manifestation of design, which existed in the watch, exists in the works of nature," according to Paley, "with the difference, on the side of nature, of being greater or more, and that in a degree which exceeds all computation."

This compelling analogy was not originated by Paley. It had been formulated by the ancient Roman Cicero fifty years before Christ, though the great orator's version was based on the technology of his time, the sundial. The proposition has been restated many times since then over the course of centuries as a powerful argument for God's existence, despite its logical flaws. (The logical conclusion from the analogy would be this: If a watch implies a watchmaker, then a shoe implies a shoemaker and a tin implies a tinsmith; likewise, if there is truly design in a bird's wing, then it implies not an all-powerful creator but simply a bird maker, as snowflakes imply a snow maker, etc. If the analogy is correct, then there would be a multitude of designers, not the one God of Abraham. Modern proponents of intelligent design overcome the logical fallacy of the watchmaker analogy by making no explicit claims about God, instead positing an unknown designer or designers.)

Paley's analogy held sway for much of the nineteenth century; *Natural Theology* was required reading at Cambridge and the young Darwin found it one of the most captivating works he encountered at the university, and a convincing and rational argument for God's existence. His thinking about nature was influenced deeply by Paley's detailed and convincing observations about how the design of each creature seemed so perfectly adapted to its environment—"contrivance" was the word Paley often used, and Darwin would later adopt it for his own observations of nature. This notion of adaptation and nature's contrivance would play an important role in Darwin's subsequent theorizing, and in his eventual decision that Paley had gotten it all wrong.

Charles Darwin was born to privilege in 1809—his maternal grandfather was Josiah Wedgwood, founder of the famous Wedgwood china works, and his father was a prominent physician. His paternal grandfather, Erasmus Darwin, was a doctor and a famous naturalist, who wrote that all life originated from the same "living filament"—an anticipation, in his grandson's words, of the theory of common descent. Exposure to his grandfather's thinking, including this verse from Erasmus's *Temple of Nature*, certainly influenced Darwin:

Organic life beneath the shoreless waves
Was born and nurs'd in ocean's pearly caves;

First forms minute, unseen by spheric glass,
    Move on the mud, or pierce the watery mass;
These, as successive generations bloom,
    New powers acquire and larger limbs assume;
Whence countless groups of vegetation spring,
    And breathing realms of fin and feet and wing.

Charles Darwin was expected to become a physician, at least by his father, but the young Darwin made it clear he was not cut out for a career in medicine. His father eventually relented, sending him to Cambridge to study theology instead. But the study of nature was his passion, and after graduating with a degree in theology, he set off in 1831 for a five-year journey aboard HMS *Beagle*, allowing him to study the geology and natural history of the Galápagos Islands, the Cape Verde Islands, Brazil, and other tropical locales. He rowed onto virgin beaches, crunching through thick layers of gleaming white shells; he tromped through rain forests and took careful notes and sketches of creatures never before studied; and he collected ancient seashells high in the Andes Mountains, evidence that the soaring peaks had once been submerged in an ancient ocean. During the long voyage, he read Lyell's pioneering works on geology and the immense stretches of time required to shape the land and oceans, and he began viewing the life and the fossils he observed and collected in the same way—as shaped over long periods of time, rather than created and designed instantaneously.

Publication of his findings about geology, botany, and zoology during the voyage established Darwin as an eminent geologist and first-rate naturalist on his return, garnering him a respected place in the world of science and academia. His ability to observe and describe nature has long set the standard. But he would spend the next two decades puzzling out the deeper significance of his findings, hesitant to make them public because he knew they would be viewed as controversial, even heretical. In one letter to a friend frequently cited by creationists anxious to discredit Darwin, he joked that he would be perceived as "the devil's chaplain." But though he seemed hesitant to publicize his theories, his convictions remained firm: Darwin's close observations of nature, particularly how similar animal populations living on separate but nearby islands seemed to develop different traits and a different appearance, led him to reject notions of creation, of intelligent design, and of fixed species. Those old ideas, however pleasing, that seemed to explain the apparent design of nature, simply did not hold up in the real world.

Darwin found that if you looked closely enough, nature conveyed a very different message. How could, for instance, the Galápagos Islands serve as home to thirteen separate species of finches, each similar to the other, yet each peculiarly adapted with different-shaped beaks for their particular island habitats? Clearly these finches had migrated over time from the mainland and from one island to another, and then, once separated, had begun to diverge and to become distinct from one another. But how? And why? Why did the giant sloths, whose bones Darwin recovered on his voyage, go extinct, while other creatures thrived in the same environment at the same time? And how was it that some animals seemed poorly designed for their environments, in defiance of Paley's perfect watchmaker—woodpeckers that lived on treeless terrain, land birds with webbed feet—yet they managed to adapt and survive through makeshift means that no divine designer would ever have intended? Why did pythons have vestigial legs, and why did the bones inside the wings of a bat parallel the bones in the human hand and arm? This was evidence not of a master design, Darwin realized, but of a slow and gradual change in existing forms, spread across the ages, inherited from remote—and shared—ancestors. The evidence he painstakingly assembled on his voyage, then presented, bit by bit, in his classic book, pointed to very slow, very gradual changes in living things over millions of years, to creatures suddenly dying out and disappearing when their forms no longer allowed them to survive in a changing climate or environment, and to new forms of life that emerged and thrived in their place.

When he finally published his findings in *The Origin of Species* in 1859, Darwin marshaled reams of evidence to accomplish two proofs. First, he laid out an airtight case that all life varies over time and across generations—in other words, evolution is a fact and notions that species are fixed and immutable are incorrect. Second, the mechanism that brings about evolution is not intelligent or divine, but a gradual, undirected, natural process that Darwin called "natural selection."

The concept as he explained it seemed then, as it does now, deceptively simple yet utterly profound in its implications. It has three parts:

1. Individual living creatures of the same species differ from one another, and those variations arise at random, without a plan or purpose.
2. Variations can be passed on to offspring through heredity.
3. Some variations—a sharper beak, a better ability to grasp, a bigger brain, a longer neck to reach high tree branches and leaves—enhance a creature's ability to adapt and survive in a given environment. Those creatures that are best adapted, then, are most likely to reproduce in greatest numbers and pass on their variations, which become concentrated in subsequent generations.

That's it, in a nutshell: Darwin's entire theory of evolution through natural selection, explainable in a few short sentences. For centuries, this had stared everyone in the face who cared about and observed nature, but no one could see it, so trapped were they by the human tendency, expectation, and desire to find design and purpose in the world—to impose, in other words, human qualities on nonhuman nature. Darwin's brilliance was in seeing beyond the appearance of design, and understanding the purposeless, merciless process of natural selection, of life and death in the wild, and how it culled all but the most successful organisms from the tree of life, thereby creating the illusion that a master intellect had designed the world. But close inspection of the watchlike "perfection" of honeybees' combs or ant trails extending unerringly from nest to food sources reveals that they are a product of random, repetitive, unconscious behaviors, not conscious design. So it is with the gradual transformation of a simple organism into a more complex one, Darwin's theory proposes.

This form of evolution, which Darwin called "descent with modification," suggests that all life on Earth could have begun with one or a few very simple organisms. Across millions of years, Darwin theorized, gradual changes built up and new species arose, just as multiple species of finches arose in the Galápagos. Darwin used as an analogy the artificial selection that animal breeders apply to give rise to new types of dogs and other domestic animals with desirable traits in a fairly short time. Given the much longer period of time natural selection has to work its magic, the changes wrought in the wild are even more dramatic than those animal breeders can accomplish, according to Darwin. And just as breeders do not allow animals with undesirable traits to reproduce, so did nature doom wild creatures with unhelpful variations, causing their numbers to dwindle and finally disappear as fitter animals survive and reproduce.

Darwin did not attempt to address where those original organisms— the common ancestor or ancestors—came from, and his theory is incapable of explaining it, as natural selection requires living creatures to work. This left plenty of room, as he saw it, for the action of a creator, if people wished to believe in one. Darwin wrote near the end of *The Origin of Species*:

> Authors of the highest eminence seem to be fully satisfied with the view that each species has been independently created. To my mind it accords better with what we know of the laws impressed on matter by the Creator, that the production and extinction of the past and present inhabitants of the world should have been due to secondary causes, like those determining the birth and death of the individual. When I view all beings not as special creations, but as the lineal descendants of some few beings which lived long before the first bed of the Silurian system was deposited, they seem to me to become ennobled. . . . And as natural selection works solely by and for the good of each being, all corporeal and mental endowments will tend to progress towards perfection.

Evolution, then, was not in his opinion inherently hostile to religion in general or to Christianity specifically—it was hostile only to a literal reading of scripture and, in particular, of the Old Testament (which was already losing ground, owing to geologic studies of the age of the Earth). It is true that in later life Darwin described himself as an agnostic, though not as an atheist, but he did not come to doubt God and religion because of his scientific research or because of his theory of evolution, as critics of evolution sometimes allege. Darwin habitually quoted scripture during the voyage of the *Beagle,* he was trained to be a pastor, and he considered God the ultimate lawgiver even as he theorized that religion evolved as a tribal survival trait. Rather, he lost faith in God when his ten-year-old daughter, Annie, with whom he had an especially close and affectionate relationship, died in 1851, two years after a bout of scarlet fever. She would come in to visit him in his study as he worked on his then-secret theory of evolution and stroke her shy and reserved father's beard and hair, and he adored her; Darwin never fully recovered from the loss of Annie.

Darwin's theory of evolution by natural selection differs in other important ways from the chaotic, random forces that the ancient Greek atomists proposed, although creationists often accuse Darwin of asserting that life—all the separate species of plants and animals, and humanity itself—arose through random chance alone. What Darwin's theory actually suggests is that while variations in organisms arise randomly, natural selection is not at all random; it consistently "selects" the most useful traits in an organism for reproduction and inheritance, and just as consistently relegates to extinction the variations that are harmful or unhelpful in the fight to survive. This is a very firm pattern, as regular as a snowflake, and it is this pattern that creates an illusion of a conscious, intelligent design in the vast panoply of life on Earth.

The idea may be simple, but its power to shape thinking about the world has been unrivaled. Darwin offers an explanation of how parasites adapt to prey on their hosts, and how those hosts evolve defenses and immunities to parasites; why the teeth of horses, tigers, men, and mice are so superficially different yet fundamentally the same; how the isolation of Australia led to the parallel evolution of marsupial dogs, mice, tigers, and moles with a strange resemblance to their placental counterparts in the rest of the world; why birds of prey have keen vision and cave-dwelling fish have no vision at all; and how the complex and beautiful yet utterly ridiculous (if you believe in intelligent design) waltz of symbiosis allows termites to digest otherwise indigestible wood fiber with the help of protozoa that live in their guts, which are in turn dependent on bacteria to produce essential digestive enzymes.

A theory that explains so much and challenges so many assumptions arrived, as Darwin anticipated, with the force of a bomb: It caused a sensation and enormous controversy. But his careful documentation of the evidence convinced most scientists that evolution was a fact—there was no denying it. Eyes were opened to the evidence that had been in front of the world's scientists all along: species were not immutable; they changed over time. Suddenly it seemed so obvious. Even the staunchest creationist has given Darwin that due (with the lines of battle shifting to the degree of change that is possible).

Darwin's proposal of natural selection as the mechanism of evolution, however, required many more years to overcome widespread skepticism. Its final, widespread acceptance by the scientific community occurred only in light of the new science of genetics, first elucidated in 1865 by the Austrian monk Gregor Mendel with his famous peas, then ignored for nearly four decades before the principles he outlined were rediscovered. With an understanding of genes, about which Darwin (like all his peers) was ignorant, the cause of variation—random genetic mutation—and the principles of inheritance finally became clear. Evolutionists referred to the combination of Darwin's theory and genetic theory as the great "synthesis," on which modern "neo-Darwinian" evolutionary theory is built, a process of merging disciplines that began in the years before the Scopes trial and concluded shortly before World War II.

The discovery of the double helix structure of DNA as the molecular building block and instructional "code" of life in the following decade, and the unraveling of the human and various animal genomes that has followed, contributed vastly to evolutionary theory. This new research has provided additional evidence of the relationships between all living creatures, which seem to share many of the same basic genes and DNA sequences—the codes that determine a creature's body plan and attributes. Darwin's belief in common descent seemed to be ratified by these breakthroughs in the new science of biochemistry and molecular biology. But even with this wealth of new information added to Darwin's original theory, the understanding of the mechanism of evolution through natural selection remained largely the same as when he first explained it.

Although Darwin steered clear of it in *The Origin of Species* in order to avoid predictable outrage from those who believed in man's divine origins, one classic, concrete example of "descent with modification" often cited by evolutionary biologists is the development of modern humans—*Homo sapiens*—from a long line of earlier hominids and apelike ancestors. The chain of what appear to be transitional species is well documented in the fossil record. There was the small-brained *Sahelanthropus tchadensis* from 6 million years ago, with its combination of chimpanzee and human qualities. The powerful, bipedal *Australopithecus afarensis* lived 4 million years ago. The bigger-brained, toolmaking *Homo habilis* arose 2.4 million years ago; skull fossils suggest a brain capable of human speech. *Homo erectus*, master of fire, a species that first appeared 1.8 million years ago and was still living just 300,000 years ago, had a brain nearly as large as a modern human's. Finally, archaic forms of modern humans began to appear a half million years ago. They assumed (more or less) modern characteristics within the last 200,000 years, though more "robust" or apelike facial features and teeth are apparent in skeletal remains as recent as 10,000 years old—evidence of continuing human evolution in the very recent past. It was once theorized that apes and humans diverged from a common ancestor as far back as 30 million years ago, but recent DNA analyses suggest that the ancestral parting of the ways happened no more than 10 million years ago, and possibly within the past 5 million years—a mere blink in the evolutionary history of the world, which stretches back 3 billion years or more.

But how does natural selection bring about such changes? How could tiny, random variations gradually transform an apish animal with no self-awareness, no ability to use fire or tools or to do anything more complicated than forage for fruit and grubs, into a thinking, sentient being capable of wielding a stone knife, painting murals on a cave wall, or writing a sonnet? It defies common sense, yet if Darwin is correct, that's exactly what happened.

One important measure of these ancient hominids is their brain size, which can be deduced from their skulls. When considered as a proportion of overall body weight, brain size is a fairly good rough measure of any species' overall intelligence, based on what we know about living animals' relative intelligence and brain size. (Bottle-nosed dolphins and humans have proportionally the largest brains and are among the most cognitively gifted of creatures; fish have proportionally minuscule brains and are considered, in general, the least intelligent vertebrates.)[1] Over time, the different species of ancient hominids seem to have developed progressively larger brains in proportion to their size, and brain size seemed to correlate with both their increasing ability to use tools and fire, and their increasing resemblance to modern humans in stature and appearance.

Natural selection enters the picture because, while every species has an average brain size, the brain of each individual *Homo habilis* or *Australopithecus* would vary at random. This is apparent in modern humans; our brains can range in size from 1,200 to 1,600 cubic centimeters. A similar spread occurred in the ancient hominids, living in the wild, trying to survive in a difficult environment with hostile weather conditions, fierce predators, and periodically scarce food. If larger-brained individuals had an advantage in intelligence, enabling them to better use a rock as a primitive tool or weapon, or to plan a better hunting strategy, or to succeed in a myriad of other ways in which being smarter poses an advantage, then natural selection would favor larger-brained hominids over their smaller-brained relatives. The larger-brained hominids would survive and reproduce in greater numbers, and their qualities would be concentrated in subsequent generations with the pattern repeated across thousands and millions of years.[2]

Another problem Darwin sought to address in advance of his critics is how natural selection could account for complex organs with many working parts, such as the human eye. Darwin confessed that it seemed impossible for an eye to develop gradually through small steps, primarily because natural selection could act in favor of only a functional eye. A classic creationist challenge, in Darwin's time and to this day, asks the question, "Of what use is half an eye?" The separate parts of the eye—lens, cornea, retina, optic nerve—all have to be present at the same time for the eye to work and for natural selection to favor it. Darwin's system of gradual variations could not cause those parts to materialize all at once—this would be a miracle, and proof that the creationists were right all along. Yet natural selection could not favor the separate parts, because they do not function on their own and therefore confer no survival advantage. This posed a potentially fatal puzzle for Darwin's theory (creationists and intelligent design proponents argue that it *is* fatal).

As Darwin put it, "If it could be demonstrated that any complex organ existed which could not possibly have been formed by numerous, successive slight modifications, my theory would absolutely break down."

In Victorian England, no part of the body seemed more impossibly complex or specialized than the eye, but Darwin said it could, in fact, have arisen by the gradual process of evolution. Pointing to extremely primitive, still living organisms with eyes that are little more than light-sensitive spots, then to animals with slightly more complex visual systems, and so on up to the complex eyes of birds and mammals, Darwin showed how each shared some of the basic parts, adding to and improving them to produce successively better organs. This was proof by analogy, not direct evidence, but it suggested a reasonable evolutionary pathway.[3] Darwin's explanation appeared to be confirmed many years later through a computer program that sought to test the plausibility of eye evolution by natural selection with a virtual organism possessed of a flat, three-layer eyespot similar to the sensory organ on many simple microscopic creatures. The simulation caused a tiny random variation in the curvature of the aperture of this primitive eyespot and in the refraction capabilities of its clear top layer. The variations that improved visual acuity were chosen as aids to "survival" and were passed on to subsequent generations, mimicking natural selection. After a thousand successive generations, the model had created a nearly perfect humanlike eye. By some calculations, this degree of complexity in an eye could be achieved in a mere 364,000 years of evolution in the natural world.

A similar evolutionary puzzle that confronted Darwin is the evolution of flight. How could natural selection account for the development of a bird's wing? The ancestors of birds had to be flightless—in evolutionary theory, wings are a specialized adaptation of limbs originally intended for walking or grasping, which in turn are adaptations of body parts on marine creatures. Since a rudimentary wing stub, the necessary precursor to a functional wing, would not enable a flightless creature to fly, limbs suitable for flying could not be favored by natural selection. So how can Darwin's theory account for functioning wings?

Darwin had an answer for this challenge as well—a principle that more recently has become known as exaptation. Darwin explained that some features could develop to serve a different purpose initially, then be adapted for a new purpose later. A small, feathery flap, for instance, too small for flight, could be used to keep a creature warm in cold weather. A bird frequently tucks its head under a wing when sleeping or when cold; what if this was the original purpose of the structure that eventually became a wing? Natural selection could favor these natural capes and select for larger and more thermally efficient variations. If the creature happened to live in trees, eventually this newly shaped limb could prove useful for jumping and gliding, at which point natural selection would begin to emphasize variations that aided flight. Bird evolution would then be well on its way.

Recently discovered fossil evidence suggests that some flightless dinosaurs—the ancestors of modern birds—had this same head-tucking behavior.

Darwin knew that his theory and—more important—the hard evidence to support it were incomplete, primarily because the fossil record was spotty. Fossils form rarely, and only under precise conditions; more than 99 percent of the creatures that have lived on Earth have become extinct, with only a relative few leaving any trace behind. As Darwin cautioned, "The crust of the earth with its embedded remains must not be looked at as a well-filled museum, but as a poor collection made at hazard and at rare intervals." So, while he offered proof of changes over time in various species, such as the Galápagos finches, he relied on analogies to explain other aspects of his theory—with the understanding that the theory would stand until some important aspect of it was disproved, for example by the discovery of an organ that could not evolve through natural selection. "But I can find no such case," he confidently wrote.

Darwin was content to let his writings speak for themselves, and to continue writing books that examined plant and human evolution, and then the evolution of emotions and human psychology, leaving his supporters in the scientific community to defend his theories. They did so, vociferously and, in the end, successfully. *The Origin of Species* was reprinted many times and was the last great work of science written both for a specialized readership of scientists and for a mass audience. It was widely read throughout the world, lauded, mocked, satirized, and discussed, until it became entrenched in the popular imagination, a cultural touchstone of enlightenment or corrupt materialism, depending on the point of view. The weight of the evidence as Darwin's century drew to a close seemed to scientists to favor evolution and its power to explain so many diverse observations and natural processes. By the time Darwin died in 1882, his ideas were being taught in universities worldwide, replacing the works of William Paley and the advocates of intelligent design.

Darwin had won the first battle. But not the war.

Chapter 7

THE WATCHMAKER RETURNS

Michael Behe never wanted or expected to become a star of any kind, much less an icon of intelligent design (which he believes in wholeheartedly) and creationism (which he does not). The affable, bearded biochemist had been content to teach college classes and putter away in his lab at Lehigh University, located, aptly enough, in Bethlehem, Pennsylvania, just over 100 miles east of Dover. On that verdant, small-town campus he continued the work on nucleotide sequences and rare "backward" DNA known as Z-DNA, which he had first researched at the National Institutes of Health after graduate school. In his spare time at home he tended the garden, enjoyed an occasional beer, and spent time with his wife and nine home-schooled children. (Home schooling was chosen not for the typical religious reasons, Behe says, but because he and his wife enjoy having the children close and are "control freaks"). Behe liked the quiet, anonymous role of professor at a respected but not elite college, the sort of teacher who lectures in a conversational manner, an advocate of scientific plain speaking instead of the more erudite constructions some academics favor; he often describes scientific ideas as "cool" or, when he's particularly excited about something, "very cool."

Like most well-trained scientists educated in the latter half of the twentieth century, he never questioned the wisdom or power of the theory of evolution. He first learned about it as a kid in Harrisburg attending Catholic schools, where the nuns and teachers had no problem at all with Darwin, their attitude being: If God wanted to create life through natural selection, who are we to say otherwise? Later, Behe took the requisite biology courses in college with their ample references to evolutionary theory; but in his particular field, which focuses on the minute world of protein sequences and chemical reactions inside living cells, Darwin wasn't really of much concern to him. Life's origins and the past weren't what he wanted to contemplate; he sought to unravel the here and now inside cells—the amazing chemical factories and microscopic "machines" that seemed almost impossibly complex beneath the microscope. When he thought about evolution at all, he always assumed that other scientists, those who specialized in evolution, had all the answers and evidence well in hand. Wasn't evolution, after all, the single most important theory in the biological sciences? Behe assumed that there was more evidence in support of it than just about any other biological theory. So everyone assumed, Behe laments.

Then he picked up a two-year-old copy of the 1985 book *Evolution: A Theory in Crisis* by the molecular biologist Michael Denton of New Zealand. Behe's eyes were opened, he recalls.

He was, as he said years later, horrified. As Denton presented it, evolution was overblown, exaggerated, and contradictory and its most profound assertions were utterly unsupported by the evidence. Denton argued that natural selection could adequately explain only the process of "microevolution"—small changes within species, such as the evolution of pesticide-resistant insects or the different appearance and capabilities that professional breeders had achieved in dogs and horses.

Where the theory went awry, according to Denton, was in extrapolating from the relatively minor variations of microevolution a grand theory of macroevolution, or common descent, that supposedly could account for the emergence of new and radically different organs, body systems, and whole species. This had never been observed to happen, Denton argued, and there was no evidence that it ever had happened— Darwin had offered clever rhetoric and plausible analogies, but the explanation did not hold up. As far as Denton was concerned, it was "inconceivable" that slow, gradual modifications across generations of living creatures could produce such distinct evolutionary changes as, for example, the highly efficient flow-through lungs of birds, which differ radically from the bellows-type lungs of mammals, reptiles, and amphibians. "Common sense tells me there must be something wrong," according to Denton.

"Ultimately, Darwin's theory implied that all evolution had come about by the interactions of two basic processes, random mutation and natural selection, and it meant that the ends arrived at were entirely the result of a succession of chance events," Denton wrote. "Evolution by natural selection is therefore, in essence, strictly analogous to problem solving by trial and error, and it leads to the immense claim that all the design in the biosphere is ultimately the fortuitous outcome of an entirely blind random process—a giant lottery."

Behe read Denton's book in less than two days, unable to put it down. It was like watching a car wreck, with Darwin at the wheel. Fifteen years later, Behe still sounds stunned by the revelatory power Denton's argument carried for him—Denton, a scientist, was challenging the great Charles Darwin not for religious reasons but for scientific ones. Evolution and the evidence in its support had always been, in Behe's mind, "as plain as the nose on my face. That assumption was shattered for me."

Evolution had been an article of faith, not science, Behe decided. For most scientists to this day, as Behe sees it, evolution is a blind belief, a sort of godless religion.

Denton, however, was roundly criticized by evolutionary biologists, who asserted that his book made a number of fundamental mistakes—including a huge one commonly made by nonscientists: He categorized the combination of variations and natural selection as random. In this he was only half correct. Natural selection may be unconscious but, as Darwin and his successors made clear, it is the opposite of a random force. It can drive changes in an organism in a very linear, persistent fashion—as had been observed in the laboratory, in nature, and in simulations such as the one that modeled eye evolution. Denton was wrong about evolution's being one big lottery. The correct analogy would be a game of darts in which the players cannot see the target. Some darts will find their mark while the majority will miss—a random process. But the rules of the game eliminate all but the best-thrown darts. Because nature tosses an immense number of darts—the mutation rate in any single gene in an organism will run in the millions—natural selection has plenty of well-targeted darts to choose from, and the march toward new and complex forms is not so difficult to understand, after all. But presenting an accurate metaphor would not have supported an attack on evolution.

As for Denton's specific argument about such organs as birds' lungs and his inability to "conceive" of them evolving in small, gradual steps, proponents of evolution mocked it as the "argument from ignorance." By this they meant that Denton's inability to imagine how evolution works is not an argument for anything other than his own limitations— and it is countered by powerful evidence of a primitive version of bird-type lungs found in certain dinosaur fossils believed to be ancestors of modern birds.

Nevertheless, Denton's book had a powerful influence on Behe. With Denton's critique in mind, he turned to the field he knew best, biochemistry, and looked for papers that might offer evidence that evolution governed the microscopic processes he taught, researched, and knew so well. He recalls finding no such papers, certainly nothing that could explain the biochemical evolution of such intricate structures as the eye. In Darwin's day, scientists knew nothing of the complex inner workings of the cell—it was, to use a term Behe especially likes, a "black box," a machine that performs a function without anyone knowing how it works. Because the biochemistry of retinas, lenses, rods, and cones was a black box in 1858, all Darwin could consider was the anatomical structure of the eye, and his argument that simple photosensitive eyespots could gradually evolve through natural selection to more complex forms therefore sounded reasonable, even undeniable.

But fourteen decades later, the complex biochemistry of eyesight had been unraveled. How, Behe began to wonder, could random mutation and natural selection even begin to assemble all that complexity at the molecular level—all those many chemical reactions, triggers, and combinations that biochemists such as Behe often refer to as molecular machines? "Each of the anatomical steps and structures that Darwin thought were so simple actually involves staggeringly complicated biochemical processes that cannot be papered over with rhetoric," Behe would later write in his popular 1996 book *Darwin's Black Box.* "Darwin's metaphorical hops from butte to butte are now revealed in many cases to be huge leaps between carefully tailored machines—instances that would require a helicopter to cross in one trip."

There was no proof that the sweeping changes of macroevolution as Darwin envisioned it had ever occurred, Behe concluded after reading Denton's book, or even that they could occur.[1] Instead of evolution, Behe began to peer at the microscopic world of cell biology and discern something else: design. Intelligent design.

William Paley had been right, Behe decided. Nature did show evidence of design. Paley's problem wasn't that he was mistaken. His problem was that he simply couldn't see inside living things to their microscopic building blocks in order to prove his point.

That would be Michael Behe's job. The evidence was there, he felt sure, that life really was too complex to have evolved without supernatural intervention. All he had to do was prove it.

The opponents of Charles Darwin's "dangerous idea," as the philosopher Daniel C. Dennett has called it in his book of the same name, had been waiting a long time for someone like Michael Behe: a real scientist with no obvious religious ax to grind, a proper academic with legitimate research credentials, who was reasonable, likable, and bright. He wasn't a Bible literalist, though he did believe in God, and he was no anti-evolution extremist: Behe acknowledged that Darwin had been right about a great many things, and that Darwin had been doing the best he could with the technology and data available to him. Behe even conceded that the evidence in living genetic structures made Darwin's idea of common descent with modification likely to be true; all creatures appeared to be related. But Darwin was wrong on one major point, as Behe saw it: the cause of modifications in living creatures could not possibly be natural selection. Anyone who understood how the amazing molecular machines worked inside living cells knew that they had to have been designed, Behe says.

If humans and other creatures share a common ancestor, Behe argues, that is because a designer made it that way. As a scientist, he could conceive of no other explanation. If that proposition had religious implications, so be it. Those implications might please Behe, he says, but they did not sway his scientific judgment one way or another.

In short, Behe was made to order for a new generation of anti-evolution warriors. And theirs had been a long wait. The evolutionists might have lagged behind in American public opinion polls by wide margins, but they had scored one victory after another in the U.S. courts and in the schools since the early 1960s. Previous attempts by anti-evolutionists to counter the teaching of evolution in the nation's public classrooms had failed spectacularly, because they couldn't do what Michael Behe did: separate the science from the religion that inspired it.

Henry Morris, a civil engineer with a preacher's heart, who would later found the Institute for Creation Research in San Diego, blazed the trail for Behe with a first attempt to establish a "creation science." He was a coauthor of the 1961 book *The Genesis Flood*, which argued that the geologic features commonly believed to show the Earth to be billions of years old could actually be best explained scientifically as a result of the biblical flood of Noah's Ark fame. Morris's book sold briskly at first, heralded by supporters as a major blow in the battle against Darwin, but it soon became clear that his conclusions and evidence read more like articles of faith than science. The assumption that the Book of Genesis is both a work of literal history and a scientific guidebook is implicit on every page and in every interpretation of geologic strata. According to Morris, since Noah's flood had to have happened 4,000 years ago and had to have inundated the Earth for forty days and forty nights, interpreting (or, in some cases, simply ignoring) the geologic and fossil record was relatively easy. Whatever agreed with Genesis was correct, whatever failed to mesh with the inspired words of King James's scholars was misunderstood or the work of Satan.

A common thread in Morris's claims—one that still appears in contemporary creationist thought as well as the pronouncements of the Discovery Institute and the Dover school board—is that scientists worldwide are engaged in a vast conspiracy to prop up evolutionary theory and to conceal evidence of divine origins. In the twenty-first century, in Dover and elsewhere, belief in this conspiracy has been reframed to become part of a perceived "war against Christians." To this day, Henry Morris's observation of four decades ago is still quoted by creationists and fundamentalists who view evolutionary theory as an assault on their values as pernicious as bans on school prayer and legalized abortion: "Evolution is . . . anti-Biblical and anti-Christian . . . the pseudo-scientific basis of atheism, agnosticism, socialism, fascism, and numerous other false and dangerous philosophies over the past century."

The problem with Morris's theories, his book, and his whole approach is simple: The Bible is not a scientific document, and so the scientific (as opposed to theological) claims of *The Genesis Flood* were quickly revealed to be not just wrong but laughably wrong. A list of the many specific creation-science claims Morris and his supporters have made that have been disproved would take up dozens of pages, but a few of these false claims include the first member of a new species would have no mate; the Earth would be covered in 182 feet of meteor dust if it were billions, rather than thousands, of years old; the moon is moving away from Earth at a rate of 3.8 centimeters a year, and so it would have begun its existence an impossible few feet above Earth's surface if the world was really billions of years old; archaeopteryx fossils, the bridge between dinosaur and bird, were faked to show feathers that did not exist; and scientists at NASA found a missing day, in keeping with biblical accounts of the sun standing still for Joshua.[2] This last claim, which would be repeated at the presentation "Evolution Is Stupid" in Dover in 2005, is nonsensical, as there is no way for NASA to detect a "missing" day through space exploration or astronomy; days are strictly an earthly phenomenon, based on the time it takes the Earth to complete one full rotation on its axis, and there is no celestial record of past days, no cosmic "tape" to rewind. (Indeed, the fact that a day is defined by the rotation of planet Earth poses a major problem for any literal reading of Genesis that features six twenty-four-hour days of creation: There was no Earth in existence until the third day of creation, according to the Bible, and so no twenty-four-hour day could exist before then. But the ancients who wrote the Old Testament did not know that the Earth was a globe, that it rotated, or that it circled the sun, and so they incorrectly believed that the time period humans call a day was relevant throughout the universe. This is one of many reasons why Genesis, like almost all ancient writings, reflects an understandably flawed description of the world, time, and space.) The list of discredited claims against evolution grew so vast that a leading creationist organization, Answers in Genesis, has published on the Internet a long catalog of these and other creationist assertions that should *not* be used in arguing against evolution.[3]

Despite its flaws, creation science enjoyed brief popularity in some public school systems following the publication of *The Genesis Flood*, and it was widely taught at the discretion of sympathetic individual teachers who were uncomfortable with the new content on evolution that appeared in many textbooks after Sputnik. But creation science went on to lose every significant court battle, and it completely broke down as a state mandate in 1981 in Arkansas. There, a federal district court heard voluminous testimony from leading creation scientists and evolutionary biologists, then struck down a state law mandating that creation science be taught alongside evolution. The decision in the this case—*McLean*— was not appealed and so did not set precedent; but its reasoning, eloquence, and scope, and its finding that creation science was nothing more than a "religious crusade," informed the debate in the courts after that, setting the stage for the decision by the Supreme Court in *Edwards* six years later, which struck down similar balanced-treatment laws nationwide. Henry Morris's *Genesis Flood* could no longer be used in the classroom; other creation science textbooks were also banished.

Yet a steady stream of controversies continued, as the underlying conflict remained as strong as ever. Each side had its victories during the next quarter century, with the evolutionists prevailing in court, but the creationists winning in the court of public opinion and, along the way, gaining political clout as religious fundamentalists became the single most important constituency for the Republican Party.

This situation was apparent in 1982, a year after a judge in Arkansas decided against creation science in *McLean*. In the gubernatorial battle in Arkansas that same year between an incumbent Republican who claimed that his time in the statehouse was "a victory for the Lord," and an up-and-coming Democrat by the name of Bill Clinton, the candidates never argued over whether creationism should be taught in public schools. Their disagreement was on how much creation should be taught.[4]

The following year in St. Louis, a junior high school earth science teacher was forbidden by his school district to show the film *Inherit the*

Wind. He had wanted to introduce to his class in an engaging way the conflicts at issue in the Scopes Monkey Trial. But school officials, keenly aware that many voters in the district were religious conservatives, decided that the 1960 film painted too unflattering a portrait of creationism. Score one for the anti-evolutionists.

In 1984, Texas repealed a ten-year-old regulation that had required all science textbooks to describe evolution as only one possible theory of life's origins—a regulation that, because of Texas's size and proportional impact on the entire textbook industry, had led to very restrictive discussions of evolutionary theory in many nationally distributed textbooks. A few years later, the restriction was imposed again by a more conservative state board. Another win, after a brief setback, for creationism.

In 1985, a state study in California revealed that, while high school textbooks had greatly increased coverage of evolution since the 1960s, publishers had systematically removed mention of evolution or portrayed it in a grossly erroneous fashion in virtually every science text used in middle and elementary school. As a result, the California Department of Education rejected as unsuitable more than twenty middle school science texts published that year, declaring them unfit for millions of schoolchildren in California. This forced the nation's textbook publishers to expand, correct, and update references to Darwin and evolution in future editions. Alabama responded that year by taking the opposite course, banning forty-four different textbooks from its public schools for "promoting" a godless and humanistic "religion"—evolution. This ban was later overturned by a federal appeals court. In the end, 1985 was a good year for the evolutionists' cause.

In 1988, the year after the *Edwards* decision permanently banned creation science from public schools, creationist textbook publishers continued to market successfully tens of thousands of books to public school districts. In these districts, school boards and individual teachers with creationist beliefs skirted the Supreme Court's ruling by introducing the topic without formal policies or curriculum statements. California, meanwhile, moved against a principal producer of such textbooks, Henry Morris's Institute for Creation Research, banning its creationist university from conferring master's degrees in science, and ordering the school to limit degrees to religion or creation.

In 1990, Ralph Reed's Christian Coalition funded and helped organize hundreds of fundamentalist candidates in local elections, with great success. In San Diego County, California, alone, sixty of eighty-eight religious-right candidates supported by the Christian Coalition won. In 1993, one such slate of candidates, which won a majority on the school board in Vista, California, ordered teachers to include scientific criticism of evolution in biology classes, and authorized the teaching of creationism in history and literature classes. The policy polarized the town for the next year. Every school board meeting became a confrontation between religious and secular, conservative and moderate; one side would shout the word "God" during the Pledge of Allegiance, while the other side, peppered with scientists from nearby universities and research institutions, would wave fossils to tweak young-Earth creationists. In 1994, voters threw the fundamentalist majority out of office, and their policies went with them.

In 1996, Pope John Paul II dealt creationists a blow when he released a formal Vatican position paper stating that the human body might not, after all, be an immediate creation of God, but the product of a gradual evolutionary process. "Fresh knowledge leads to recognition of the theory of evolution as more than just a hypothesis," the Pontifical Academy of Sciences informed the world. This was a first. The Vatican had never opposed evolutionary theory outright, but there had been a history of discomfort with Darwin; in 1950 a papal encyclical had warned that the theory could easily be used by atheists and materialists who sought to remove God from creation. Now, however, the pope had deemed evolution to be no threat to faith, as the critical element of creation was that God created the human soul, not the human body, in his image. In other words, science and religion occupied two separate spheres; in the pope's view, Darwin might have explained where the human body came from, but that had nothing to do with the spiritual and divine aspects of human existence.

The following year, this victory for evolutionists was offset by a major victory for creationists: the National Association of Biology Teachers, under pressure from religious fundamentalists, altered its formal platform on teaching evolution. The policy had previously read: "The diversity of life on earth is the outcome of evolution: an unsupervised, impersonal, unpredictable, and natural process." Although scientifically accurate, the language seemed provocative to fundamentalists, who considered it atheistic; the association agreed to drop the words "impersonal" and "unsupervised."

That same year proved a watershed for intelligent design, which entered the public consciousness in a big way, featured in the nationally televised debate on the Public Broadcasting System show *Firing Line*, followed by extensive publicity. Also, a "Book of the Year" award was given by the evangelical magazine *Christianity Today* to a new best seller about intelligent design: *Darwin's Black Box,* by Michael Behe.

When Michael Behe first came to the conclusion that there was a crisis in evolutionary theory and he began considering this new idea of intelligent design, he kept it quiet until he had secured tenure at Lehigh University. "This subject is radioactive," Behe says ruefully. "If you don't have tenure, stay away, because you will be eaten alive."

By then, Behe had the germ of the idea he would eventually call irreducible complexity—a restatement of the watchmaker analogy at the molecular level. He had several "irreducibly complex" examples in mind, but his "Exhibit A" was the bacterial flagellum, a complex whiplike appendage that operates like a small motor to propel simple microscopic creatures through water. Under high magnification, it resembles a machine, with parts that look roughly like wheels and flanges. The flagellum is a standard feature in biochemistry texts, presented with drawings that depict these complex cells as actual machines with gears and driveshafts—more a matter of artistic license to illustrate a point than an accurate rendering of what these "machines" look like under a microscope. What the machine actually consists of is a complex series of separate chemical components that must be assembled in total for the system to function. If any one of the chemical substances is removed, then the flagellum doesn't work. Therefore, Behe argues, none of the individual pieces offers an adaptive advantage for natural selection to act upon, and no gradual, Darwinian assembly of the flagellum could ever take place. It was the perfect illustration of both Paley's age-old notion and Denton's New Age criticism, Behe decided. He likened it to an old-fashioned spring-loaded mousetrap: Take away any one part and the trap won't work.

Irreducible complexity was the silver bullet anti-evolutionists had been waiting for, the answer to Darwin's statement: "If it could be demonstrated that any complex organ existed which could not possibly have been formed by numerous, successive slight modifications, my theory would absolutely break down." Behe believed he had found the monkey wrench that would break down Darwin's theory.

He was not sure what to do with this idea, or if it represented something truly powerful and original, until he read Phillip Johnson's *Darwin on Trial* in 1991, and then wrote a letter to the editor of *Science* magazine to complain about a dismissive and, in Behe's view, unfair review of Johnson's book in that prestigious scientific journal. The letter was published, and a short time later Behe received a thank-you note from Johnson. They began a correspondence that culminated when Johnson invited Behe to the retreat at the Pajaro Dunes resort where intelligent design shifted from a vague idea among disaffected academics to a movement with a plan—and money. Behe recalls strolling on the warm sands with Howard and Roberta Ahmanson, the philanthropist couple from Orange County with a long history of underwriting conservative religious causes. Irreducible complexity was a big hit at the gathering, and Behe found himself thinking, for the first time, that he actually might have something new to say, something no one had thought of before.

In 1996, Behe's book appeared, selling more than 200,000 copies. It was roundly panned by the scientific community, which not only ripped into Behe's ideas but criticized him for going to the general public with them rather than first publishing in a peer-reviewed science journal.

But Behe was unrepentant, arguing that he could never get a fair hearing from a scientific community unwilling to reconsider its mistaken embrace of all things Darwinian. And he went further than merely arguing that irreducible complexity required an intelligent designer. He asserted that fossil evidence, analogies drawn from living creatures, and supposed transitional forms found in the fossil record—all the evidence Darwin and his proponents relied on to support evolutionary theory—proved nothing about evolution. Unless it could be documented at the molecular level, in the biochemical processes of life, there was no proof that macroevolution had ever even occurred.

"Where's the evidence?" Behe asks. "It's just not there."

Behe's work created a sensation with extensive press coverage portraying him as posing a new challenge to Darwinian orthodoxy. In 2000, he joined a delegation from the Discovery Institute, led by Phillip Johnson, that went to Congress to brief Republican leaders for more than three hours, with more than fifty senators and congressmen in attendance. While Behe explained his scientific attack on evolution, and described how Darwin's "creaking" nineteenth-century rhetoric could not begin to explain the complexity now detected in nature, Johnson told the receptive gathering of national leaders that scientists had become a secular priesthood, jealously safeguarding their secular "creation story"— evolution. They have ousted God from science, Johnson complained, and the only cure amounts to ousting the scientific priesthood. Senator Sam Brownback, the Kansas conservative, whose home state was in the midst of its own evolution war at the time, likened this battle to John Brown's battle to abolish slavery.

Shortly after the meeting, Senator Rick Santorum of Pennsylvania offered an amendment to the No Child Left Behind Act that called for teaching the gaps in evolution and such alternatives as intelligent design. Johnson had suggested the wording Santorum used. Santorum's addition was ultimately removed from the law itself, but it was included in a legislative report outlining the Senate's goals, and proponents of intelligent design touted the language as a major victory. "The Darwinian monopoly on public science education, and perhaps on the biological sciences in general, is ending," the Discovery Institute exulted in a press release.

The rise of intelligent design had an immediate impact—an acceleration in challenges to evolution around the country because at last there was another option besides biblical creationism. Behe's work and Johnson's rhetoric, along with the other efforts by the intelligent design movement, helped spark a new round of evolution wars—battles fought less in the halls of academe than in the court of public opinion, where polls had shown that the proponents of ID held sway. And so teachers in Alabama had to pass out textbooks with a state-mandated disclaimer stamped on them: "Evolution is a controversial theory some scientists present as scientific explanation for the origin of living things, such as plants, animals and humans. No one was present when life first appeared on earth. Therefore, any statement about life's origins should be considered as theory, not fact." A principal in Louisville, Kentucky, meanwhile, announced over the public address system at his high school that Darwinism had been disproved by new scientific evidence and that evolution no longer would be taught. A survey of biology teachers in Louisiana found that a quarter of them believed in creationism and nearly a third thought it an appropriate subject for high school biology courses. Tenth-grade students in a public school in Oregon told their biology teacher he was the devil incarnate for trying to "convert" them to a belief in evolution. The Institute for Creation Research could no longer keep up with the demand for its "young Earth" raft trips through the Grand Canyon, featured in the slick full-color guide "Grand Canyon: Monument to Catastrophe," explaining how Noah's flood had formed the canyon. The Republican Party's platforms in at least seven states— Texas, Kansas, Oklahoma, Oregon, Missouri, Iowa, and Alaska—called for teaching challenges to evolutionary theory, effectively making opposition to evolution a political litmus test for Republican candidates. No other major political party in the world has done this.

By 2001, nearly half the states began considering legislation mandating changes in how schools handled—and questioned—the science of evolution and its new competitor, ID. A benchmark of sorts was taken from a pro-evolution perspective, and it suggested that evolutionists had a big problem on their hands: The gains of the 1960s in the teaching of evolution had eroded in American public schools. A national study by the Thomas Fordham Foundation found that more than a third of states largely ignored evolution or actively undermined students' understanding of it in kindergarten through twelfth grade. The foundation, which asserted that evolutionary theory was a critical part of any solid science education, reviewed every state's standards for teaching science and its specific references to evolution and gave an unsatisfactory grade—D or worse—to nineteen of them. Twelve states received an F for science standards that omitted evolution or contained only "useless" references to it; most of these states avoided even the word "evolution." Kansas ranked at the bottom, according to the report; it was described as "disgraceful" and given a grade of F-minus for "shunning biological evolution while also deleting all references, direct or indirect, to the age of the earth or the universe, including even radioactive decay."[5] Even the big bang was banished from Kansas as unworthy of being taught.

For the first time, the U.S. government also entered the evolution wars under the leadership of President George W. Bush. While campaigning for office, the new president had expressed a belief that creationism should be taught in public schools, and his Justice Department reflected his philosophy by creating a new "religious discrimination" legal team whose charter seemed to blur the line between church and state. Its cases were unlike any the government had championed before. When the Salvation Army was accused of imposing a religious test on employees in publicly funded aid programs, the Justice Department unit got involved, too—to defend the Salvation Army's right to discriminate. The unit also attacked cities that refused to give zoning preferences to religious organizations; it investigated a school in Plano, Texas, for barring a fourth-grader from giving out candy canes with religious messages on campus; and it threatened to sue a community college in Florida until it lifted a ban of R-rated movies that had prevented on-campus showings of Mel Gibson's ultraviolent religious film *The Passion of the Christ.*

The religious rights unit's most sensational case, however, put it in the middle of the evolution wars—and not on the side of science. Federal prosecutors began an investigation of a biology professor in Texas for requiring students to embrace a scientific theory of human origins in order to get a letter of recommendation. This was not a university grading policy or a restriction on admissions, but simply one individual professor's personal policy for the voluntary practice of writing recommendations. Yet when a creationist student with ambitions to become a doctor filed a complaint about the recommendation policy, the Justice Department unit used its investigative powers and the threat of a lawsuit to pressure Professor Michael Dini of Texas Tech University into changing his approach. Now he simply requires students to "explain" evolutionary theory. Previously he had required students who wanted his recommendation to "truthfully and forthrightly affirm a scientific answer" to the question of human origins. Ralph F. Boyd, Jr., the head of the civil rights division of the Justice Department, a man charged with battling racial injustice, police brutality, employment discrimination, and hate crimes, was sufficiently outraged to get involved personally in the Dini case on behalf of the student, arguing that the professor's policy amounted to a public university dictating religious beliefs.

"A state-run university has no business telling students what they should or should not believe in," Boyd said. "If the separation of church and state is to mean anything, it must surely mean that such matters of conscience are beyond the reach of government inquiry."

This was an extraordinary and aggressively "pro-faith" position for the federal government to take; it equated evolutionary theory with a religious belief and cast the theory specifically as an enemy of Christianity. In short, the Justice Department of the United States under George Bush had adopted a classic creationist philosophy and a primary principle of the wedge strategy: that evolution was inherently hostile to religion. And the department clearly indicated that it would use its powers and resources to enforce that view, even if it meant compelling a professor to write letters of recommendation for Christian students who rejected evolution and embraced what the professor considered pseudo-science.

By then, the evolution war had become part of a larger political campaign: the Republican leadership's controversial "Culture of Life," in which religion, politics, and government at times seemed all but indistinguishable. Abortion rights (and death threats against abortionists), the extraordinary congressional intervention in Terri Schiavo's right-to-die case (and the death threats against her husband and the judge who sought to remove life support for the vegetative Schiavo), opposition to embryonic stem-cell research (and death threats against researchers who advocate it), and other fundamentalist casus belli were knit together in the sometimes incendiary "Culture of Life." The term was first coined by Pope John Paul II as a precept of the Catholic Church that included opposition to the death penalty, but it was appropriated by President Bush and in 2004 was made a part of the Republican Party's platform (minus the objections to capital punishment). Republican leaders, including the disgraced former House Majority Leader Tom DeLay of Texas, made opposition to evolutionary theory a major theme of the Culture of Life, expressing the ardent belief that teaching evolution was a root cause of immorality, materialism, and godlessness. Descendants of animals have no qualms about aborting their unborn or ending Terri Schiavo's sad life, this line of reasoning went, whereas the children of God could never countenance such acts. In this view, the theory of evolution became the devil's greatest weapon—and a political bludgeon to wield against liberals, judges, teachers' unions, and others. It is no surprise, then, that proponents of evolution and opponents of creationism in classrooms in Dover and elsewhere in the country, like the other perceived enemies of the "Culture of Life," have also received death threats. When even the Justice Department investigates scientists, academics, and teachers for declaiming about evolution, it would seem that the time for a new paradigm—perhaps one such as intelligent design—has arrived.

Scientists, long the heroes of popular culture, the swashbuckling adventurers in the science-fiction novels Michael Behe devoured as a kid, were suddenly recast as the bad guys, the truth hiders, the God killers—the evil priesthood of Phillip Johnson's imagination. In one of the most extraordinary moments in the history of the U.S. Senate, Americans watched the majority leader, Dr. Bill Frist, diagnose the comatose Terri Schiavo by examining a brief video that had been edited in the most biased and misleading way possible, gainsaying the word of her own doctors and substituting his own opinion that she had a real chance of recovery, when in fact she had none. If a man like Frist, a physician, had no faith in his fellow men of science, if he believed that they were so evil and so blind that they would end a life unnecessarily, why should anyone else have faith in science? Why believe them when they say evolution is the bulwark of all modern biology and medical research, from new antibiotics to the fight against cancer? Just ask Tom DeLay: "Give me one example that proves evolution. One example! You can't." And why not doubt global warming, too? After all, Senator James Inhofe of Oklahoma, the Christian dominionist who chaired the Senate's environmental committee and who believes America should be transformed into a Bible-based theocracy, assured the nation, "Man-made global warming is the greatest hoax ever perpetrated on the American people." Maybe it was true, also, that vaccinations really hurt kids more than they help—hadn't the chair of the House Government Reform Committee, Dan Burton of Indiana, said so, holding four years of hearings to challenge the wisdom of mandatory inoculations? Maybe HIV really doesn't cause AIDs, as Phillip Johnson has argued. Maybe the Earth really is only 6,000 years old and scientists are conspiring to cover up this information, as Kent Hovind and a battalion of like-minded lecturers have insisted for decades. The senior senator from Kansas agrees with him, as do many of his colleagues—educated, powerful men, who say that evolution is nonsense and scientists are lying. They say there is a "war on Christians," and they are merely coming to the rescue of the faithful, but in their public statements these national leaders seem to be talking more about a war on science.

By 2005, it no longer mattered if Behe's irreducible complexity was a valid scientific alternative, or a shibboleth drawn from Aristotle, William Paley, and old creationist tracts citing the bacterial flagellum as evidence of God years before Behe came along. Whatever the truth of the matter, Michael Behe's scientific critique of Darwin had already made its contribution to the wedge strategy. It had helped spark the battle in Dover, skirmishes across the nation, and what looked to be a looming, all-out war on the practice of science itself.

Chapter 8

THE WATERS OF KANSAS PART

By 2005, numerous fronts had opened up in the new evolution wars. There was a battle in Cobb County, Georgia, over a book sticker reading "Evolution is a theory, not a fact"; this case was still on appeal. Citizens had gone to war over curriculum standards in Ohio (another "F" grade from the Fordham Foundation). Proposals worked their way through legislatures in twenty-one states to limit evolution, criticize Darwin, or open up classes to alternatives such as intelligent design.

Even with this flurry of activity, however, the opposing camps understood in 2005 that there were two main fronts in the war, two clashes that would in all probability decide the future course for evolution and design in America, either one of which, if not both, could end up on a path to the U.S. Supreme Court. There was Dover, and there was Kansas.

The issues were basically the same in each place, but the fields of battle were distinctly different. In Kansas, unlike Dover, the conflict would not move to the neutral ground of courtroom. Instead, it would be fought (at least in the opening rounds) in the political arena of the elected Kansas Board of Education, where one side controlled a majority of votes, and therefore appeared to dictate the rules of engagement, if not the eventual outcome. In Pennsylvania, neither side would have a tactical advantage with the judge when it came time to stride into court and lay the evidence on the table. In Kansas, however, the politics of the moment meant that the anti-evolutionists and creationists held the upper hand.

And the fighting in Kansas would come first.

———

There is a large geologic formation beneath the broad, flat plains of Kansas, a curiosity that stretches many miles north, well into Canada, called the Niobrara Chalk. Mostly it lies buried deep in the Earth. But in the badlands of western Kansas, scattered through the Smoky Hills, lumps and swaths of the Niobrara lie exposed, the soil and rocks that once covered the great chalk beds blown and scratched and washed away by nature's tides across the Great Plains. Massive crumbly pyramids and towers of chalk rise up from the plain, stark monuments to ancient life long departed. The formation reveals the secret of Kansas's past, for chalk beds are formed in only one way—slowly, across the ages, always in shallow waters, the result of the slow accumulation of calcite in layer after layer, the fallen shells of countless single-cell marine organisms. Chalk is a thing of "deep time," to use the evocative phrase from John McPhee's fine 1981 book *Basin and Range.*

Kansas is as flat as a griddle for a reason. Long ago it was a seafloor, 600 feet below the surface of a tropical ocean that warmed and covered the Great Plains, dividing North America in half, east and west. The sea is long gone, but it left its legacy beneath the fields of wheat and corn, the Niobrara. This Kansas sea is much younger than the swampland that once covered Dover, and the shallow waters of the Midwest were filled with giant, fearsome, wondrous creatures. This ocean was known not for the tiny, harmless trilobites of Pennsylvania, but for giant prehistoric sharks, snakelike fanged mosasaurs, and thirty-foot behemoths with six-foot skulls filled with jagged teeth—the deadly *Tylosaurus*—their fossilized skeletons still swimming through the chalks, some of the most spectacular marine fossils in the world. Hundreds of them were uncovered, sold, and displayed during the "bone wars" of the nineteenth century, a time when Kansas became one of the hottest spots in the world for fossil hunters and a geologic wonder for scientists. In Hays, Kansas, the Sternberg Museum of Natural History has many of these monsters on display, along with a re-creation of their long-ago shoreline with models of the gigantic inhabitants in their habitat—ancient land reptiles that had returned to the sea and then adapted, with limbs initially designed for land evolving into a swimmer's paddle-like appendages. The museum informs the public that these sea creatures lived some

70 million to 85 million years ago, during the Cretaceous period. The Niobrara is presented as a vast laboratory for evolution and as evidence of a very, very old and very changeable Earth.

In August 1999, the state of Kansas decided that the Niobrara Chalk and all its fossils simply did not exist—at least, not as the vast majority of scientists and a beloved museum conceived of them. In that year the state's independent, elected board of education adopted science teaching standards that conspicuously omitted most references to evolution or other principles of cosmology, biology, and geology that might contradict creationist views or that suggested the Earth might be older than 10,000 years. It didn't ban evolutionary theory from the classroom—in Kansas, school standards are voluntary and local school districts have the final say—but the standards do dictate what will be included on annual statewide tests. In the new standardized test-driven environment of public schools, where measures of achievement are critical to budgets, promotions, and enrollments, and where the label "failing school" can be one bad test score away, removing a subject from the standards almost guarantees it will be given short shrift in the classroom, if not dropped altogether.

The reaction was instantaneous. Within a month, the Nevada-based publisher of a new textbook for seventh- and eighth-graders in Kansas— *Kansas: The Prairie Spirit Lives*—deleted an entire chapter on geology, natural history, and paleontology, omitting any reference to Kansas's ancient sea, the fossils recovered from it, and the mosasaurs so proudly displayed at the Sternberg museum, a destination of school field trips for generations. "We want to put out a better book . . . but we don't want schools to reject it," the head of the publishing company told reporters. Critics of the new standards described this as the scientific equivalent of Holocaust denial.

The very meaning of the word "science" was altered that year in Kansas. Once, it had been defined by the board's own science curriculum experts as "the human activity of seeking natural explanations for what we observe in the world around us." The newly adopted standards altered that definition to read "seeking *logical* explanations." The creationists who pushed through this seemingly trivial alteration knew that it was potentially the most momentous change of all: It opened the door to supernatural explanations as the most "logical."

It opened the door—the public school door—to miracles, to purpose and design in the universe, and to God.

Under those standards, if a science teacher in Kansas believed that the best explanation for the Niobrara Chalk and the extinct animals fossilized within it lies in forty days and forty nights of rain from the time of Noah, then that teacher—or school or school district—not only was free to instruct public school students accordingly, but would be justified in feeling duty-bound to do so. This is something that a majority of scientists adamantly oppose, but that the American public, in large numbers, considers a good idea. The state board of education also altered the definition of a scientific theory (not just evolution, but any theory) from "a well-substantiated explanation of some aspects of the natural world" to simply "an explanation of some aspects of the natural world." The nation's scientists overwhelmingly considered evolution to be well substantiated, but Kansas would make it clear that it most decidedly did not agree.

Despite the strong creationist and fundamentalists sentiments of large numbers of Kansans, their state came surprisingly late to the evolution wars.

Before 1999, there were no pitched battles about Darwin or creationism here, and no big lawsuits or controversial legislation as in nearby states. Instead, other hot-button cultural issues dominated—sex education, prayer, an emphasis on local control. In Kansas the most contentious political battles are not between Republicans and Democrats, but between the moderate and conservative wings of the state Republican Party, and these factions quarreled over even the most minor school issues. Even the use of calculators in math classes led to a knock-down, drag-out fight about math literacy and technology in the classroom—a silly yet ultimately telling clash between the traditional and the modern. When the state science standards came up for their regular quadrennial review, with a mandate to bring them up-to-date for the needs of a new millennium, a similar clash over evolution became inevitable.

The school board of ten elected officials, each representing a geographic district of the state, first established a committee of twenty-seven teachers and scientists to write state-of-the-art science standards for the board. They needed a committee of experts to do this because the board members in Kansas, as is typical nationwide, are not required to possess any particular qualifications beyond living in the district that elects them. They are not scientists, and most are not educators. So the committee was to do the heavy lifting of researching and crafting science standards, then the board was supposed to provide "citizen oversight." The timing of rewriting the standards coincided with a national movement toward a more standards-based education in all public schools begun during the first President Bush's term in office. The idea was to create a well-defined path for each grade that (unlike the old standards) focused less on what teachers should "cover" and instead specified the knowledge and skills each student should master. The new approach was intended to provide a measurable, testable set of benchmarks to gauge students' achievement. An eminent panel of Kansan educators and scientists recommended to the board in 1999 what appeared to be a solid set of standards, grounded in mainstream science and based on a model curriculum supported by the National Academy of Science.

But statewide elections a few years earlier, during the tumultuous years of Republican ascendancy nationally and the party's "Contract With America," had brought an extremely conservative majority to the Kansas Board of Education. Steve Abrams, a veterinarian in Arkansas City—he proudly describes his small-town practice as including "dogs, cats, pigs, cows, horses, everything but skunks"—was one of these conservative members. He came to office without a thought about evolution, or science education in general, but with a huge concern about literacy, kids' lack of interest in reading, and the atrociously written job applications he'd get from high school and even college graduates interested in working at his veterinary practice. His focus shifted when he first read the science committee's proposed curriculum standards. He recalls being surprised, even taken aback, at how thoroughly "dogmatic" they seemed to him in presenting the theory of evolution. "It wasn't presented as a theory," the big, bluff animal doctor with the booming voice and thick mustache recalls. "It was presented as fact. . . . Well, having had quite a bit of science background, I know science is never fact. . . . Science is provisional."

Abrams also happens to be a young-Earth creationist who accepts the literal biblical account of cosmic and biological origins as a matter of faith. That belief, he says, is his "bias"—something he readily discusses, in part because he thinks everyone ought to acknowledge having a bias, particularly when it comes to the topic of origins and evolution. Those proposed science standards reflected a bias as well, Abrams asserts—a bias that, in his estimation, leads evolutionists to see facts and hard evidence where there really are none. That said, Abrams insists he would never try to block or modify science standards for public schools because of his understanding of Christian scriptures. Throughout the long debate in 1999, he repeatedly argued that his objections to the proposed science standards were strictly scientific. His stated goal was to make certain that all the science taught to students in Kansas—whether it was the theory of gravity or electromagnetism or evolution—adhered to five basic and universal principles: "It has to be observable, measurable, testable, repeatable, and falsifiable." That's been Abrams's mantra for years, he says, and the concern it reflects is about science, not religion.

"Evolution is not good science," Abrams told the press in 1999. "And as such, we don't believe it should be presented."

As reasonable and scientific as Abrams's "five principles" sounded, they created quite a stir among a number of teachers, scientists, and voters, who formed a group called Kansas Citizens for Science to provide organized opposition to his approach. A strict application of Abrams's five principles, they said, would make impossible whole bodies of scientific thought and theory. Scientists can't observe the formation of the continents hundreds of millions of years ago; nor can they repeat those events. Yet there are solid, tested, widely accepted geologic theories on how the continents formed—just one of many examples of "historical" science that do not fit neatly with Abrams's views. Interpreted as narrowly as possible, these principles left much of evolutionary theory subject to challenge by Abrams and like-minded critics of Darwin, while many alternative explanations, such as intelligent design or creation science, could well qualify as suitable subjects for the classroom. But to open up the classrooms to such new ideas, Abrams and his allies on the board knew they also had to excavate one significant "bias" from the school standards—they had to define science as a search for "logical" explanations, rather than "natural" ones.

So Abrams led the charge to dump the proposed standards recom-

mended by the science committee in favor of the anti-evolution standards ultimately adopted by the board in Kansas during the summer of
1999. These were drafted in secret by a group of citizen volunteers dominated by a local creationist organization, were later reworked by Abrams
to attract sufficient votes, then introduced without warning in the midst
of a board meeting.

The standards ultimately adopted contained a long list of creationist
principles couched in the numbingly bureaucratic language of curriculum guidelines, though their provenance was undeniable. When the
standards addressed evolution, it was to contend, as creationists have
long argued, that evolutionary theory adequately explains only small
changes and adaptations—microevolution. The school board decided
that the emergence of new species and the theory of common descent,
the heart of evolutionary theory, would not be addressed at all. The standards went on to contradict a mainstay of evolutionary theory by asserting, "Natural selection can maintain or deplete genetic variation but
does not add new information to the existing genetic code." No new information meant no new types of animals, no new abilities, no evolution
of wings or flight. If this one statement were true, the root of Darwin's
theory—the literal "origin of species" he highlighted in the title of his
book—could never happen. Common descent and all the science built
around it would have been proved false. This is one of the main arguments of the Discovery Institute and the wedge strategy, as well as a
frequent claim of creationists.

The new standards suggested a path to the truth for students. Schools
were told they could engage in numerous student activities in line with
creationist beliefs, such as exploring the weaknesses in popular conceptions of dinosaurs—specifically, how long ago they lived and what made
them become extinct. The creationist belief in Noah's flood as the cause
of the dinosaurs' extinction in the recent past was not mentioned explicitly, but observers on both sides of the issue interpreted this part of the
standards as an invitation to consider creation science and the notion
that man and dinosaur walked the Earth at the same time, not many
millions of years apart, as mainstream science holds. Another part of the
standards invited students and teachers to explore alternative explanations for the geologic record, such as the hypothesis that some rock strata

were "laid down quickly." This is a reference to the geologic theories first put forth in 1961 in the creationist Henry Morris's book *The Genesis Flood,* and the standards invited students to consider how the Grand Canyon might have been formed by a sudden cataclysm, such as a volcanic explosion—a theory that young-Earth creationists have long advanced, but that mainstream scientists (and the National Park Service, which operates Grand Canyon National Park) consider ridiculous.

Unsurprisingly, the nation's leading scientific organizations and members of the science faculties at the major universities in Kansas condemned the standards as perpetrating a lie. But the board majority seemed to take strength from such criticism, using it to continue to argue that evolutionists—or, more often, "Darwinists," a term that in some quarters in Kansas was just as bad as calling someone a "Marxist" or "abortionist"—were hopelessly biased and untrustworthy on such important matters.

There was a fail-safe: Unlike most states, the curriculum standards adopted by the Kansas state board would remain voluntary. Individual schools and districts could choose to teach evolution or its alternatives as they saw fit. And because the main objection scientists had to the new standards was the deletion of key concepts (common descent, the big bang, evidence of a 4.5-billion-year-old Earth) rather than an overt mandate to teach creationism, a direct constitutional challenge, as in *Edwards*, the balanced-treatment case in Arkansas a dozen years earlier, seemed unlikely. Of course, any school district that adopted the standards and added a creationist program to its science courses would be at risk of a lawsuit.

Kansas was roundly criticized for "dumbing down" its science standards in 1999, and not just through the "disgraceful" grade from the Fordham Foundation. The state's Republican governor suggested that the school board ought to be abolished, and he called the board's action "a terrible, tragic, embarrassing solution to a problem that did not exist." The lieutenant governor complained that the state would lose high-tech businesses and jobs because the Board of Education was earning a reputation for being antiscience. The chancellor of the University of Kansas predicted that National Merit scholars would go elsewhere. The National Academy of Sciences expressed its disapproval and issued a statement to educators nationwide, cautioning them that the inclusion of creationist (or any religious) teachings in science classes "compromises the objectives of public education."[1] *The New York Times*, meanwhile, editorialized, "The real losers here will be the very schoolchildren the board members thought they were protecting." Public reaction in Kansas was decidedly mixed, with a substantial measure of confusion in the middle. In an interview on National Public Radio, a Kansas high school student summed up the generally sour sentiment—and presented a sobering reminder that both sides had difficulty getting their message out: "No one was there that's still alive today that actually witnessed creation or evolution. It's just what a person believes. I mean, we have no right to say what exactly is true."

After the creationist standards were adopted in 1999, the divisiveness of this issue and the sudden interest of major news organizations put the Kansas Board of Education in the public eye as never before. The once-sleepy board elections became high-spending affairs as moderates and conservatives fought for control in 2000. For the first time, the national media covered these elections, one candidate raised an unprecedented $90,000 in campaign funds, and television spots were purchased by a candidate. In the end, three of the incumbents who had voted for the anti-evolution science standards were ousted (Abrams, however, enduringly popular in his district in southern Kansas, was reelected) and the board majority shifted from conservative creationists to moderates who supported teaching evolutionary theory. The standards, which many school districts had ignored, were revamped in 2001 to match the original recommendations of the committee of experts. Evolution returned to the schools of Kansas, and the Niobrara once again could have 80 million birthdays instead of being described simply as "really old."

In the next election cycle, however, the balance of power shifted yet again. One of the moderate Republicans on the board, Val DeFever, who had accused Abrams and his allies of injecting "pseudoscience" into Kansas's schools, was upset by a virtually unknown conservative who rarely appeared in public, did not talk to the press, and was a graduate of Kentucky Mountain Bible College (class of 1956) with a degree in Christian Education. Iris Van Meter had one other winning quality: She was the mother of the executive director of the Kansas Republican Assembly, a new but potent organization of Kansas's conservative Republican wing, dedicated to antiabortion, pro-Christian policies, and supporting George W. Bush as an exemplary conservative president. Ten days before the vote, a mailer went out in DeFever's district asserting that the incumbent had ties to an atheist organization that supported her candidacy, and letters to the editor of a local newspaper and radio ads made similar claims. This was denounced as a smear—and the claims appear to have been false—but atheism is the kiss of death to a Kansas politician. It cost DeFever the election.

By 2004, other incumbents fell and an even split on the board was overturned as the conservatives won enough seats to form a majority voting bloc. Abrams took charge again as chairman. And the science standards that would affect 450,000 public school students once again came up for review. The state braced itself for yet another war over evolution, and the hosts of late-night talk shows dusted off their "What's the matter with Kansas?" monologues.

This time around, five years later, the creationists would be staying out of the limelight. Now it was intelligent design's turn at bat, the new masters of anti-evolution spin, sophisticated and well financed. And their first task was to frame the issue not as creationists going for the gold, but as an upstart school board interested only in the highest-quality education possible, taking on a science establishment that brooked no criticism and wanted to crush all dissent from the "gospel" of Darwin. In this formulation of the issue, the Kansas Board of Education simply wanted to raise questions about evolutionary orthodoxy and to consider real scientific evidence that evolution might not be all it was cracked up to be. This was brilliant public relations, as it allowed the Discovery Institute to portray the inevitable outcry from scientists as a kind of Scopes Monkey Trial in reverse. This time it was not a state ban on teaching evolution that was being challenged, but the Darwinists' ban on even questioning evolution. Students from anti-evolutionist families throughout Kansas had been coached for the past year with handouts from the Discovery Institute on how to stand up in class and object to evolutionary theory, posing the "ten questions" that Discovery recommends children ask their science teachers—and bringing biology classes to a standstill, at least temporarily, as fourteen-year-olds rattled off questions about the "Cambrian Explosion." Kansas would be a dream come true for the "wedge" strategists.

As in 1999, a panel of twenty-five science teachers and other experts was formed to draft new science standards. This time, however, Abrams and his fellow anti-evolutionists on the school board appointed to the standards committee eight similarly minded critics of evolutionary theory. The committee split along predictable lines, with most of the members crafting detailed standards that reflected mainstream scientific thought on evolution—"The Majority Report"—and the eight hand-picked dissenters producing their own anti-evolution "Minority Report." The minority's recommendations once more redefined science to include supernatural causes and contained many criticisms of evolutionary theory that had been proposed by Phillip Johnson, Michael Behe, and other lions of the intelligent design movement. The minority committee report, which immediately captured the school board's favor, did not mandate the teaching of ID, nor did it seek to banish evolution and other "old-Earth" and "old-universe" theories, as the 1999 standards had done. The eight minority authors said they simply wished to provide students with more information about evolution, not less—and to give them the tools and knowledge to critically analyze this grand, sweeping theory of all life, "to follow the evidence wherever it leads."

The rest of the standards committee, however, felt that the minority approach was even more radical than the 1999 standards in its criticism of Darwin's theory and of "methodological naturalism"—a term creationists consider derogatory but scientists consider synonymous with the "scientific method." The minority report advocated teaching and *testing* students on the notion that science is not necessarily limited to natural explanations. The mention of testing meant that local school districts would probably feel compelled to include nonnatural explanations such as intelligent design in the science curriculum, so their students would be prepared for the all-important annual statewide tests. The standards also stated as fact a variety of scientific assertions that the vast majority of mainstream scientists vehemently reject: that evolutionary theory is controversial in the scientific community, that it cannot explain the diversity of life or the existence of DNA, and that teachers should not discourage students from exploring nonnatural explanations.

The difference between the two approaches can be seen in one telling detail. The majority report featured a suggestion for how teachers should react when students raise nonscientific or religious issues and questions: "Treat the question with respect. . . . Explain why the question is outside the domain of natural science and encourage the student to discuss the question further with his or her family and other appropriate sources." This is a common approach in many states, and it was once taken in Dover, too: treat religious issues respectfully, but as outside the scope of a biology class.

That advice was stricken from the minority report. Instead, with science defined as including supernatural causes, teachers would be empowered (not required or mandated, but definitely permitted) to engage in classroom discussions of supernatural alternatives to evolution.

Four public forums were then held in communities around the state to seek public input on the majority and minority recommendations, a chance for ordinary Kansans to add their voices to the process. The events drew huge crowds by school board standards; in the past, such meetings had attracted very few spectators, even on the most controversial issues. This time, more than 1,200 members of the public attended the forums, and more than 200 comments were offered. Opponents and supporters of the minority report seemed about even in numbers. Scientists, professors, and science teachers in attendance almost all opposed the minority standards, and a number of them took pains to explain the scientific principles at issue, and how acceptance of evolutionary theory did not conflict with their own religious beliefs. On the other side, most of the speakers who praised the minority standards cited religious and creationist reasons, rather than scientific reasons, for their support. The overall impression was exactly the opposite of what the school board and the Discovery Institute wanted. It sounded very little like the battle between scientific ideas that the intelligent design movement sought, and very much like the old format of Darwinism versus creationism, which was a sure constitutional loser should there ever be a court challenge.

This situation led a retired attorney, John Calvert, head of the Kansas-based Intelligent Design Network, to write in the Discovery

Institute's Internet blog that the forums weren't having the desired effect, and that a different approach was needed to inform the public:

"It would have helped to have more scientists on our side. If that had been the case we would have won the debate hands down. As it was, the objective observer would leave scratching his head. . . . One thing is obvious. This is not the proper process for deciding this issue. Focused hearings from experts are desperately needed to cut through the misinformation, ridicule and half truths."

Calvert, who had given up a successful career in business law to champion intelligent design in school board battles around the country, carried considerable weight with the board of education and its chairman—at times, he seemed almost like a shadow member of the board. A week later, at the next school board meeting in February 2005, Steve Abrams proposed exactly what the leading proponent of intelligent design in Kansas had advocated: focused public hearings structured to pit the curriculum committee's majority and minority views against each other—Darwinists versus anti-evolutionists. In a trial-like approach, both sides would present their scientific cases over three days, using any experts chosen, with the opportunity to cross-examine the witnesses and attempt to shred opposing arguments. Calvert was appointed to represent the minority view, happily cast in the role of William Jennings Bryan as a prosecutor of evolution, putting Darwin on trial in real life as Phillip Johnson had done in his book.

A three-person panel of school board members, including Abrams, would preside over the hearings and act as judges. All three were creationists, so that the comparison to the Scopes Monkey Trial and its creationist presiding judge became a bit close for comfort. But Abrams was undeterred by claims that he had stacked the deck: "This is a real opportunity to discuss the real scientific issues. . . . Not straw men. Not who has the best credentials. But the actual science."

The second "judge" at the hearings would be board member Connie Morris, a former elementary school teacher who owns a ranch and a plumbing business with her husband, and who has written an autobiography detailing her "recovery through Christ from incest, rape, domestic violence, substance abuse, and poverty during her early years in the Appalachian Mountains region." She had been on the board for two years.

The third "judge" was Kathy Martin, a retired teacher who had been a board member for only a year and who was up-front about her creationist beliefs and her disdain for evolution. A few weeks before the hearings began, the newspaper in her hometown, Clay Center, Kansas, quoted her as saying that she had already made up her mind about the issues, and that America, as a "Christian nation," ought to teach prayer, creationism, and intelligent design in public schools. She later said her remarks had been taken out of context and denied wanting ID or creationism taught in science classes.

A month after the hearings were proposed, the Kansas Citizens for Science decided to boycott them, calling on scientists everywhere to do the same and refuse to take part by "defending" evolution. The math teacher Jack Krebs, a future president of this volunteer organization— and a member of the majority on the curriculum writing committee— recalls agonizing over the boycott. Abrams and Calvert had put the evolutionists in a no-win situation. If they participated in the hearings, they would give credibility to the twenty-four witnesses Calvert had lined up to slam Darwin and extol intelligent design—it would appear that there really was a scientific controversy, just as the Discovery Institute had long asserted. But if the evolutionists stayed away, they would at best appear sullen and unwilling to engage in the time-honored democratic exercise of debate; at worst, they would appear to be afraid, as if their ideas were too fragile to stand up to a bit of give-and-take. After much discussion, the group settled on what it decided was the lesser of two evils—to stay away from the "rigged hearing" and deny "the antievolution members of the board the veneer of respectability."

Abrams and Calvert accused the scientific community in Kansas of attempting to stifle scientific debate. They were clearly frustrated—they had carefully scripted questions and had envisioned a withering interrogation of the experts on evolution, a breakthrough that they hoped would greatly advance the anti-evolution cause. The boycott had caught them flat-footed, in the vaguely ridiculous position of staging a war with no one to fight. They searched for months to find an evolutionary scientist willing to play the role of opposition in the hearings. But no one raised his or her hand—the boycott was being supported by the nation's major scientific organizations. Finally, a Topeka attorney, Pedro

Irigonegaray, volunteered for a limited role. He would "represent" mainstream science at the hearings—not by calling any witnesses but by going after the antievolutionist experts.

The national media assembled in a packed auditorium in Topeka, the aisles bristling with cameras and microphone booms, ready for a second Scopes Monkey Trial. In Dover, both sides eagerly watched what amounted to a dry run of their own case.

Chapter 9

WHAT WILL WE TELL
THE CHILDREN?

The hearings in Kansas began on May 5, 2005, at Memorial Hall in the state capitol area in Topeka. It was National Prayer Day. In the capitol building, evangelical choirs sang hymns and the notion of separating church and state was derided in angry speeches. In the hearings, what *The New York Times* described as a "parade of PhDs" began, scientists testifying one after another, offering a barrage of criticism of mainstream evolutionary theory and the orthodoxy behind it. Never before had so many advocates of intelligent design and critics of evolution been assembled in one place and given a national platform. It was intended as a graphic demonstration that there truly was a scientific controversy about evolution.

"According to Darwinists, these scientists aren't even supposed to exist," wrote Discovery Institute senior fellow John West on the institute's website after the first day of the hearings. "But they are here nonetheless, and they deserve to be heard."

The first witness was William H. Harris, a University of Missouri nutritional biochemist who does research on heart disease and omega-3 fatty acids in fish. He was the leader of the curriculum minority group, and a managing director with John Calvert of the Intelligent Design Network. Although macroevolution is not his area of scientific expertise, Harris testified under friendly questioning from Calvert about its inadequacies as a theory and argued that a "reasonable interpretation of the data" led to the conclusion than an intelligent designer had created

DNA. Finally, he asserted that Kansas's current science standards amounted to indoctrinating children into a godless, naturalistic religion called evolution. "Part of our overall goal is to remove the bias of religion that is currently in schools," Harris said.

With Harris's testimony, a four-part pattern emerged that would be repeated throughout the hearings with nearly every witness.

First, Calvert would gently question his well-prepared witness.

Part two would be a cross-examination by Pedro Irigonegaray, who had developed a set line of questions. He asked the witness to examine the majority curriculum standards and show him where there was any mention of religion, naturalism, or atheism, or a ban on questioning evolutionary theory. Harris, like the witnesses that followed, had to admit that there was no such language in the majority standards, though he did at one point assert that he saw it "between the lines." Then the attorney filling the Clarence Darrow role would ask each witness how old the Earth was, which in Harris's case led to considerable equivocation. Eventually, he grudgingly conceded that his best guess was 4 billion years old, though he first said many believed 10,000 years was about right.

In part three, Abrams and the other board members would praise the witness, pile up additional criticism of evolutionists, and ask follow-up questions intended to undo any damage from Irigonegaray's pointed queries—as when Abrams asked Harris why evolutionists were so "unwilling" to acknowledge the evidence of design in DNA. It was during this stage that the board members were at greatest risk of revealing their sometimes startlingly poor understanding of the science they were supposed to judge, and their lack of impartiality. Connie Morris, for example, who had received a deluge of e-mails questioning her about the standards, heartily thanked Harris for doing such a great job testifying, and then added, "I wish you had been in my office this morning to answer all the e-mails."

In part four, reporters would flock during breaks in the testimony to the information booth that the Kansas Citizens for Science had set up in the hallway to give the responses scientists had declined to provide inside the hearing room. This was another risky move that exposed them to accusations of hypocrisy from Abrams, who said the evolutionists were "sniping from the hallways" instead of engaging in honest debate and sitting for questions from the board. Jack Krebs stood his ground, however, gamely defending his organization's boycott and dismissing Harris and the other witnesses as charlatans. "It's clear from the beginning that this is not a real science discussion. This is a showcase for intelligent design," Krebs told reporters on the first day. "They have created a straw man. They are trying to make science stand for atheism so they can fight atheism."

Charles Thaxton, a retired chemist and Discovery Institute fellow who has researched and written on the origins of life, came next. Thaxton does not believe natural processes provide a plausible explanation for the beginning of life, and he spoke about the "fine-tuning" of the universe—how all conditions are improbably but exactly right to support life as we know it—as evidence of purposeful design. (This argument for design has been dismissed as circular by mainstream scientists and philosophers of science as the "if things were different, then they would be different" argument for design. It's not that conditions are just right to support life as we know it; it's that life exists as we know it because of the conditions that exist in the universe. Different conditions could support different forms of life, or no life at all.)

Thaxton went on to provide one of the oddest moments of the hearings. The science standards proposed in the majority report did not bring up the chemical origins of life; the curriculum committee, like most mainstream scientists, considered hypotheses on the origins of life extremely speculative and their implications controversial. Indeed, the origin of life is a topic that creationists usually want kept out of public school classrooms, as in Dover, because the notion of life arising in a primitive Earth through a complex series of chemical reactions is even more challenging to religious viewpoints than evolutionary theory. But in Kansas, it was the authors of the *minority report* and their witnesses who wished to introduce the topic into the standards, so that they could then criticize attempts to explain the origins of life through natural, chemical means.

Though Thaxton was giving them exactly what they wanted, two of the three board members seemed utterly confused by him, particularly by his references to "prebiotic soup"—the theoretical environment in which life might have begun on the early Earth. The composition of that environment is highly speculative and has been the subject of lively scientific debate and of innumerable experiments dating back to the 1950s and the famous experiments by Stanley Miller.[1] But Connie Morris wanted to know, yes or no, if there was "evidence of the soup" (no—the soup was gone billions of years ago) and if the soup posed a "critical problem" for Darwin's theory (curiously, Thaxton said yes, but this is untrue, since Darwin never addressed the origins of life or the prebiotic soup; evolutionary theory assumes that life is already present). Kathy Martin became so flummoxed by this "soup idea"— the board members had begun to speak of it as if it were a pot on a primordial stove—that she began to assert, "You have to use imperialism when you deal with science," apparently meaning empiricism. When he was cross-examined, Thaxton expressed disbelief in the general theory of common descent and specifically rejected the notion that man had evolved from more primitive hominids, revealing a creationist bent in his scientific beliefs.

The Discovery Institute was well represented, and one of its most famous fellows took the stand, embryologist and theologian Jonathan Wells, author of *Icons of Evolution*. He explained how his book and the video based on it—which had so entranced William Buckingham in Dover but had left the science teachers there cold—reveal that many illustrations of evolution in school textbooks are incorrect. These range, he said, from staged photographs of peppered moths that had adapted over time from predominately white to black wings because of sooty trees in Britain, to discredited nineteenth-century illustrations of supposedly common features in the embryos of diverse creatures. (The first point was a poor one—adaptation of peppered moths has been proved to have occurred across multiple generations, regardless of the staged illustrations. But Wells was correct about the embryos: The illustrations by Ernst Haeckel continued to appear in modern textbooks long after his nineteenth-century work was shown to be inaccurate. However, the general notion they depict—that common features appear and then disappear in the embryos of birds, reptiles, and mammals—is well accepted in mainstream science.)

Wells also spent a great deal of time suggesting that molecular evidence—a topic outside his area of expertise—does not support Darwin's key theory of common descent. The tree of life collapses because the relationships do not proceed at the genetic level as evolutionary theory predicts, he asserted. On cross-examination, he was asked if that belief put him in the minority of scientists.

"I enjoy being in the minority," Wells replied. "I'm more comfortable."

"More than being right?" Irigonegaray asked.

Out in the hallway, Krebs and his colleagues pointed out that Wells was disregarding decades of accumulated genetic research and human genome analysis showing that people and other creatures shared very large amounts of DNA and similar genetic materials—genes that had been passed down for millions of years. Then a handout made the rounds that seemed to undermine Wells's claim that his objections to Darwin were grounded in science, not religion. The handout referred to Wells's involvement with the Unification Church and its leader, Sun Myung Moon, whom followers call "father" and revere as the messiah. The handout included an old Internet posting Wells had written: "Father's words, my studies, and my prayers convinced me that I should devote my life to destroying Darwinism. . . . When Father chose me (along with about a dozen other seminary graduates) to enter a Ph.D. program in 1978, I welcomed the opportunity to prepare myself for battle."

Abrams bristled at this attack, accusing the science group of engaging in character assassination and dwelling on personal beliefs and credentials instead of science, but both sides used the same brutal tactics. The opening testimony of the first witness called by the board's prosecutor, John Calvert, in a well-rehearsed set piece, concerned an Internet posting pulled from an online forum of the Kansas Citizens for Science. It was supposed to illustrate the science group's willingness "to smear and demonize" opponents. The posting by a Kansas science advocate read in part:

> My strategy at this point is the same as it was in 1999;
> notify the national and local media about what's going on and
> portray them [the school board and its allies] in the harshest
> light possible, as political opportunists, evangelical activists,

ignoramuses, breakers of rules, unprincipled bullies, etc. There
may be no way to head off another science standards debacle,
but we can sure make them look like asses.

The chilly climate of the hearings became downright hostile, and
any pretense of evenhandedness on the part of the board members or
restraint on behalf of the their opponents pretty much fled the building.

Most of the out-of-town reporters fled, too, filing their stories after
the first day of hearings, then skipping the next two days, leaving the
locals to cover the rest. But even the Kansas media covered only part of
the proceedings and reported only a small fraction of the testimony, to
the board's lasting disappointment.

On the second day of the hearings, Bryan Leonard, a high school
biology teacher, was brought in from suburban Columbus, Ohio. He
had developed a model curriculum for his school, "Critical Analysis of
Evolution," that the state of Ohio eventually adopted for statewide use.
The program proved highly controversial, as scientists complained that
the specific pieces of evidence "against" evolution used in Leonard's lesson plan were either incomplete, misleading, or simply wrong. His approach, however, was exactly what the Discovery Institute advocated—a
"teach the controversy" strategy that did not address intelligent design
explicitly, but took up only criticisms of evolution, drawing heavily on
exactly the sorts of objections that Michael Behe raised in *Darwin's
Black Box* and that Jonathan Wells cited in *Icons of Evolution*. Up to six
weeks of class time would be spent going over the evidence for and
against evolution, Leonard explained—a far cry from ninety-minute
lesson plan of Dover, Pennsylvania. The members of the Kansas board
immediately asked if they could get a copy of Leonard's plan for their
schools, and Leonard happily agreed to provide it. (Months later,
Leonard's national exposure would not serve him so well, as questions
soon were raised about the composition of his dissertation committee—
it included two ID supporters, one of whom was a fellow witness in
Kansas.)

Leonard's cross-examination was unpleasant. Irigonegaray asked his
standard question about the age of the Earth, and Leonard was the first
witness (though there would be many others) who would not agree that

the Earth was billions of years old. A curious exchange followed, at times hilarious, at others almost painful to witness, as Leonard simply refused to answer the question directly, then finally conceded beliefs that put him firmly in the creationist spectrum of thinking.

Q: First, what is your opinion as to what the age of the world is?

A. I really don't have an opinion.

Q. You have no opinion as to what the age of the world is?

A. Four to four-point-five billion years is what I teach my students.

Q. I'm asking what is your opinion as to what the age of the world is.

A. Um, I was asked to come out here to talk about my experiences as a high school biology teacher.

Q. I'm asking you, sir—

A. I was not under the impression that I was asked to come out here—

Q. I'm asking you—

A. —talking about—

Q. —Sir, what is your personal opinion as to what the age of the world is?

A. Four—four to four-point-five billion years is what I teach my students, sir.

Q. That's not my question. My question is, What is your personal opinion as to what the age of the world is?

A. Again, I was under the impression to come out here and talk about my professional experience—

Q. Is there a difference?

A. —more of—

Q. Is there a difference between your personal opinion and what you teach students the age of the world is?

A. Four to four-point-five billion years is what I teach my students, sir.

Q. Is—my question is, Is there a difference between your personal opinion and what you teach your students?

A. Again, you're putting a spin on the question. You know, now I'll spin my answer, sir, to say that my opinion is irrelevant. Four to four-point-five billion years is what I teach my students.

Q. The record will reflect your answer. Do you—do you accept the general principle of common descent, that all of life was biologically related to the beginning of life? Yes or no?

A. No.

Q. Do you accept that human beings are related by common descent to prehominid ancestors? Yes or no?

A. No.

Next, Irigonegaray coaxed a surprising admission from the biology teacher. Though Leonard had come here to recommend the minority curriculum standards over the majority report, he had not actually read the majority recommendation on the science curriculum standards. Even the attorney seemed surprised at this answer. "You have been brought to Kansas to challenge the majority opinion and you have not taken the time to read it?"

"I read the part of the minority report that . . ."

"I didn't ask you about the minority," Irigonegaray snapped.

"No, I have not read the majority opinion."

After that, one witness after another was forced to admit that he had not read the majority curriculum standards he had been called to Kansas to testify against. In all, eight proponents of intelligent design who testified on the second day of the hearings reluctantly confessed that, no, they had not actually read the proposed standards they were characterizing as inadequate. They had read only the twenty-eight-page minority report, with its emphasis on criticizing evolution, not the 107-page majority standards. Krebs later said that this proved beyond any doubt that the proceedings were a "kangaroo court."

Kathy Martin, the board member who had already expressed her support for the minority standards and her opposition to the majority, next elicited gasps from the audience when she attempted to comfort one of the witnesses for being unfamiliar with the majority proposed science standards—by admitting a similar ignorance on her own part. "Please don't feel bad that you haven't gotten to read the whole thing," she said, "because I've not read it word for word myself."

The board next gave a warm welcome to another teacher, Roger De-Hart, the former science instructor from Washington state who taught intelligent design for ten years to his students at public high schools, then was forced to stop, and eventually moved to a Christian academy in California where he could resume "teaching the controversy." He told the board that kids learned better when they were given both sides and allowed to debate the merits of evolution, that it energized them and got them involved in the subject, but that the ACLU and Darwinian extremists had finally succeeded in "censoring" him. It got so bad, DeHart recalled, that when he transferred to a different school in Washington, one fellow teacher, who was affiliated with the National Center for Science Education, warned him that she'd been told by the organization to keep an eye on him.

"Like the security police in the biology classroom?" Calvert asked.

"More or less," DeHart responded sadly.

Under cross-examination, however, DeHart said that he believed in a young Earth, that the way evolution was taught in the United States is atheistic, and that humans did not evolve from a common ancestor shared with other species.

"What is the alternative explanation for how the human species came into existence if you do not accept common descent?"

"Design."

"Who was the designer?"

"Science cannot answer that. When I'm teaching my class I do not answer that."

Kathy Martin, however, was of the opinion that schools should answer exactly this question, even as her chairman continued to insist that the hearings were not about religion: "Evolution is a great theory. . . . There are alternatives. Children need to hear them," Martin had said before the hearings began. "We can't ignore that our country is built on Christianity, not science."

Those sentiments are the kind that the Discovery Institute assiduously avoids in public—the sort of overt religious appeal that had left the proponents of ID more than a bit nervous about Dover, and that could mean the difference between a policy that passes constitutional muster and one that gets shot down in court.

The next witness, Stephen Meyer, who testified by phone, quickly got things back on a properly secular track. Meyer was one of Discovery's cofounders and one of its most polished public speakers. He commended the minority standards and asserted that they would require teaching *more* evolutionary theory, not less, by revealing the theory's shortcomings as well as its accomplishments. Such a method would inevitably show that design was a better explanation for much of biology, such as the software-like information encoded within DNA. And it was this, Meyer suggested, that many evolutionists wished to avoid, out of fear that their whole Darwinian house of cards would collapse if the subject of origins was taught honestly.

Under cross-examination, Meyer refused to be baited by Irigonegaray's stock questions. He readily agreed that Earth was billions of years old, and conceded that there was some evidence of limited common descent. And though he "was skeptical" about the idea that humans had a common ancestor with all other creatures, it would not bother him if this turned out to be true. "It would not affect my conviction that life is designed."

The exchange between Meyer and Irigonegaray grew testy when the philosopher refused to concede that his views on evolution put him in the minority of scientific thought. "I don't concede that. . . . We know that 400 scientists at least have signed a statement of dissent. . . . There is significant scientific dissent from Darwinism."[2]

Evolutionists have derided this number—400—as paltry when compared with the hundreds of thousands of working scientists in the world, and they suggest that some of the signers lack expertise in evolution or related fields, while others were unaware of exactly what they were signing.[3] But Irigonegaray did not raise these issues and simply cut Meyer off, saying there was no question on the floor, then ended his cross-examination. If anyone got the better of the trial attorney during the hearings, it was Meyer.

Michael Behe was the last witness called, and he made his standard presentation about irreducible complexity. At John Calvert's prompting, he also described a research paper he had recently published with another scientist, David Snoke, a physicist at the University of Pittsburgh. Behe said the paper showed that the evolutionary mechanism of gene duplication and mutation took impossibly long to evolve even the simplest new protein features—features that are essential for biochemical evolution to take place. "The amount of time that it would take . . . becomes prohibitive, over a hundred million generations."

Behe described this as a huge problem for evolutionary theory, one that was underappreciated by both the general public and the scientific community. It meant that complex features would take too long to evolve, if they ever did, and that modern biochemistry undermined Darwin's ideas about natural selection driving large-scale changes in species. As he often does, he used the analogy of a mousetrap to explain how hard it would be for evolution to assemble even one set of parts, assuming the mousetrap was alive.

> If you had that spring in the mousetrap and you wanted it to attach to, say, the catch, and you didn't want an intelligent agent to put it together, how likely would it be if it mutated in a way similar to the way proteins do in the cell? In order to do that, you need a long time and a very large population size— prohibitively large—just to get those two things together, not even worrying whether their shapes were the right shapes for the purpose—just to get them to stick to each other. It's a problem.

The lack of evolutionary scientists on hand to raise questions about this remarkable claim essentially let Behe off the hook, because his testimony did not tell the whole story. Irigonegaray is no scientist, however, and he had no way to know what was being left unsaid. But the researchers in the Dover case were taking careful notes, and this bit of testimony would eventually come back to haunt Professor Behe.

One of the most genuine and engaging witnesses—one whom Steve Abrams later said he found particularly inspiring—was, perhaps unsurprisingly, the least partisan, the least ideological, and the only one who expressed no interest in movements and causes, unless a teacher advocating for her students could be considered a cause. Jill Gonzalez-Bravo, an eighth-grade science teacher at Rose Hill Middle School in southeastern Kansas, told the board that she had been stymied time after time when her students would pose questions about evolution and human origins. Many of her pupils seemed uninterested and unmotivated much of the time, but they would get excited about evolution, starting heated discussions and debates. But Gonzalez-Bravo had been trained to believe that there was no controversy about evolution. The science standards were clear about what to cover, and the National Academy of Sciences had cautioned teachers against doing anything to undermine the teaching of evolution.

Yet as far as her students were concerned, evolution was hugely controversial.

"Students cannot comprehend how a process largely founded on chance could be so specialized," she told the board. "When I presented evolution to them, the contents somehow impacted their conscience. . . . It took from them the idea that they were born for a purpose. I was telling them . . . something completely counter to their mind-set and beliefs. And that troubled me."

As she struggled with how to respond to her students' concerns and interests while remaining true to the teaching standards and her role as a public school teacher, things got even more complicated for Gonzalez-Bravo. She gave birth to her own children, heightening her concerns as she contemplated how her kids would face these issues; she became a Christian after years as a "secular humanist"; and she began researching criticisms of evolutionary theory. In particular, she felt persuaded by evidence cited in such works as *Icons of Evolution*, notably the staged photographs of peppered moths used in some older science textbooks. She became convinced that there was a legitimate controversy and a properly scientific debate about evolution that she felt she should teach.

The board members took this to mean that she then began to "teach the controversy" to her students—they applauded that—but Gonzalez-Bravo corrected them. "I don't teach the controversy. I am teaching evolution in the classroom . . . because the standards don't allow for the controversy. So I would not do that. I teach to the standards."

This brought her to why she was there to testify. She wanted to explain why she felt that Kansas's science standards should be modified, as the minority report would do, so that she would have the freedom to teach what her students yearned to discuss and learn and explore. Not to proselytize. Not to attack evolution. But to have a full discussion.

There is no overestimating the effect that a reasonable person can have on an essentially unreasonable process. No amount of fulminating by the Discovery Institute or bluster by the board or harrumphing by the evolutionists could come close to equaling the appeal and reason displayed by this sweet young teacher, who was clearly conflicted about testifying and had twice refused to be a witness. If there were a poster child for "teach the controversy," it would be Jill Gonzalez-Bravo, a former Peace Corps volunteer who played by the rules and followed the standards even when she disagreed with them, and for whom the issue was not her own beliefs, but the real needs and questions of the kids she taught.

The press did not mention Jill Gonzalez-Bravo, however, and her testimony made no headlines or blogs or newscasts. She came late in the hearings, and by then the reporters were bored, and she really did not fit into the neat, tidy narrative of creationists and pseudoscientists versus arrogant evolutionists who refused to debate. She did not think academic freedom entitled her to teach as she pleased regardless of state standards, as Roger DeHart and his supporters at Discovery had argued; nor did she agree with mainstream scientists that evolutionary theory was religiously neutral. In the ivory tower, among men and women who had read the great philosophers and who practiced science for a living, perhaps that religious neutrality existed. But in her world, down in the trenches with the eighth-graders at Rose Hill Middle School, there was nothing neutral about Charles Darwin's dangerous idea.

There was, of course, a reasonable, rational response to Jill Gonzalez-Bravo's reasonable, rational plea. First, had both sides been represented with experts at the hearings, any number of scientists could have pointed out that much of the evidence against evolution she considered so persuasive might, in fact, be baloney—such as the claims about the peppered moths. Yes, those photos in the textbooks were staged—but they were staged to illustrate real data, findings, and evidence of the evolution of the moths. Michael Behe uses similarly "staged" drawings of the bacterial flagellum that makes its parts look like an assemblage of real machines, not the bits of protoplasm they really are, but no one suggests that his use of such illustrations disproves his theories.

Second, it might have been suggested that there was a valid alternative to introducing a discussion of design, creation, or other potentially religious ideas in a public school classrooms when a student asks, "Do you believe in evolution or creation?" She could tell her students that science explains only the natural world, not the spiritual world, and that questions of God and creation are best discussed at home or at church, not in science class. This is what the majority report called for—as part of the standards the school board had already decided to reject.

But the hearings, by the second day, were no longer about reasoned discourse. There was no room for counseling a teacher reaching out for guidance and help. By then, it was all about scoring points; it was all about winning. Pedro Irigonegaray did the only thing he could do under the circumstances, the nicest thing he could do. As the teacher steeled herself for the cross-examination she had dreaded, he said, "I have no questions."

"You are merciful," Gonzalez-Bravo blurted.

Chairman Abrams barked, "That's not appropriate." And the mood was broken, partisanship flooding back into the room, the momentary truce evaporating like dew in the desert.

"I think you did a great job," the board member Connie Morris told Gonzalez-Bravo. "Someone used the phrase critical mass of data coming forth that refutes Darwinian evolution and eventually that will catch up with the evolutionists. And you, you dear, wonderful teacher, will be on the leading edge."

The proceedings ended as they had begun: with rancor.

Instead of presenting days of pro-evolution testimony, Pedro Irigonegaray made a two-hour closing argument, heaping scorn on the board and the proceedings:

> These hearings have been an unjustified waste of taxpayer
> money intended first to justify the Board's support for

> inserting creationist claims into the science standards and
> to provide a showcase for the National Intelligent Design
> Movement. . . . We reject the show trial hearings whose
> purpose is to make it appear that intelligent design creationism
> and the well established science of evolution are on equal
> footing.

He then took the board members to task for falsely linking evolution
with atheism. Irigonegaray said the charge of atheism was a favorite bit
of rhetoric among proponents of intelligent design, not because it was
true but because they perceived their greatest enemy to be "theistic
evolution"—the belief, held by many scientists and endorsed by the Vatican, that there is an all-powerful God, but that he has chosen to act
through natural laws, such as evolution. The ID movement disdains this
belief, Irigonegaray said, because if more people understood and accepted theistic evolution, they'd have no need for intelligent design. To
illustrate, he read a passage by a prominent Kansan critic of evolution:

> Evolutionists start with the bias that everything must have
> a natural explanation, i.e., God does not, cannot, be part of
> the answer. . . . Creationists on the other hand start with the
> bias that God did indeed create all animal and plant types.
> They believe it is their responsibility to study and explain how
> he did it. The two worldviews are diametrically opposed and
> mutually exclusive. There's another group that tries to meld
> the two views together, they are the theistic evolutionists. They
> usually take the tack that God created something and then left
> it to evolution to work it out. If these people are talking about
> the God of the Bible then they do not understand what is
> written in the Bible or they do not understand the philosophy
> of evolutionary theory.

The person being quoted was the school board chairman, Steve
Abrams.

After Irigonegaray delivered his parting shots, Abrams and John
Calvert leaned forward in their chairs, anxious to turn the tables. They

had long anticipated the opportunity to cross-examine Irigonegaray, who had been skewering their witnesses for days. But Irigonegaray demurred, saying that he was representing science as an attorney, not a witness.

The chairman and his "prosecutor" became livid, saying that Irigonegaray's behavior was a "breach" of their agreement, but he calmly said no, the board had presented evidence for three days, and he had taken but two hours to argue his case, and as far as he was concerned, that was it.

Board member Kathy Martin tearfully complained that she had been up all night filling a notepad with questions that would now remain unanswered—questions that she believed would address misrepresentations by the press, by scientists, and by Irigonegaray. "The Board has been accused of being closed minded and these hearings as being a rip. I guess we'll leave it up to the public. . . . Based on what we've heard these four days, which side of the issue is being closed minded? Why are some scientists tenaciously holding onto the evolutionary tenets that are unproven, as we have heard, and are often disproven?"

Connie Morris, the recovered abuse victim, was more sorrowful than angry. "I believe your behavior here was abusive," she told Irigonegaray. "I do understand abuse, and I just want you to know that I forgive you, truly."

And so the hearings ended, shedding little light on the subject of evolutionary theory, changing few minds, and hardening more than a few hearts. Those inclined to doubt Darwin now had reason to be angry at the members of the scientific community for considering themselves "too good" to mix it up with the rabble, "unwilling to take part in the democratic process," as Steve Abrams put it. "What are they so afraid of?" was the triumphant refrain at the Discovery Institute, announcing victory in Kansas by default.

On the other side, the evolutionists railed about a school board that cared little for the welfare of its students or their scientific literacy. The popular Kansas blog RedStateRabble said that the Discovery Institute was deluding itself, that this had been a victory for evolution: "The (boycott) strategy was high risk, but it has paid off handsomely. The defense of science is in good hands in Kansas." Eugenie Scott, director of the

National Center for Science Education, even compared the Kansas school board to the rulers of Stalinist Russia:

"The Kansas hearings are a show trial, like in the sense of the Soviet Union back in the fifties. The board already has its conclusion. They're just going through these motions, making a big show for the public, to get an idea out."

After the hearings, the creationist majority on the Kansas board voted, as everyone knew it would, to adopt the minority report standards almost verbatim, to redefine science to include nonnatural causes, and to create testable standards for 450,000 Kansas students that required them to learn as much about the criticisms of evolution as about the theory itself, if not more. No other theory was singled out in the standards for such treatment—not atomic theory or relativity or plate tectonics. Only Darwin would suffer the fate of constant attack, if the Kansas board had its way.

As an exercise in science, the hearings were less than useless, and any lasting victory claimed by either side could well prove illusory. There was no real exchange of ideas, no peer review, no attempt to share and compare and replicate data and findings, and certainly no attempt to compromise. Only the elections for the board of education the following year would determine who had the public's approval—and which faction would control the state's schools in the future. The standards could disappear as quickly as they arrived if the incumbents were ousted; or they could become the law of the land for many a year if the creationists held on.

The timing of the elections would prove critical, as the newly adopted standards could not go into effect until the start of the following school year. Then each school district in the state would have the option of interpreting them as it pleased, ignoring them if it wished, or embracing them as it saw fit. The evolution war in Kansas had its visceral aspects, the in-your-face confrontations, the accusations, and bluster, and they had been on display in no small quantity at the hearings in Topeka. But the war also has a less obvious chess-like quality, and the hearings were not only an end in themselves, but also an opening gambit in a larger battle plan. Both sides knew that somewhere in Kansas, and probably in numerous places, school districts would eagerly adopt the new standards and begin to denigrate evolution while promoting alternatives. It appeared inevitable. This was the goal of the wedge strategy, and if the standards held past the next election, it would be achieved.

And then, one or two or a group of parents who did not want religious explanations offered in their public school science classes would pick up the phone and call the ACLU, and the Discovery Institute and Phillip Johnson would have the perfect test case to bring up to the Supreme Court. Not like the one in Dover, where Discovery's playbook had not been followed, where creationism had been on the table before ID, and where a Christian law firm with its own agenda was calling the shots. In Dover, ID had become mixed in the public's mind with creationism and religion, exactly what Discovery wanted to avoid at all costs. Here in Kansas, however, Discovery's playbook had been followed nearly to the letter.

Here, it was hoped, Charles Darwin's last stand would begin.

Unless, of course, Dover became the Supreme Court test case first.

PART III

UNNATURAL SELECTIONS

WE CAN ALLOW SATELLITES, PLANETS,
SUNS, UNIVERSE, NAY WHOLE SYSTEMS OF
UNIVERSES, TO BE GOVERNED BY LAWS,
BUT THE SMALLEST INSECT, WE WISH
TO BE CREATED AT ONCE BY SPECIAL ACT.

—Charles Darwin

WE SHOULD AFFIRM THE REALITY OF GOD
BY CHALLENGING THE DOMINATION OF MATERIALISM
AND NATURALISM IN THE WORLD OF THE MIND.
WITH THE ASSISTANCE OF MANY FRIENDS
I HAVE DEVELOPED A STRATEGY FOR DOING THIS. . . .
WE CALL OUR STRATEGY THE "WEDGE."

—Phillip E. Johnson

SCIENCE IS WHAT YOU KNOW.
PHILOSOPHY IS WHAT YOU DON'T KNOW.

—Bertrand Russell

Chapter 10

SEND LAWYERS, GEEKS, AND MONEY

To TRO or not to TRO? That was the question—or, rather, the irritating Shakespearean pun—that kept bouncing around inside Vic Walczak's head during the waning months of 2004. It was a little mantra with big implications for the nascent lawsuit against the Dover school board, for the parents and kids who opposed the new intelligent design policy, and for the future of the nation's evolution wars. The aura of the Big Case had settled around the battle in Dover—and this was before the suit had even been filed. The press had begun circling, the science and evolution bloggers in both camps were all over it (and each other), the school board and its lawyers seemed to exude confidence, and the usually straightforward TRO decision had become fraught with meaning and peril.

At the American Civil Liberties Union, Walczak was sometimes referred to as "Mr. TRO," so great was his fondness for the gloriously risky proposition of the temporary restraining order. The legal director of the Pennsylvania division of the ACLU, Walczak had asked judges for more TROs than anyone else in the organization, going after schools for censoring students' free speech, local governments for mistreating protesters, university officials for discriminating against gays and lesbians, and police departments for brutalizing minorities—causes both popular and unpopular (though mostly the latter, defending the Constitution being a fairly lonely business). He filed for them all the time, making himself something of a riverboat gambler among the courtroom set.

This time, however, Mr. TRO wasn't so sure his usual strategy was the right move.

A TRO is a judicial stop sign—a temporary order from a judge that immediately "restrains" a person, a corporation, or a government agency from doing something that would cause someone else irreparable harm. The justification for such orders is simple: Sometimes a judge has to "freeze" events, in advance of the long-drawn-out process of a trial, forestalling potential damage before it can happen. This could involve stopping a factory about to dump wastewater into sensitive wetlands, barring a school from expelling a student for mocking the principal on his MySpace website (one of Walczak's actual and more absurd cases), or stopping a school district about to begin a program to criticize the theory of evolution and introduce intelligent design to its students. Whatever the cause, when waiting for a trial means waiting until it is too late, a TRO can provide the solution.

That's why Vic Walczak liked these judicial shortcuts: TROs are the courtroom equivalent of shooting the moon in a game of "hearts" or sinking the eight ball on the break. Legally, you still have to prove a case and win a trial to ultimately prevail and make the restraint permanent—that's what the law says—but in the real world, once you have a TRO, it is often decisive. If the Dover school board was blocked from carrying out its policy, the trial's outcome would be perceived as a forgone conclusion, and the school district would be under enormous pressure to settle out of court, saving money, time, and the divisiveness of a pitched legal battle. But as in cards or billiards, going for a win in one shot carries the risk of backfire. One bad play, an errant cue ball falling into the corner pocket—or, in the legal world, a rushed and incomplete presentation before a cautious or intransigent judge, or a better presentation from the other side—and you are dead. Once you go for a TRO and the judge says no, that is a huge, sometimes unrecoverable, setback, an indicator of a potentially weak case. Any motive the other side might have to settle would evaporate. And the policy would go into effect as scheduled, the lawsuit's teeth unceremoniously pulled.

So that was the critical calculation before Walczak. Option one, no TRO, allowed for a long investigation and the construction of the most airtight case possible, followed by a trial nine months or a year down the line. But this also meant that the plaintiffs would have to suck it up and take their lumps out in the real world. The board would be able to carry out its new ID policy unfettered.

Then there was option two, the TRO, which responded to the immediate needs of parents who didn't want their kids to have to take any lumps at all—they wanted that judicial stop sign. In a matter of weeks, their boys and girls would receive lessons in ID and criticism of evolution—or, more likely, they'd be forced to opt out, with the attendant risk of stigma, of being branded "ape boy" and "monkey girl" by the other students. The TRO could prevent all that.

Nothing appeared straightforward about this calculation, least of all the evidence at the outset of the case. There was the creationist talk of the school board members, which seemed to Walczak an obvious winner on constitutional grounds, an argument in favor of a TRO application. But then there was the policy the board ultimately adopted, which sounded much more secular and, to many ears, reasonable—what's wrong with looking at gaps and problems in a theory? What's wrong with considering alternative explanations? There might be good and persuasive counterarguments, but a brief TRO hearing, which would last at most a couple of hours, might not provide the opportunity to convey such nuances. Then there was the dispute over the Dragonfly Book. The board members had railed about Darwin and Christian nations, all of which would make for a great argument in favor of a TRO—except that the board ended up buying the Dragonfly Book, evolutionary chapters and all. One of the main critics of evolution in Dover, the board president, Alan Bonsell, had ended up voting for the Dragonfly Book before championing the new ID policy. So that angle of attack might sound ridiculous and churlish. And, finally, there was intelligent design itself. Was it religious? Was it scientific? Walczak knew next to nothing about this growing movement, and he assumed that whichever judge drew the assignment of hearing the case would know nothing about it at all. A ninety-minute TRO hearing would not provide a very good forum to flesh out that subject.

The decision was complicated further by the murky and volatile situation in Dover. The staff of the ACLU had to arrange clandestine meetings with potential clients and witnesses, as if they were exchanging state secrets rather than discussing a school board's venture into modern biology. Many families didn't want to be linked publicly to a potential lawsuit until it was actually filed, and even then, there was trepidation. Teachers at Dover High—Bertha Spahr, for one—had taken to locking up documents in a safe for fear that crucial evidence might somehow disappear.

Within weeks of the board's adoption in October of the new ID policy, it had become common knowledge around Dover that a lawsuit was in the offing. To no one's surprise, the community did not roll out the red carpet for the representatives of the ACLU. Even some critics of the school board seemed uncomfortable with the idea of "outsiders" intervening in local affairs. Supporters of the school board, and the board members themselves, were quick to fuel the resentment over this "interference" and to appeal to many conservatives' distaste for the ACLU. Bill Buckingham would later miss few opportunities to refer to the ACLU as the "American Communists Legal Union," a reliable laugh line with friendly audiences. Many assumed that the ACLU had stirred up trouble by prompting the parents to sue, although that was not the case. Most of those outside the small circle of parents who had turned to the ACLU for help had little idea what the organization was really about, except for some vague notions that it seemed hostile to religion and had defended the free speech rights of Nazis and other unsavory characters, nor did they know who this Walczak character was, other than he and his colleagues were invaders from the big city, talking tough on the evening news.

That Walczak would be confused with a communist or a Nazi sympathizer exemplifies one of the great ironies facing those who work for the frequently demonized ACLU, for nothing could be further from the truth. Witold "Vic" Walczak is a career ACLU man, which, with typical deadpan humor, he likens to "joining the priesthood." Despite the organization's inflexible opposition to the slightest breach in the separation of church and state, the analogy holds. In terms of personal commitment, separateness, the vow of poverty (there are few areas of the law guaranteed to pay a lawyer less), and the dedication to belief (in the Constitution and its Bill of Bights), career ACLU lawyers are a kind of secular priesthood. Since the organization was founded in 1920, its staff attorneys have played a role in many of the critical and landmark court cases and causes that for generations have shaped the cultural zeitgeist and the rights of Americans: *Scopes*, *Gitlow* (in which the Supreme Court ruled that the Bill of Rights applied to the states as well as the federal government), *Brown v. Board of Education* (ACLU had a supporting role in the famous school desegregation case), *Gideon* (which guaranteed legal representation in criminal cases regardless of ability to pay), *Miranda* (of the famous Miranda warning given to everyone arrested in the United States), and *Roe v. Wade* (in which the constitutional right of privacy was extended to abortion rights—something for which cultural conservatives have never forgiven the ACLU).

Walczak's choice of a career is intimately tied to his family history. His mother was a child in the Warsaw ghetto during World War II—a Holocaust survivor. His grandmother died in the Treblinka concentration camp in Poland; his grandfather was one of the few who escaped and lived to testify at the Nuremberg war crimes trials about Nazi atrocities. Both Walczak's parents were Polish refugees who took asylum in Sweden, where he was born. The family moved to the United States when Vic was three years old, spending time in Tennessee and Oklahoma before settling in suburban New Jersey. This afforded him the opportunity to grow into an inveterate Bruce Springsteen fan and an equally obsessive detractor of the New York Yankees, but more than anything, to be part of a family that treasured liberty, having known its terrible opposite in a way few others could claim.

While earning a degree in philosophy, Walczak took time off from college to work as a public defender's investigator and a juvenile delinquency caseworker in Washington, D.C., eye-opening experiences that led him to consider a career in the law. During his senior year at Colgate University he worked as a volunteer translator assisting refugees forced to leave Poland during the tumultuous rise in the 1980s of the Solidarity movement, which would eventually force the peaceful overthrow of communist rule and the crumbling of the Warsaw Pact alliance. Walczak's relationship with the activists in Solidarity inspired him to travel to Poland to see the situation for himself and to visit and attempt to assist the relatives of the refugees he had come to know. He began gathering evidence of misconduct by the Polish government, naively oblivious of

the dangers of such work until one night he found himself photograph-
ing policemen beating protesters—at which point the police started
beating him, too. A group of dockworkers then appeared out of nowhere, knocking the policemen down and yelling at Walczak, "*FRUNĄĆ!*
Flee!"—sending him on a late-night headlong race through the cobbled
streets of Gdansk, "my own little 'Midnight Run,' " Walczak recalls.

When he visited a friend at the U.S. consulate the next day, the diplomat brought out a huge, quarter-century-old portable radio and
cranked up the volume. "We know the consulate's bugged," his friend
said, leaning close and murmuring under the music. He instructed Walczak to bring all his papers and film and any other pieces of evidence to
the consulate as soon as possible, so that they could be sent home in the
diplomatic pouch, out of reach of Communist Party officials—and before Walczak could get into hot water over the material. "If they find the
stuff on you, you'll go to jail, and it could take us four weeks to get you
out," his friend said. "Kraków Prison is not a good place to be."

Walczak did as he was told. And, sure enough, when he left for home
a few days later, he was strip-searched at the airport and his suitcase was
cut open. The authorities at the airport were contemptuous and un-
touchable but ultimately unable to hold a U.S. citizen without any real
evidence of "subversion."

By the time Walczak got to law school in Boston a few months later,
he knew exactly the sort of law he wanted to practice: civil rights. What
separated America from Poland in 1984 was nothing intrinsic in the
people, the places, or the cultures, Walczak decided; it was the United
States' Bill of Rights, the envy of the world, a part of the national fabric
that sets America apart from all other countries, though it has long been
under siege on the home front and is in daily need of defense. Polls for
the last half century have consistently found that a majority of voters,
after having parts of the Bill of Rights read to them aloud without the
source being identified, would vote against such radical ideas as free
speech and limits on police powers if given the chance. In 2000, 53 percent of Americans polled favored a ban on public speech that criticized
religion; the same poll found that 37 percent could not name the five
freedoms guaranteed by the First Amendment (freedom of religion,
speech, public assembly, and the press, and the freedom to raise griev-

ances with the government). After 9/11 in particular, as is common during national crises, support for the Bill of Rights waned, to the point that
a clear majority of Americans polled asserted that the entire First
Amendment "goes too far in the rights it guarantees." The Founders'
vision of America rests on a razor's edge, as they knew it would, and
Walczak wanted to be part of defending that legacy.

After a stint as a legal aid lawyer in Maryland while his wife attended medical school in Washington, he joined the ACLU in Pennsylvania in the mid-1980s, holding a variety of positions, most recently as
the organization's legal director based in Pittsburgh.

Dover would not be his first church-state case—he had filed numerous suits involving school prayer over the years. Many people perceived
those suits as attacks on religion and often didn't understand when Walczak tried to explain how the suits were just the opposite—they were, as
he saw it, in defense of religion. He used to give talks at high schools
about this, as part of the ACLU's community outreach. To drive home
his point, he would start off a class by saying he'd like to dignify the
proceedings by beginning with a little prayer. This would surprise the
kids, but at first no one would object. "Now I'd like everyone to step out
from their desks and kneel down and face east over there, and we're going to say a prayer for Allah," Walczak would say. His expression would
be completely serious, and the kids would stare at him as if he were a
madman; they were all or almost all Christians, and finally, a kid would
say, "I'm not doing that. This is a public school; you can't make me pray
to Allah."

"Why not?" Walczak would ask. "Would it be any different if I said,
'Let's say the Lord's Prayer'?"

You could really get kids thinking that way, as Walczak recalls—the
lightbulb would go on, just like that. Adults, however, were another
matter; they tended to be more attached to their opinions, in Walczak's
experience, and such rhetorical exercises seldom changed long-held
views about church and state. In this regard, the ACLU was a target of
particular vehemence. Richard Thompson, for instance, the head of the
Thomas More Law Center and lead defense attorney for the Dover
school board, had spoken many times about his view that the ACLU,
federal judges, and the "homosexual lobby" were "at war" with Chris-

tians, engaging in a "concerted effort to destroy the religious foundations of our nation." Even Walczak, who thought himself impervious to invective, had been startled by some of the things Thompson had said in public, such as this description of the religious "cleansing" he felt was under way in America: "It's almost like a genocide. It's a sophisticated genocide." Walczak, who saw himself as a defender of religious freedom, whose family had been the victim of true religious and racial genocide, was more than slightly appalled at this sort of talk. Finding common ground, or even achieving common courtesy, is not easy when such worldviews collide; hostility is never far from the surface, and more often than not is laid bare.

Aside from the school prayer suits, Walczak had also handled a couple of cases involving the teaching of creationism in public schools. One case had been nipped in the bud with a simple warning letter to a school district, in which Walczak announced that the ACLU would sue if the district went forward with plans to introduce creationism in classrooms. The letter included a concise explanation of why the Supreme Court case law on the issue meant the ACLU would inevitably prevail in court, at great potential cost to the school district; the district immediately canceled its plans. This was Walczak's preferred method of resolving church-state cases—informally, without filing suit, and without school districts' running up crippling legal bills to be paid with money better spent in the classroom. His other case involving creationism, set in a high school in suburban Pittsburgh during the late 1990s, had been more complicated, but it too was quickly resolved. The high school had staged an assembly in which a creationist lecturer informed the students that man and dinosaurs had lived simultaneously and that there was scientific evidence to show that the Earth was exactly as old as the Bible described it—in short, a young-Earth creationism presentation, of exactly the sort the U.S. Supreme Court had forbidden a decade earlier. Attendance was compulsory. After the assembly, one teacher asked the students in her classroom to break up into groups, with those who believed in creationism lining up on one side of the room, and those who believed in evolution on the other. Only two students—one Jewish, one Catholic— stood up in support of evolution. The others students made fun of them while the teacher looked on approvingly. The two outcasts became Walczak's clients.

"You can imagine," he recalls drily, "that one settled rather quickly."

The "mock the monkeys" case had been simple, though: the constitutional questions clear, the conduct clearly reprehensible, the eventual apology heartfelt. It had been a slam-dunk winner the moment Walczak reviewed the facts; the outcome was never in doubt, and there was no need for a full-blown trial.

Dover would be different. Walczak felt no less certain about the constitutional questions and the importance of this case—he believed unequivocally that what the Dover school board had done was wrong—but proving it in court would be another thing entirely. Dover had layer on layer of complex scientific questions, months of contradictory conduct by a recalcitrant school board, a complicated paper trail, erased recordings of meetings, think tanks offering glossy brochures and curriculum guides, and a national strategy by the opponents of evolution to create a path into public schools designed to pass constitutional muster by avoiding the pitfalls creation science had encountered. Clearly, the law firm advising the school board and the center of intelligent design advocacy, the Discovery Institute, felt that path might be blazed through the quiet streets of Dover with a new idea, ID, that they insisted was science, not religion. The top experts in intelligent design would line up to testify that being God-friendly did not make their discipline *about* God, and their well-funded presentations undoubtedly would be practiced, articulate, digestible, and appealing. They would play well on the witness stand.

The simple truth was that the ACLU could not handle this one alone. Despite its fearsome reputation and its long list of landmark cases, the organization's state-by-state budgets have to be carefully nursed along, with no room for single-handedly taking on monster lawsuits like the one shaping up in Dover. A single big case such as this one, with its likely legion of depositions, pretrial investigation, expert witnesses, and inevitable layers of legal motions, could use up the entire litigation budget for years to come. The money just wasn't there to do it right; nor was the manpower.

Walczak knew he needed three things: money, experts, and a cadre of private lawyers willing to take on a big constitutional case not only for zero pay but at considerable cost to their firm's charitable works program—its pro bono budget. There was an incentive, though:

Walczak could offer takers the chance to lead a high-profile, potentially landmark case. This is how the ACLU handles many of its big cases—in partnership with private firms with deep pockets and a social conscience.

There was only one thing for Walczak to do, then: He called the National Center for Science Education (NCSE) in Oakland, California. The NCSE is the only organization in the United States with the full-time, single-issue purpose of defending the teaching of evolutionary theory. For lawyers working on the side of evolution in a case like Dover's, NCSE could provide biologists, paleontologists, philosophers of science, mathematicians, and a host of other consultants and expert witnesses— "one-stop shopping," as Walczak puts it.

Founded twenty-five years earlier as a citizens' group to counter the legislative push for creation science during the 1980s, NCSE has evolved into the Darwinian answer to ID's Discovery Institute and to the creationists' Answers in Genesis outreach organization. It has amassed a quarter century of experience in the evolution wars, serving as a clearinghouse for scientists and teachers concerned with improving science education and with the defense of evolutionary theory—which NCSE equates with a defense of science in general. The NCSE is also the repository of documents and archives collected across more than two decades on creationism cases and tactics, ID arguments, and the points where they intersect. The paleontologist Kevin Padian, a leading researcher on dinosaur to bird evolution and the evolution of flight, served as president of the organization during the Dover case, and the list of NCSE's supporters and consultants in the science community was a veritable who's who: the evolutionary biologist Richard Lewontin; the philosopher of science Michael Ruse; the cellular biologist Ken Miller, coauthor of the Dragonfly Book; the biochemist Bruce Alberts; Niles Eldridge, curator of the celebrated Darwin exhibit at the American Museum of Natural History in New York; and "Bill Nye the Science Guy," star of public television's current incarnation of the old "Mr. Wizard" science series and a familiar face to high school students nationwide. For every expert the proponents of intelligent design could produce—from the mathematics guru William Dembski to the biochemist Michael Behe to the philosopher of science Stephen Meyer—NCSE asserted that it could produce two, three, or more experts to dissect and, more often than not, oppose their claims about ID.

In that initial call, Walczak received a first, quick lesson on intelligent design from NCSE's executive director, the anthropologist Eugenie Scott. The reading list she provided for him was daunting, particularly to a guy who did not consider himself adept at science, but he gamely took down the titles, most of which seemed to have "Darwin" in them. Then he asked her the question that was of greatest concern to him early in the case: Did she know any lawyers with deep pockets who might be interested in signing on? Scott thought maybe she did, and she promised to contact one member of her center's advisory board, an attorney with the Pepper Hamilton firm in Philadelphia, to see how he felt about it. Walczak was intrigued, but he tried not to get his hopes up: Pepper Hamilton was an international law firm with ten offices, 400 lawyers, major clients in business, finance, and government, and the sorts of pro bono resources that could turn Dover into a dream case.

When Scott e-mailed the lawyer she had in mind, she apologized to him in advance, saying she wasn't sure he would want to be involved in a case such as Dover with the huge commitment it would entail, but if he had any free time, maybe he could help out a bit. An hour later, Eric Rothschild e-mailed back to say that he would do a lot more than a bit. "I've been waiting my whole life for this case," he told Scott.

Rothschild was a young partner at Pepper, a successful specialist in the somewhat rarefied form of commercial law known as reinsurance litigation—the contract disputes that arise between insurers and the companies that insure them. Reinsurance spreads the risk around the industry, but when a major disaster occurs, such as the destruction of the World Trade Center on 9/11, disputes about who owes what are inevitable, as are the lawsuits to settle those disputes as debts mount and companies go belly-up. Rothschild's passion for science, which had prompted him to volunteer for the NCSE advisory board, dated back to his early legal career. He had immersed himself in another big science case then, serving on the legal team that represented the insurance companies sued by cancer patients who attributed their illnesses to the infamous nuclear meltdown at Three Mile Island (which, by coincidence, was thirty miles from Dover). Rothschild's clients prevailed, as he mustered reams of scientific evidence indicating that the radiation released in the meltdown had been too inconsequential to have caused cancer rates to rise in the region.

Rothschild sensed right away that Dover would be a bigger and more personally rewarding science case to work on, and he immediately recruited another partner at Pepper, Steve Harvey. A former lawyer in the Justice Department under the first President Bush, Harvey had made a name for himself defending the government against claims of constitutional violations before entering private practice. The two attorneys worked well together, having teamed up earlier in the successful defense of the manufacturer of a flu vaccine accused of causing a rare neurological disorder. For the Dover case, with its religious overtones, the combination of Rothschild and Harvey provided an extra bonus: Rothschild was a practicing Jew and Harvey a practicing Catholic; both considered themselves believers in theistic evolution. Rothschild, who leans more to the liberal side of the political spectrum, and Harvey, who considers himself a traditional Pennsylvania conservative, both felt this case hewed to their own principles. Keeping government out of religion and preventing school districts from usurping the rights of parents, they say, could be considered both conservative and liberal causes.

Before a final decision on entering the case could be made, however, the two lawyers wanted to meet the clients, and they began with Tammy Kitzmiller, who would be listed as the first named plaintiff when the lawsuit was filed because her daughter would be in the first biology class under Dover's new ID policy. The two lawyers drove out to Harrisburg to meet her for lunch on the Sunday after the e-mail from Eugenie Scott, and they gathered at an Irish pub. Kitzmiller was nervous. The idea of a lawsuit, of being in the public eye, was not something she cared for in the least. She was a single mom, a private person, protective of her kids—but she also seemed to the attorneys to be a woman of determination, possessed of a steeliness belied by her slight frame and her quietness. The school board had crossed the line, she told the lawyers. Intelligent design is religion, plain and simple. And when it comes to God, or "the designer," or religious notions of creation, "it's not the school board's job to tell my daughters what to believe," she said. "That's what families are for."

Accomplished lawyers know a good witness when they see one, just as they know a likely disaster within minutes. Being a good witness is not a matter of polish, eloquence, or poise, though those qualities certainly are helpful. Fundamentally, it's a matter of heart. It's a matter of raw, heartfelt conviction. The lawyers could see—anyone could see—that this was not a plaintiff who had been chased down and recruited by lawyers. This was a woman who, against her natural inclination to avoid the spotlight, had come forward to do something difficult: to say no to authority, to risk the disapproval of neighbors and strangers, to have her name transformed into a thing, a court case, a shorthand term for the war between evolution and creationism—all because she believed it had to be done.

Before the drive back to Philadelphia had ended, Rothschild and Harvey had decided they wanted the case. They would join the ACLU and Americans United for Separation of Church and State, the Washington-based public interest group that had agreed to provide the services of the appellate attorney Richard Katskee—whose elegantly written briefs are legendary, veritable beach reading in a field better known for its ponderous prose. A short time later, Pepper Hamilton's pro bono panel enthusiastically endorsed entering the case, sealing the deal and ensuring that both sides of the Dover dispute would have capable, well-funded legal teams. In return, Rothschild would act as lead counsel, with Harvey as his co-lead, a gesture of appreciation for the money and support staff Pepper Hamilton was going to pump into the case. Doing it right would cost the firm millions of dollars, but if you're going to take on a public interest case, this was the sort most lawyers dreamed of: huge potential impact, extremely high profile, and a rare chance for a major corporate law firm to align itself with a cause that simultaneously championed civil rights, religious freedom, the foundations of science, and the "little guy" in the form of the parents of Dover.

With that, the legal team for the second Scopes Monkey Trial was complete, with one exception: The lawyers still needed a science geek.

For the plaintiffs in this case, the strategy would boil down to two separate but related issues. The first was the more clear-cut, or so it seemed: the conduct of the Dover Area School Board. This wasn't just about the evolution–intelligent design policy itself, considered in a vacuum. Through a long-standing legal test intended to detect breaches of the constitutional "wall" separating church and state, the lawyers would present evidence concerning the motives and "legislative intent" of the school board that helped shape the new policy, and they could present evidence of how that policy would affect students and citizens in Dover. If the intent of the board, or the principal effect of its policy, was to advance religion or to favor one faith over another, then the policy would be deemed unconstitutional, a violation of the First Amendment establishment clause, and the parents would win. If the evidence on this score was powerful and clear, it also would suggest a temporary restraining order as the best route to take.

If, on the other hand, the board's attorneys were right, and the primary intent of the policy was the improvement of science education, the Dover school board would win the case. Even if the board members had also harbored religious motives, or if the new policy had a real religious impact on citizens and students, that would be okay, so long as the effect was merely secondary to the legitimate purpose of improving science education. And, in an added twist, it wouldn't necessarily matter if the board's efforts failed to improve science education in Dover; so long as the road to the new policy was paved with good intentions, then the school board still would prevail.

The second half of the case was far more complex and involved the nature of intelligent design itself. Was it at root a religious cause, or was it a scientific idea with religious implications? The former would be unconstitutional, the latter constitutional. Was it a valid alternative to the theory of evolution, with criticisms of Darwin that were grounded in science and logic? Or was it faith dressed up in scientific clothes, a masquerade that resurrected old creationist criticisms grounded not in the scientific method but in Genesis and the gospel of John? If the lawyers could show, through expert testimony, that intelligent design itself was an unscientific religious proposition, then the board's intent would be irrelevant. The case could be won on the science. If the school board could show that ID was scientific—even if it wasn't good science—the policy would be upheld.

Winning on the first set of grounds would serve the parents well, but it would be a "small" victory, in the sense that it would be specific to the Dover school board and the explicitly religious comments the board members had made in adopting their policy. A case decided on those grounds alone would have little or no value as a larger precedent and would do little to settle questions about the constitutionality of teaching intelligent design itself. At most, the case would become a guide for other school boards interested in bringing intelligent design into the classroom. Rule one: Don't talk about creationism and Christian nations when designing your new policy.

Winning on the second set of grounds, however, would be an entirely different matter. Having a judge declare intelligent design to be a religious rather than scientific idea would be to win big. "That," Harvey observed, "would drive a stake through the intelligent design movement."

The lawyers would initially concentrate on the first grounds. The evidence for that part of the case would be key to the TRO decision, which had to be made as soon as possible. Time was running out, as the statement on evolution and ID would be read to students in January 2005.

At the same time, Scott would begin recruiting a team of scientific experts for the second part of the case strategy, the scientific grounds and the history of intelligent design. Research to support that part of the case would begin right at NCSE, with an obsessive young scientist just hired by the center as a public information specialist: Nick Matzke, a boyish self-described "science nerd" in his late twenties. No one quite understood at the time—least of all Matzke—that his involvement would prove pivotal, and that he would end up uncovering the single most decisive piece of evidence in the case. For the moment, all he could see was the months of work that lay ahead, and he immediately plunged into the musty archives of the center, disappearing for days at a time. Like others on both sides of the case, he sensed that he had become part of something important, and he began spending extra hours at work and on his own time at home doing research for the case. "I have a reputation for being thorough," Matzke would say many months later. "Or you could just call it completely insane."

Matzke grew up in Oregon and Indiana. He was fascinated, as a child, by questions of creationism and evolution, ever since his beloved grandmother, a young-Earth creationist, introduced him to her Bible-based beliefs about the world. When he was nine, she began regularly sending him creationist literature to counteract the influence of the dinosaur books he devoured. By the time he was ten, however, he recalls deciding that creationism didn't add up for him. "I wasn't buying it," he says now. "Even at that age, I knew something wasn't right."

His ninth-grade science teacher proved to be a star, the sort who routinely inspired a handful of his students to choose science as a career, and Matzke became one of them. He choose biology and chemistry as his double majors at Valparaiso University, a Lutheran campus in Indiana. He then went on to pursue a master's degree in geography, though his most intense interest was in the related field of biogeography, with its strong connection to evolutionary theory. He managed to pursue that interest in another way, however, by becoming an early and prolific Internet blogger on the subject of evolution, creationism, and intelligent design. He soon found himself part of an international network of scientists, teachers, and science-literate laypeople whose hobby led them, to use Matzke's term, to "cyberstalk" the claims of creationists and ID proponents. They traded online barbs with creationists and advocates of ID, attempted to debunk their scientific claims, and defended attacks on evolutionary theory from ID theorists, who had their own highly motivated posters and bloggers acting as counterweights to the evolutionists. For Matzke, this was strictly a hobby—albeit one that consumed many hours of his time as he sat hunched over a computer keyboard. The blogging consisted of one part civil discourse, one part making highly technical concepts palatable to both a scientific and a lay audience, one part stream-of-consciousness rant, and one part plain old-fashioned name-calling—his side would hurl around terms like "IDiots," then steam when a response dismissed them as "blind Darwinian dogmatists." The various science and ID blogs Matzke prowls have some of the liveliest exchanges and communities on the Internet.

In early 2004, having received his master's degree and casting about for a job until he could afford to pursue his doctorate, Matzke landed at NCSE, where, to his astonishment, he found he could get paid for doing what he liked to do in his spare time: write and blog about science, evolution, and criticism of intelligent design. "I still can't believe I get paid for this," he muttered more than two years after coming to NCSE. "I'm just a science nerd. This is so cool."

As the Dover school board began crafting its new policy—in the months before the policy became national news—Matzke was focused on the highly publicized evolution battles in Kansas and Ohio, and on attacking a new paper by the Discovery Institute's star, Stephen Meyer. Proponents of ID considered Meyer's article a breakthrough—the first paper on intelligent design published in a peer-reviewed science journal of the Smithsonian Institute, *Proceedings of the Biological Society of Washington*. Online critics of ID were quick to point out that in their opinion the paper wasn't a breakthrough at all: It was a "literature review" article that discussed previously published scientific works by others, rather than an account of new scientific research by Meyer.

Meyer's article sought to advance the argument that Darwinian evolutionary theory simply cannot explain the creation of new genetic information, leaving the action of an intelligent designer as the better explanation. DNA contains a sort of digital code, much like computer software, that is the programming language for all of life. All our experience, according to Meyer, tells us that it takes an intelligence to write software—all the random mutations in the world could not construct Linux or Apple's OS X, and the same holds true for DNA. The primary example in the fossil record of this theoretical shortcoming, according to Meyer, can be found in the unprecedented profusion of new life-forms that emerged in the distant past during a (relatively) short span of time—the so-called Cambrian Explosion more than 500 million years ago. From simple microscopic life-forms in the pre-Cambrian period, entirely new types of organisms arose with diverse body plans and organs, seemingly with no predecessors in the fossil record—not the sort of evidence gradual descent with modification is supposed to leave behind, according to Meyer. Reams of new genetic information had to arise out of nowhere to account for all the new forms of life that emerged during the mere 30 million years this explosion encompassed, far too short a span for Darwinian mechanisms, or so Meyer argued. And that, he said, left only design as an adequate explanation.

Matzke cowrote a scathing attack on Meyer's paper that was published on the leading Internet site for writing about evolution and criticizing intelligent design, PandasThumb.org. (The website is named for a famous essay by the late evolutionary biologist Stephen Jay Gould, who wrote about the panda's unusual adaptation for stripping the bark from bamboo shoots—a rather clumsy extra thumb—as a proof of evolution in action.) "Meyer's Hopeless Monster," as Matzke's posting was called, asserted that the paper was utterly lacking in scientific merit; that it was rife with bad science and uninformed conclusions that ignored well-known and experimentally observed instances of new genes and genetic information evolving; and that it falsely equated arguments *against* evolution with arguments *for* design. The Cambrian Explosion had been cited for decades by creationists as supposed proof of "gaps" in evolutionary evidence, but more recent discoveries had uncovered predecessor organisms in the fossil record that had previously been overlooked because of their fragility and small size, Matzke observed. And 30 million years (or 50 million, according to some researchers) for an evolutionary explosion of new types of life was not such a short time for natural selection to work wonders under the right conditions—*Homo sapiens*, after all, had only 5 million or 6 million years to evolve from a chimplike creature into a modern human, if recently obtained evidence from the chimp and human genomes is correct. The review concluded dismissively that Meyer and his fellow intelligent design theorists had not really advanced beyond invoking miracles, and would not be taken seriously by scientists until they did so: "They need a positive research program showing scientists that ID has more to offer than 'Poof, ID did it.'"

Meyer did not flinch. Indeed, he used the criticism to advance the Discovery Institute's arguments. He fired back on the Discovery Institute's blog that Matzke's article unfairly led to a retaliation against the editor who had approved the ID article for publication. The mainstream scientific establishment was so defensive about evolution's shortcomings that it would even engage in academic witch hunts against journal editors in order to protect the status quo and conceal the fatal flaws in Darwinian evolution, Meyer wrote:

> The National Center for Science Education and others have repeatedly insisted that the theory of intelligent design is not scientific because it has not been published in peer-reviewed science journals. Yet when an article proposing design does appear in a peer-reviewed science journal we are told that it should not have been published because it isn't scientific. . . . Unfortunately, this tidy circularity does nothing to address the mounting evidential difficulties facing neo-Darwinism. Nor does it get the unwanted elephant out of the room. If all living systems look as though they were "designed for a purpose" as neo-Darwinists have long acknowledged, and if neither neo-Darwinism nor any other materialistic evolutionary theory accounts for the most striking appearances of design in living systems . . . then perhaps livings systems look designed because they really were.

The posting on PandasThumb.org attracted several hundred comments and a nearly endless series of follow-ups, parries, and angry attacks from each side. It was in this overheated climate that, in October 2004, after the Dover school board policy was adopted, Matzke was assigned to work full time on the case. He began by pulling together boxes of files NCSE had accumulated over the years, piecing together a bibliography for the lawyers, developing lines of testimony for the scientific experts who would be recruited to testify for evolution, and preparing the attorneys to depose and, later, cross-examine the experts for intelligent design that the defense would bring in. Matzke was already something of an expert on the theories of Michael Behe and the bacterial flagellum, but it was still a huge job—especially as he was expected to continue with his other duties at NCSE as well.

Matzke, however, was in heaven. The old files proved a gold mine and would eventually yield important evidence. One of the most valuable bits Matzke uncovered in the archives: documents from the early 1980s concerning *Of Pandas and People*, the textbook advocating intelligent design that was about to be introduced into the biology classrooms of Dover. Matzke noticed that some old promotional material for the book, written years before publication and somehow squirreled away at NCSE, didn't mention ID as the subject of the book.

It mentioned creationism.

One promotional letter was dated before the Supreme Court's decision in the Edwards case that teaching creation science in public schools was unconstitutional. And the letter referred to an early draft of the book that was circulated at the time to prospective customers.

Matzke shot off an e-mail to Eric Rothschild, saying he felt confident that, if the team could get hold of the old draft of *Pandas*, it would prove invaluable. "I am reasonably sure that the word 'creation' would be substituted for . . . 'intelligent design' at many points within that manuscript. This would prove our point in many ways."

Of course, he added, if the publisher had any sense at all, it would have destroyed those drafts long ago.

Rothschild needed little convincing. After the lawsuit was filed and the judge had begun the process of legal "discovery"—during which each side was entitled to demand information and depositions from the other—Rothschild sent a subpoena to the publisher of *Pandas*, the Texas-based Foundation for Thought and Ethics. To everyone's surprise, the publisher had not destroyed the original manuscript pages of the textbook. The foundation had saved them—not one draft of the book, but five.

And suddenly they were in Nick Matzke's hands—6,000 pages of what would turn out to be the single most important body of evidence in the case, at least from the plaintiffs' point of view. First, though, someone had to go through each and every one of those pages.

"It was no problem," Matzke says happily. "That's what science nerds do."

Chapter 11

MONKEY SUIT

The lawyers, their clients, and a supporting cast made up of a scientist, a clergyman, family members, and the inevitable anti-evolution protesters gathered within the imposing granite façade of the Pennsylvania state capitol at Harrisburg on December 14, 2004, to announce their long-anticipated lawsuit against the Dover Area School District. Fifty-seven days had passed since the intelligent design policy had been adopted, and a press conference had been called to accuse the school board of wholesale constitutional violations. According to the lawsuit, the board had adopted a policy "to compel public school science teachers to present to their students in biology class information that is inherently religious, not scientific, in nature."

A group of print and broadcast journalists crowded around the lawyers and parents as copies of the lawsuit were passed out. This material featured a compact, acerbic history of intelligent design, the wedge strategy, and the movement's common ground with creationism—an unusually detailed pleading for such a preliminary document, thanks in large part to Nick Matzke's ferreting through the NCSE archives.

"Teaching students about religion's role in world history and culture is proper, but disguising a particular religious belief as science is not," Vic Walczak told the reporters, the ACLU man working hard to cast the suit as pro-science, not antireligion. "Intelligent design is a Trojan horse for bringing religious creationism back into public school science classes."

Reverend Barry Lynn, head of Americans United for Separation of Church and State, took the microphone as well, lamenting that the problem was not with Christianity but with the Dover school board's "misguided crusade." He complained, "They want our public schools to operate like Sunday schools."

The final tally of plaintiffs was as follows: eleven parents from eight Dover families had decided to join the suit—a diverse group in terms of background, income, and politics. Most were strangers to one another, and none were close (other than those married to each other), at least at the outset. Each seemed to represent a different part of Dover's culture.

There were Bryan and Christy Rehm, both teachers. They had four children, one of whom was in eighth grade and facing the biology course in the following school year. They represented, if unofficially, the voice of their friends, Dover High's estranged science teachers, who in a few weeks would face reading a statement to their students about evolution and ID with which they vehemently disagreed. Christy stepped up to the microphone to say, "I have faith in ourselves as parents. I have faith in our church community. And I have faith in our community as a whole. I don't believe it is the job of the school to teach my children a religious viewpoint." Biographical materials provided by the ACLU made certain that press accounts dutifully reported the Rehms' volunteer work as Sunday school teachers and Bryan's status as an Eagle Scout—inoculation against the charges of atheism that, sooner or later, were bound to arise in the "spin" war that would be fought during the yearlong wait from lawsuit to trial.

Joining the Rehms as plaintiffs were Deborah Fenimore and Joel Leib, who had a son in middle school, and whose ties to Dover's past and its roots as a bastion of religious tolerance were among the region's strongest. Leib, a teacher in a local private school, is sixth-generation Dover, part of a clan that has lived there since before the Revolution. "If somebody dropped a bomb in Dover," he quipped, "that would be the end of the line."

Cynthia Sneath was the co-owner, with her husband, of an appliance repair company. With a first-grader and a preschooler, she faced no immediate crisis, but she felt too strongly to stand aside and let others fight the board on her behalf. She described her own education as, "Graduated high school, diploma, life lessons, hopefully a dose of common sense," and her bottom line was her desire for her children to go farther in life and in their education, particularly in the sciences. She also voiced a desire for compromise and healing, telling the press she hoped the school board would back down and reverse its policy so they could drop the lawsuit.

The middle school science teacher Steven Stough, whose daughter was in eighth grade, came to the press conference, too, his buzz cut perfect and his opinion that ID was not science firmer than ever. He brought the insights of a specialist in science combined with the practical plain talk of an athlete: "I'm a coach, track and field. And I have an expression for that sort of thing: You can't shine mud."

The other plaintiffs included Julie Smith, a medical technologist whose tenth-grade daughter felt, as a result of the school board's actions, that she could no longer believe in evolution and still be a Christian. There were also Aralene "Barrie" Callahan, the former school board member, and her husband, Frederick, president of a paper company in York, who felt that the board was trying to intimidate teachers and stymie science. Beth Eveland, a legal assistant with two young daughters, experienced the board's actions as an outright attack on her values, a "slap in the face."

And finally there was Tammy Kitzmiller, who worked for a landscape company and struggled each day to make a life for her kids. She had one daughter who was scheduled to hear the new presentation on intelligent design in a few weeks. Tammy leaned forward and spoke into the microphones, plain and direct. "The Dover school board created this policy for religious reasons," she said. "It's wrong."

The staged scene had a surreal aspect to it, beginning with a measure of control, then spinning off into unexpected directions. The attorneys had recruited a prominent paleoanthropologist from Penn State University, and he spoke to the press about the scientific issues at stake in the lawsuit. A pair of protesters inched ever closer, listening with disdainful expressions and waving signs that read, "ACLU Censors Truth" and "Evolution: Unscientific and Untrue." One of the demonstrators passed out literature promoting the creationist views of Kent "Dr. Dino" Hovind—probably the last person with whom the advocates of design at the Discovery Institute would wish to see their cause associated. Meanwhile, the professor from Penn State, Robert Eckhardt, patiently explained the mainstream scientific views of evolution and ID, which definitely did not accommodate the notion that Adam and Eve saddled up a few friendly dinosaurs for a quick spin around the garden.

Contrary to the assurances of the Dover school board and the arguments emanating from Discovery, Eckhardt doggedly informed reporters that there was no controversy about evolution in the scientific community. Disagreement about details? Sure. Uncertainty about how (though not *if*) certain features evolved? Absolutely. But those were healthy, productive differences between working scientists, he explained, which certain proponents of ID misleadingly exploited to suggest a larger problem with evolution itself. Not so, he insisted. The notion that intelligent design represented a powerful new challenge to science or that evolution was mired in controversy was "absolute nonsense."

Eckhardt had led a research team whose careful, recently published study of a fossil thighbone 6 million years old provided new evidence of the earliest known human ancestor to walk fully upright. The chimp-sized hominid strode through what were then the rain forests of Kenya, and Eckhardt's findings suggested that the advent of bipedalism occurred about 2 million years earlier than had previously been thought, adding new evidence for the gradual evolution of man from early, ape-like primates. Eckhardt's Laboratory of Comparative Morphology and Mechanics at Penn State was a leader in this sort of analysis, using medical CAT scans and other high-tech methods to unravel evolutionary mysteries right in Dover's backyard. Yet, he complained, no one on the school board had even thought to consult with Eckhardt or any of his colleagues at Penn State before opining on Darwin's failings and crafting a new curriculum for evolution and intelligent design.

Eckhardt's eloquent pronouncements triggered two unintended but predictable events. First, one of the protesters sneered at the scientist, in his professorial tweed jacket and heavy glasses, proclaiming Eckhardt a "screaming leftist unbiblical liberal." Second, Eckhardt's statement guaranteed that every reporter present would do his or her best to quote an advocate of ID taking the contrary position, thereby creating a portrait for news readers and viewers that was the exact opposite of what Eckhardt had hoped to convey: scientists disagreeing, which surely meant, in the popular imagination, that there really must be a controversy. The news media's eternal quest for a simpleminded form of "balance," the "he said, she said" convention of daily journalism that for the most part abdicates responsibility for gauging the relative truthfulness of opposing sides, made this result inevitable. Therein lies one of the most ingenious techniques of the Discovery Institute: Simply saying that there is a controversy (or baiting evolutionists into denying that a controversy exists) guarantees that there will be a controversy in the next edition of the newspaper. In biology, this is called an autocatalytic reaction—a self-sustaining replication of a particular molecule or protein. The lawsuit, then, represented not so much a setback for the ID movement as an opportunity to replicate controversies almost endlessly. And the existence of a genuine scientific controversy would be valuable evidence indeed once the lawsuit came to trial—perhaps all that was needed for ID to find a constitutionally protected place in the public school classroom.

As Eckhardt finished, the ever-present Reverend Jim Grove, the anti-abortion pastor who would become a fixture at the trial the following year, materialized on cue in the capitol rotunda to challenge the professor to a debate. Eckhardt laughed and shook his head, which Grove gleefully took to be a concession of defeat for evolutionary theory, uttering what would soon become his common refrain: They don't have "the facts." In Grove's estimation, the missing facts consisted of the unsolved and unsolvable mystery of how a fish could turn into a bird—his vision of what evolution purports to be.

There was an even odder turn of events at this press conference than the image of the bristly Grove pronouncing victory by default over evolution. Amid the reporters, gawkers, and protesters, a diminutive woman with wavy blond hair scribbled notes intently as the lawyers and plaintiffs spoke. She was Angie Yingling, a member of the Dover school board. As the rest of the board declined comment in the wake of the lawsuit, Yingling had decided to observe the event in Harrisburg and to announce that her goal was to use what she learned this day to achieve the demise of the ID policy she had helped usher in.

Yingling, the malleable real estate agent whose inconsistent voting record and cluelessness about both evolution and intelligent design proved emblematic of the school board's approach to the issue, then made this revealing comment about the ID proposition she had voted for in October: "Anyone with half a brain should have known we were going to be sued."

There was a certain weird logic to Yingling's presence at an event reviled by her friends and colleagues on the school board. It was Yingling whose change of heart and tie-breaking vote had finally allowed the Dragonfly Book into the high school and enraged Bill Buckingham, although Yingling had returned to his voting bloc a few months later in approving the new intelligent design policy. This vote came not, she would later admit, because she understood much about ID, but because she was afraid of being branded an atheist. Now, as the lawsuit loomed, she had changed her mind yet again, saying: "I just want it to go away."

Reaction to the lawsuit hit Dover like a tidal wave, as national media, from *The New York Times* and the *Washington Post* to *Salon* and *Wired,* arrived on the scene to write profiles of a divided community, of science versus religion, and of Red versus Blue (Pennsylvania having elements of both). The media sensed a trend: President Bush had just been reelected five weeks earlier, and one of his most important and energized constituencies, evangelical America, was riding high and ready to push for the reforms it had long sought. An emboldened school board in Grantsburg, Wisconsin, followed Dover's lead and imposed a policy to criticize evolution in line with the Discovery Institute's claims, while South Carolina, Tennessee, Michigan, Oklahoma, and Texas quickly introduced similar statewide legislation. In California, a coalition of Christian schools sued the University of California for allegedly violating the constitutional rights of evangelical students. The offense: refusing to give credit to university applicants who had taken high school biology courses that taught creationism instead of evolution. In Florida, legislators introduced an "academic bill of rights" that would empower students to sue their professors for offending their beliefs in class—such as by teaching that evolution is a fact and that creationism is unscientific. In Kansas a clear creationist majority was swept into power on the state school board when George Bush was reelected, starting a process that would lead to the epic hearings in the spring of 2005. And in Missouri, a far-right state representative, Cynthia Davis, introduced legislation requiring all biology textbooks used in the state to include at least one chapter on alternatives to evolution, such as intelligent design. Davis called this a "commonsense" proposal that had nothing to do with injecting religion into the classroom, echoing comments by the Dover school board members as she said, "This bill is for the purpose of allowing the students to learn about science without the influence of those who are hostile toward God. The purpose of the biology class is to learn about science, not how we could have created ourselves as if atheism were the only option." In addition to linking evolution to atheism, Davis offered a chilling comment likening anyone who thought differently to the murderous terrorists of 9/11.

"It's like when the hijackers took over those four planes on Sept. 11 and took people to a place where they didn't want to go," Davis told *The New York Times* two days before the Dover lawsuit was filed. "I think a lot of people feel that liberals have taken our country somewhere we don't want to go. I think a lot more people realize this is our country and we're going to take it back."

Dover, by popular acclaim in the news media and among many of the players in the unfolding drama, had become ground zero for that sort of political calculus—the newest front in the nation's culture wars. Typical were the comments of Steve Farrell, owner of a local nursery, who wholeheartedly supported and prayed for the school board. He had found God, he said, when his newborn daughter died of severe birth defects, and since then he has seen his most ardent prayers answered time and again: prayer, he told one reporter, had cured his mother of cancer, rescued his failing business, healed his hernia, and ended a local drought. And his prayers were answered again by the school board's decision to "bring God back into public school." He even prayed for the parents who opposed the new anti-evolution policy in Dover—because, he said, they need prayers most of all.

"We just want the information to go out that ninety-five percent of us believe: that God created the earth," Farrell pleaded. "Why can't the teachers say that?"

This was, of course, the message that resonated most, and Farrell's plaintive remarks were reported across the country, sounding both spiritual and reasonable. Farrell asserted that he just wanted balance—there should be room for both evolution and God.

But there was plenty of opposition, too. The *York Dispatch* published a stinging newspaper editorial attacking the board for "angling" for publicity and abdicating its responsibilities. The biology faculty at York College wrote the school board a blistering and widely reported letter accusing it of ignorance of evolutionary theory and of having "a profound misunderstanding of the scientific process and an equally profound disregard for the science educators and students." Letters to the editor both praised and attacked the college faculty members; one letter bore the caption, "Love God or leave America, professors." Angie Yingling offered her resignation when the board would not reverse its policy (her resignation was refused for the time being, when fellow board members asked her to think about it a bit longer). And the *York Daily Record* editorialized, "Everyone who might help stop the Intelligent Design Express is jumping off. . . . [It's] like watching a train wreck in slow motion."

Richard Thompson, chief legal counsel for the Thomas More Law Center, acted as spokesperson immediately after the lawsuit was filed, standing in for the silent board members even before he was officially retained. He confidently predicted victory in court for the school board and expressed a surprising eagerness to join the battle. "There is a difference between teaching science that happens to be harmonious with the Book of Genesis, as opposed to being purely motivated by teaching Genesis," he said, previewing one of the principal arguments he'd bring to trial. "Intelligent design may have religious implications, but that does not make it religion." After all, he added, evolution has religious implications, too.

In previewing another likely trial strategy, Thompson soon began laying the groundwork for making Buckingham a fall guy—a lone voice who did not speak for the whole board—and he promised to "tear apart" the notion that the new policy on evolution was religiously motivated. The lawsuit had pointedly quoted Buckingham's reported remark— "This country wasn't founded on Muslim beliefs or evolution. This country was founded on Christianity and our students should be taught as such." But Thompson told reporters that statement meant nothing as far as the legal issues were concerned. "You cannot shackle the rest of the school board with that one statement. . . . There is nothing wrong with having a religious intent with legislation having a secular purpose."

While the board's future legal counsel came out swinging, the Discovery Institute—which was expected to provide the scientific firepower to stave off claims that ID was merely creationism disguised as science—remained curiously unenthusiastic. It greeted the news of the lawsuit more with foreboding than with its characteristic combativeness, which was then on display in Kansas. Discovery representatives went so far as to brand Dover's new policy "misguided." Although two of its senior fellows, including its cofounder Stephen Meyer, had previously published a proposed curriculum guide for teaching ID in public schools, the institute's position had changed. Now it downplayed the teaching of design because ID remained, in the words of one Discovery press release, "a growing and emerging theory." The institute's proponents of design recommended that schools focus more on incorporating criticisms of evolution into the curriculum rather than introducing ID textbooks and lesson plans.

This was an evolving position at Discovery, and one that appears to have had as much to do with legal strategy as with good science or pedagogy. This was particularly true in the case of Dover, where published reports of the board's initial discussions suggested that creationism and ID had been presented as interchangeable, and that for at least some board members there was a religious motivation behind the curriculum policy. Discovery's attorneys undoubtedly reached the same conclusion as the ACLU: Those early discussions by the school board could have loaded the case with so much creationist baggage that it could be transformed from an ideal test case for proponents of intelligent design into an ideal challenge for opponents.

These doubts first surfaced in Seattle in October 2004, two days after the Dover school board had accepted the sixty donated copies of *Of Pandas and People* and had begun formulating its curriculum changes. Discovery issued a press release asserting that it had never endorsed the adoption of *Pandas* or any other ID book as a supplemental text in public schools, though the institute still defended the school board's right to do so and to teach ID. Quoting Seth Cooper, the in-house attorney who made the initial contacts with Bill Buckingham, the press release stated:

Our recommendation is that students receive a full and fair disclosure of the facts surrounding Darwin's theory and that the leading scientific criticisms of the theory not be censored from classroom discussion.... Of course it is perfectly acceptable for teachers to discuss the full range of scientific theories of biological origins and that includes the theory of Intelligent Design, but our recommendation is that Darwinian evolution be completely and fully covered, including it's [*sic*] strengths and weaknesses.

Two months later, Discovery began to backpedal. Regardless of how "perfectly acceptable" ID was in theory as a classroom topic, Discovery apparently didn't want Dover to cross that line, and the institute's actions began to border on the desperate. In the days before the lawsuit was filed, it dispatched two representatives to Dover for secret meetings with the board, offering to take over legal representation in lieu of the Thomas More Law Center. But there was a catch: the school board would have to alter its policy, focus on advocating a "critical analysis" of evolution, and omit any requirement for teaching intelligent design—the course Discovery had most recently begun to push in Kansas and Ohio. This approach, in Discovery's view, would withstand any constitutional challenge. (The ACLU, among others, disputes that assessment, but agrees that Discovery's critical-analysis approach would be a much easier suit for school districts to defend.)[1] Mandating ID, however, was too risky, given all the creationist talk that had reportedly preceded it. "From the start, we just disagreed that this was a good time and place to have this battle, which is risky, in the sense that there's a potential for rulings that this is somehow unconstitutional," according to Mark Ryland, director of Discovery's Washington office, who flew to Dover for a last-minute meeting. But the Dover board was not interested in backing down—"I fight to win," Buckingham said afterward. It didn't hurt that there was already an offer on the table from the Thomas More Law Center to defend the existing policy. Richard Thompson was there the same day as Ryland, not speaking of caution but instead urging the board to take the case all the way to the Supreme Court if necessary. He expressed disgust with Discovery's attitude: "As soon as there's a conflict, they will back away."[2]

Two days before the lawsuit was filed, Phillip Johnson, in an interview with the *San Francisco Chronicle*, added to the distancing of the intelligent design movement from Dover, saying that he did not recommend the course taken by the Dover school board. Instead, he, too, recommended a classic wedge strategy: "Just teach evolution with a recognition that it's controversial. A huge percentage of the American public is skeptical of it. This is a problem that education ought to address."

Finally, on the day when the lawsuit was filed, a new press release appeared on the Discovery website criticizing the Dover policy for its "incoherence" and dubious constitutionality, and calling for it to be "withdrawn and rewritten." This was a far stronger statement than Discovery's previous release. Quoting John G. West, associate director of the institute's Center for Science and Culture, the new release said:

> While the Dover board is to be commended for trying to teach Darwinian theory in a more open-minded manner, this is the wrong way to go about it. . . . We don't think Intelligent Design should be required in public schools. What should be required is full disclosure of the scientific evidence for and against Darwin's theory, which is the approach supported by the overwhelming majority of the public.

There was only one way to interpret that statement, Nick Matzke and the parents' lawyers concluded gleefully: The Discovery Institute seemed to be running scared.

Many residents of Dover began to feel a similar urge to run. They had liked seeing their neighbors quoted in *The New York Times* and seeing Dover featured as some sort of buckle on the Bible belt on Bill O'Reilly's talk show on Fox, but once the novelty wore off, it was hard to feel good about where all this notoriety was taking the township. The citizens were not necessarily opposed to ID—one poll of Dover's registered voters found that 54 percent supported ID in the classroom, with 36 percent opposed and the rest undecided. If philosophy and science were the only considerations in assessing intelligent design—as the Discovery Institute argued in its wedge strategy—then ID was a huge winner with the public. But there were other factors. Many residents of Dover dreaded having their town cast as backward and as an object of national derision. Even more important, significant numbers feared that the lawsuit would prove far more costly than it was worth to the school district and the taxpayers. This is why the same poll that showed support for ID also revealed that nearly half the town, 44 percent, thought the school district was headed in the wrong direction. Noel Wenrich, one of the board's former members, calculated the odds this way: An overwhelming percentage of residents liked the ID policy, as long as it cost them nothing. "But not if taxes go up. Then it's 30 percent."

The members of Dover's school board, once so voluble, had nothing to say about the poll or intelligent design, refusing to comment or simply avoiding reporters' phone calls and questions once the new policy was in place. They preferred to talk about academic progress and lower taxes— the subjects they wanted to emphasize when all but one board member would face a primary election contest. Even Bill Buckingham, who had filled the role of quotation machine for the press for so many months, clammed up.

To an extraordinary degree, the Dover school board had insulated itself from any opposing points of view well before the lawsuit was filed. Even the former board member Wenrich had been warned to be quiet when, a few weeks after his resignation, he came back to demand an apology for having his Christian faith challenged during his last board meeting. He stormed out without getting it. Sheila Harkins, a fifty-five-year-old homemaker who was a proponent of ID, took over as board president from Alan Bonsell, and she severely limited public comment, requiring all questions to be submitted in writing in advance. There would be no more freewheeling meetings, no more shouting matches. The new intelligent design policy all but disappeared from the public discussions of the board.

The board itself was also made completely homogeneous, with not even a semblance of balance or opposing opinions. Four proponents of intelligent design, out of a dozen applicants, were appointed to fill the vacancies left by the resignations of Casey and Jeff Brown, Noel Wenrich, and, later, Angie Yingling, whose attempts to alter the board's course went nowhere. Any applicants who had experience in education or who did not voice support for ID—Bryan Rehm among them—were all rejected by the board. None of the replacements had any experience in government or education, but two of them had stood up at public meetings to praise the board for its stand on intelligent design. Another, Ed Rowand, the pastor of a local evangelical church, won his appointment after saying, "If the Bible is right, God created us. If God did it, it's history and it's also science." Another appointee, Eric Riddle, did not send his own children to the Dover schools he was appointed to oversee; he home-schooled them for religious reasons.

Rehm sought appointment to the board before joining the lawsuit and after leaving Dover High to teach at his wife's school in York, avoiding any conflict of interest at the time. Nonetheless, he had a particularly bitter experience when questioned by the board members in a kind of public job interview, in which his views on evolution were treated with disdain and Bill Buckingham asked him—alone among the applicants—if he had ever been accused of child abuse or molestation. Rehm left with his face burning.

That same month, Rehm had asked board members for permission to listen to the audiotape of the October board meeting. Although the tape of each meeting was eventually destroyed once the official minutes had been written up (invariably stripped of any controversy), that process took a month or more, and the district still had the tapes of the key meeting in which the new policy on evolution and ID was adopted. In the past, members of the public had been allowed to listen to the tapes, but this time, board members turned away Rehm and others interested in reviewing them, flatly refusing to give access to what, even under Pennsylvania's notoriously weak public access laws, were clearly public records. The explanation for the denial was given by Bonsell: On the recommendation of its attorney, the board was going to keep the tapes sealed and then destroy them in order to protect board members from potential legal problems. The *York Daily Record* responded with a blistering editorial:

> It must be nice to be William Buckingham—or even just a member of the Dover school board "majority." It's almost like having your own little kingdom and castle—Buckingham's Palace. . . . Want to make like Henry VIII and declare your

own little state religion? Go right ahead. . . . And when your
subjects raise objections, you can put them down by questioning
their faith or their patriotism. . . . If someone dares to take
offense at your insults, you can simply order that the evidence
of your royal meanness be destroyed. Rewrite history to your
own liking.

An already poor relationship between the local press and the school
board went downhill from there.

A week after the lawsuit was filed, the board officially retained
Richard Thompson and the Thomas More Law Center as legal counsel
in the case. The district administration, meanwhile, continued perfect-
ing its new ID policy by producing a final version of the "one-minute
statement" on intelligent design and the shortcomings of evolution,
which the teachers were supposed to read to their biology classes beginning January 18, 2005. All language even remotely respectful to evolution had been removed from the statement (such as a reference to
evolution as the dominant theory of biological origins); and additional
language forbidding students to ask questions about intelligent design
was added, making this curriculum not only the first to mandate the
introduction of ID in public schools, but also the first to tell students
about a new idea which they were then instructed not to discuss or
question in class. A letter would be sent home to parents a week in advance of the statement's debut, explaining the policy, including students'
and parents' right to opt out.

When the final statement was drawn up and passed out to the faculty, the science teachers at Dover High, who had already demanded the
retraction of a press release that had misleadingly implied their support
for the curriculum change, rebelled in earnest. After attempting unsuccessfully to persuade the board to modify the statement, they decided as
a group that they would have no part of it. With the assistance of their
union's attorney, the teachers wrote a letter to the superintendent declaring that, like students who did not wish to sit in class while the statement on ID was read, they, too, wished to exercise their right to opt out.
If they agreed to read the statement approved by the board, the teachers
said, they would be violating the state code of professional conduct for

educators, which, among many other responsibilities and duties, forbade teachers to misrepresent subject matter in class. After pointing this out, the letter from the teachers, using words written in all capitals, stated, "INTELLIGENT DESIGN IS NOT SCIENCE. INTELLIGENT DESIGN IS NOT BIOLOGY. INTELLIGENT DESIGN IS NOT AN ACCEPTED SCIENTIFIC THEORY." Telling students that ID was legitimate science, or even implying this by announcing that it was an alternative and referring kids to *Of Pandas and People*, would constitute a breach of ethical obligations under state law, the teachers maintained.

The science teacher Robert Eschbach explained it this way after reviewing and rejecting the board's statement, which he likened to a "warning" about the dangers of evolution: "Kids are smart enough to understand what intelligent design means. The first question they will ask is, 'Well, who's the designer? Do you mean God?'"

Board members were furious at this turn of events. There were mutterings about insubordination and possible firings, but Thompson cautioned the school board against doing anything rash or overly harsh. The board did not want a job action or additional litigation at this point, and there were other ways to achieve its goals. It was decided that the administrators would read the statement instead, and Assistant Superintendent Mike Baksa, as usual, drew the task of making the first round of classroom visits. "The Dover faculty have no right to opt out of a legal directive," Thompson told the press. "Having said that, because there is pending litigation . . . we are going to accommodate their request."

The final version of the statement was crafted as a complete script for the administrators, that even included alternative word choices for the appropriate time of day:

1. Good morning/afternoon. I am Mr. Baksa and standing over there is Dr. Richard Nilsen . . .
2. You will soon begin to study Evolution in the class and the Board of School Directors has directed that the following statement be read. The statement is currently under litigation; therefore, the administration is reading the statement, not the teacher.
3. A letter went home asking if anyone had a problem with the

statement and I would like to make sure at this time that everyone who would prefer not to hear the statement is now out of the classroom. Anyone else?

4. OK, the statement.

> The Pennsylvania Academic Standards require students to learn about Darwin's Theory of Evolution and eventually to take a standardized test of which evolution is a part.
>
> Because Darwin's Theory is a theory, it continues to be tested as new evidence is discovered. The Theory is not a fact. Gaps in the Theory exist for which there is no evidence. A theory is defined as a well-tested explanation that unifies a broad range of observations.
>
> Intelligent Design is an explanation of the origin of life that differs from Darwin's view. The reference book, *Of Pandas and People,* is available for students who might be interested in gaining an understanding of what Intelligent Design actually involves.
>
> With respect to any theory, students are encouraged to keep an open mind. The school leaves the discussion of the Origins of Life to individual students and their families. As a Standards-driven district, class instruction focuses upon preparing students to achieve proficiency on Standards-based assessments.

5. As noted in the last paragraph of the statement there will be no other discussion of the issue and your teachers will not answer any questions on this issue. If you or your parents have any questions, they can contact Dr. Nilsen, Mr. Basksa or Mr. Riedel.

6. Thank you and have a nice day.

Before the statement could be read, however, one last obstacle remained. The parents suing the district had asked the federal judge assigned to the case to order immediate depositions of key players in the case—Buckingham, Bonsell, Harkins, and Superintendent Nilsen—so the lawyers could develop the evidence they needed in order to request (or forgo) a temporary restraining order before the statement was presented to the biology classes. So, on January 3, 2005, the first Monday after the New Year holiday, Eric Rothschild, Steve Harvey, and Vic Walczak, along with the attorney Pat Gillen from the Michigan headquarters of the Thomas More Law Center, gathered in Dover to depose the four school officials about their policy and the meetings that had led up to it.

To the surprise of the plaintiffs' lawyers, all four school officials denied ever mentioning creationism at any meeting.

Nor could they recall hearing anyone else bring up the subject of creationism. They had never sought to introduce religion into the classroom, each of them assured the lawyers, and they had no religious agenda. Anyone who said otherwise, including the two newspapers in York County and the two reporters who had covered those pivotal board meetings, had it wrong, the witnesses said.

"Never said it," Buckingham swore. ". . . I've never had anyone ever talk about looking for a book of creationism and evolution."

"I was part of the curriculum committee, and I've never had anyone ever talk about looking for a book of creationism and evolution," Harkins said.

"I didn't say creationism," Bonsell said. "I never have."

Two board members did recall Buckingham's saying at a board meeting: "Two thousand years ago, someone died on a cross. Can't someone take a stand for him?" But both Harkins and Buckingham denied that this statement was made in June 2004 during the battle over the biology textbook, as reported in York's two competing newspapers. Instead, each board member made the identical assertion (in separate sworn testimony) that the incendiary remark was actually made eight months earlier, in November 2003, during a board debate concerning the phrase "under God" in the Pledge of Allegiance. Somehow, if the board members were to be believed, two reporters for two different papers had pulled the same dated quotation out of their notebooks and then lied in print about when the statement had been made, out of malice or bias or as part of some sort of conspiracy. Bonsell, however, said he could not remember when this statement was made.

The board members contradicted not only the press reports but also the recollections of numerous other witnesses, including Jeff Brown, who remembered the remark about Christ on the cross occurring during the textbook debate in June because, he recalled, "It kind of made me want to crawl under the table." And a highly regarded local pastor, Warren Eschbach—whose son Robert taught science at Dover High— found the board's testimony about never using the word "creationism" perplexing. Eschbach had been present during those early debates and was moved to ask board members, "Are you sure you want to mandate the teaching of creationism?" It was their word choice, he recalled, not his.

The board members also insisted that they repeatedly told the reporters covering their meetings that they had been misquoted and that the board's actions were misrepresented in the pages of both newspapers. Buckingham simultaneously maintained that he never actually read the newspapers, except for the obituaries. And he and his colleagues could not explain why no one at the district had ever written a letter to either newspaper formally requesting a correction or a retraction for any of these alleged inaccuracies, including the supposed confusion over the remark beginning "Two thousand years ago." Both newspapers stood by the accuracy of their reports and reporters.

As for having a religious motivation in adopting the new biology curriculum, Bonsell testified at the deposition that he had none; his only motive was to have students "critically think . . . to search for the truth." He also insisted that the policy required no actual teaching of intelligent design. Kids simply were being informed, not taught, he said.

Eric Rothschild began to get impatient as this parsing of the word "teach," which continued throughout the depositions. Rothschild finally suggested that when a school official delivers a long statement to a class, that is in fact "teaching." The students would be learning something. Otherwise, what would be the point? The only alternative, Rothschild pointed out, would be *not* learning.

"Will they be learning, Mr. Bonsell?" Rothschild asked repeatedly.

But Bonsell would not answer yes or no, saying only, "They will be made aware." He repeated this phrasing several times, apparently unaware that the dictionary definition of "to make aware" is "to give knowledge," which also happens to be a pretty good definition of the verb "to teach." Pressed further for a yes or no answer to the question "Will they be learning?" Bonsell added, "If you define learning as teaching, then no."

Rothschild snapped, "No one defines learning as teaching," but he still got nowhere.

The board members also insisted in their sworn depositions that they believed intelligent design was not in itself religious. Buckingham was adamant on this point, although he could provide no coherent explanation of what ID was. In trying to do so, he talked about molecules and amoebas, but his attempt to define ID ended up sounding more like a half-baked description of evolution. His attempt to describe evolution and common ancestry, in contrast, sounded a great deal like religion: "It can be what scientists considered two tiny amoebas way back zillions of years ago if you want it to be. It can be, for some people, it can be Buddha. For somebody else it can be Allah. For a Christian it can be a Christian God. Whatever." This was his explanation not of intelligent design or creationism but of Darwin's idea of descent with modification.

Later he added, "My faith is founded on the Book of Genesis. Again, I'm not a scientist, but it's my understanding that the theory of evolution, where it goes back to the beginning of man, it's a happenstance. It just happened. And that is inconsistent with my faith."

All the officials insisted that intelligent design did not necessarily imply that God created anything. When asked who else the "intelligent designer"—or, as *Of Pandas and People* phrased it in one passage, the "master intellect"—could possibly be other than God or some divine being, Superintendent Nilsen had a ready answer: space aliens.

When the depositions were through, the lawyers representing the eleven parents had a problem. With the board members denying any talk of creationism or religion in forming their policy, the main grounds for seeking a TRO were in dispute. Had the board members admitted a religious motivation—for which there seemed to be substantial evidence—the decision to request a TRO would have been clear-cut. But now it would require a substantial investigation, interviews of witnesses, and the complicated process of getting working journalists to testify in order to determine the truth of the matter. There would be no time to accomplish all this before the statement was read in class in two weeks.

For better or worse, Vic Walczak's and the others attorneys' TRO dilemma had been resolved. This had been one of the few points of disagreement among members of the legal team, with Walczak pushing hard to forgo the TRO and others more willing to seek it. After the depositions, however, the members of the legal team found themselves in agreement. With its flat-out denials, the school board had left them no choice but to tell the court that the parents would not seek a temporary restraining order. The district could go ahead and read the statement to students as planned.

Publicly, this was perceived as a setback for the parents who had filed the suit. The lawyers told news reporters that the battle was far from over and left it at that, content to start interviewing witnesses about their recollections of those meetings. In truth, they saw this not as a setback but as a huge opportunity. The board members had gone from being forced to admit engaging in some loose talk about religion—which appeared constitutionally questionable though not necessarily fatal—to putting their credibility on the line by denying statements dozens of other people had heard, remembered, and read in newspaper accounts. Instead of having to pillory devout board members for being honest about their Christian beliefs—a difficult task guaranteed to generate sympathy for the school board in and out of court—the plaintiffs' attorneys had just received the greatest gift possible: a chance to prove the board members liars about both God and government, a position with which no one would sympathize. And the parallel details in some of the board members' sworn testimony—particularly when it came to the identical assertions by Harkins and Buckingham that the incendiary remarks about Christ had been made at a much earlier meeting—left the attorneys very suspicious about what sort of preparations and discussions the board had undertaken to get ready for those critical depositions. Indeed, a memo written by Superintendent Nilsen to the board members after the depositions described how "gratified" he was that the ACLU had been foiled because of "the time and effort put in over the holidays." The board members "did a great job," he wrote. "The plaintiffs/ACLU could not find anything to file an injunction on our Biology curriculum."

Richard Thompson was unwaveringly exuberant when the plaintiffs decided against a TRO, and the Thomas More Law Center issued a triumphant press release captioned, "ACLU Abandons Early Effort to Stop School District from Making Students Aware of Controversy Surrounding Evolution." Thompson said, "It became clear that they simply did not have a strong enough case." The ACLU was inflating a simple one-minute statement, a reasonable exercise in fairness and balance, into a cause it could not support, Thompson maintained.

Vic Walczak angrily took exception to this, saying that the length of the statement was irrelevant. "The parallel I would draw would be, if a social studies teacher teaching World War II would talk about the Holocaust and make a statement—just a couple of paragraphs—that there are gaps in the historical records of the Holocaust, and you should know an alternative theory that the Holocaust never happened."

On January 18, 2005, Assistant Superintendent Mike Baksa walked into the biology classes and read the carefully prepared script. About 160 students heard him read the statement over the next two days, and some fifteen students opted out, along with all the science teachers, who stood in the hallway with their excused students. Afterward, the sentiments most commonly expressed by the students seemed to be: "I'm confused," "What did that mean?" and "The whole thing is kind of dumb." In keeping with the board's mandate, no discussion or questions were permitted.

One student who listened to the statement told a reporter she was baffled by what intelligent design was all about and why it was even being brought up. Many kids had questions, Danielle Yagodich recalled, but though the administrators said they were the only ones who could answer such questions, they vanished before a single kid could raise a hand. "Pretty much on the last word they were headed for the doors," she recalled.

"Half of us don't even care," another biology student observed later. "We think it's kind of stupid that people are going to argue over it."

A few students—some of those who opted out—were angry that they had to be singled out. "This stuff belongs in church, not school," one said. "It's stupid."

Many people in the community wondered what all the fuss was about. If confusion was the main result of the statement, were constitutional questions really at stake here?

Tammy Kitzmiller thought so. It was her daughter who had to trudge out into the hallway with that first group of opt-outs. It was her daughter who had to contend with the taunt, "Hey, Monkey Girl!" And it was Tammy who had to see the look on her daughter's face when she came home, feeling at once proud and forlorn, because just about the worst thing for a fourteen-year-old is to be deemed different from everyone else.

That same week, three sets of Dover High parents formally petitioned the federal court in Harrisburg to become "interveners" in the case of *Kitzmiller v. Dover*—they wanted to join the suit to represent the interests of parents who agreed with the school board's policy. One of the parents, businessman James Cashman, would a few weeks later apply for and receive appointment to the school board to take Angie Yingling's place. These parents maintained that they would feel cheated if Kitzmiller and her fellow plaintiffs prevailed.

Their legal papers argued that the lawsuit against the district, if successful, "will have the effect of censoring . . . and shielding ninth graders from all criticism of the theory of biological evolution," which would violate their "First Amendment right to information and ideas in an academic setting." The district was looking out for its own interests in the case, but the interveners argued that their presence would ensure that the parents who supported the new curriculum would also have a say. This group of parents was represented in part by the Rutherford Institute, a public interest law firm based in Virginia with a colorful past, known for supporting conservative Christian causes, for its connections to the Discovery Institute's multimillionaire benefactor Howard Ahmanson, and for bankrolling Paula Jones's dismissed (and then settled) sexual harassment suit against President Bill Clinton.

The lawyers for the plaintiffs responded that there is no such thing as a "right" to be taught a particular subject—evolution, for instance, was taught because it was mandated by state education standards approved by the legislature, not because it was demanded by parents. If the court were to manufacture such a "right to receive information," the ability of any school district to establish a proper curriculum would be eviscerated. Irate parents could sue a school district not just to have intelligent design taught, but to demand a host of other fringe, controversial, and extra subjects—astrology, Marxism, or ufology. Such a policy would be madness.

The judge agreed, and he barred the "intervening" parents from the case. Cashman took his place on the school board instead, and assumed the same public stance as the rest of the school board: that intelligent design wasn't religious, and that the new policy statement, all one minute of it, had no religious purpose or effect.

But there was a problem with this stance: No one in or outside Dover really seemed to believe it. Yes, the members of the school board had achieved at least a temporary victory by denying in their depositions that they ever spoke of bringing creationism and Christ into the classroom. Yet even their staunchest defenders saw through this or, to be more precise, ignored it as a political necessity, one that protected the bottom line of the new policy, which was God. Whenever supporters spoke up for the board and its new policy, the defense was invariably based on religious ideas. When the professors at York College wrote to the board to complain about the ID policy, letters to the editor from community members responded, "Love God or Leave America, Professors." David Sproull, pastor of the Dover Assembly of God Church, told reporters that the drive to keep ID out of school was part of a marginalization of faith in America: "Everyone in the country seems to have freedom of speech but those who talk about religion and God." School board member John Rowand said, "If God did it, it's history and it's also science." Even Cashman's daughter, Rebecca, a student at Dover High, was quoted in the *San Francisco Chronicle* about the religious implications of the new policy. "There's only one creator, and that has to be God," she said. ". . . Evolution—is that the Darwin Theory? I don't know just what he was thinking!" And nursery owner Steve Farrell had his prayers answered by the new ID policy—it was bringing God back into the schools, he joyously observed. And if this wasn't a Christian cause, why was the Rutherford Institute trying to join the case, and why was the Thomas More Law Center, the "sword and shield for people of faith," representing the board for no fee? The law center's mission statement read:

Protect Christians and their beliefs in the public square. . . .
Our ministry was inspired by the recognition that the issues

of the *cultural war* being waged across America, issues such as abortion, pornography, school prayer, and the removal of the Ten Commandments from municipal and school buildings, are not being decided by elected legislatures, but by the courts. These court decisions, largely insulated from the democratic process, have been inordinately influenced by legal advocacy groups such as the American Civil Liberties Union (ACLU) which seek to systematically subvert the religious and moral foundations of our nation.

Thomas More was a ministry, it said so right on its own website, and Richard Thompson was its chief culture warrior. Dover was his chosen cause. Yet, he argued, no one should take Dover's school policy as religious in nature.

This would prove to be an uphill battle, as the confusion—if that's what it was—was national in scope. Within days after the lawsuit was filed, Bill O'Reilly, Fox's most popular and highest-paid pundit, dedicated a segment to the controversy in Dover. He likened the ACLU to the Taliban, "infringing on the rights of all American students" by trying to keep ID out of schools. And he told his national audience—much to the dismay of the Discovery Institute—that intelligent design was a scientific statement of what many Americans already believed: "There's a deity and the deity formed the universe and things progressed from there." Then he turned to a professor from the University of Colorado—who did not understand that, on this sort of show, he was supposed to serve as a prop, not an expert—and asked, "What's wrong with that, professor?"

Before the guest could complete his answer—"What's wrong with that is, it's not science . . ."—O'Reilly had interrupted and exclaimed, "What if it turns out there *is* a God and he did create the universe and you die and then you figure that out? Aren't you going to feel bad that you didn't address that in your biology class?"

The professor was struck momentarily speechless by this (missing the chance to say, *Why, no, Bill, I won't feel bad, I'll be dead*), and O'Reilly continued with one of the most remarkable displays of nationally televised scientific ignorance ever, matched only by the host's utter certitude:

"If I were a professor of biology . . . I would say, 'Look, there are a lot of very brilliant scholars who believe the reason we have incomplete science on evolution is that there is a higher power involved in this and you should consider it as a scientist.' "

So there it was, intelligent design, broadcast into millions of homes as a religious theory of a "higher power," and there was nothing that the Dover school board, or Richard Thompson, or the Discovery Institute could say to change that. This was how intelligent design was being perceived by a mass audience, and it was a two-edged sword. The "God connection" that made ID popular was exactly the aspect its proponents needed to downplay in order to win the lawsuit. The wedge strategy wanted to save God for later, but in Dover that was not working out as planned. The Discovery Institute might coyly maintain that its explanation of the biological universe did not identify the designer, but the public was making the identification for it.

So was a former member of the Dover school board, Angie Yingling, whose turn to be deposed came next. And the unraveling of the board's denials of religious intent began.

Questioned by Eric Rothschild during a sworn deposition, Angie Yingling would provide unique insights into the petty, haphazard inner workings of the Dover school board, while portraying herself as an inattentive board member irritated by complaints, easily coerced, and unable to sustain a decision in the face of opposition or unpleasantness. She recalled that she first voted for the anti-evolution position on the textbook issue simply to "help out Bill," who often yelled at her, only to reverse herself because so many other people were yelling and upset at the board meeting.

Why, Rothschild wanted to know, did she want to help out Bill Buckingham in the first place? "Well, you know, he wants kids to understand that you might not be descended from apes, you know, something against Darwin's theory." Her understanding of intelligent design was just as uncertain as her ideas about evolution, and it flatly contradicted the school board's official position that ID was science, not religion: "It just means life is so complicated, it couldn't have possibly been, you know, just done by accident. It would be an intelligent creator. That's how I view it. Just my opinion."

Firm opinions in the absence of knowledge were the rule, not the exception, among the board members, as Yingling tells it. She swore that other members agreed with this definition of ID, at least in executive sessions where it was discussed in apparent violation of open-meeting laws, and where—as she recalled—Buckingham and Bonsell both said that the designer in intelligent design was God. They told her that bringing God back into public school classes through ID was their way of "giving back" to him, and of "trying to be a good Christian."

When she began to have doubts about the propriety of this policy, Yingling said, she held her tongue for Buckingham's sake: "I didn't want to upset him because he has enough problems." And, she added, "If you don't go along with Bill, he thinks you are an atheist and questions your religion. . . . With him, with this intelligent design, Bill thinks you don't believe in God." Yingling recalled that Buckingham and the board member Noel Wenrich nearly came to blows during one executive session when Buckingham questioned Wenrich's faith over the ID issue. They almost flew out of their chairs at one another, with Yingling caught it the middle, she recalled.

When Yingling was asked to explain how she reached her own understanding of intelligent design, her testimony became almost comical; even the lawyers seemed at a loss for words in the face of some of her comments. Yingling initially recalled learning about intelligent design from various books. But on reflection (and after realizing that she could not name a single book she had read concerning ID), she recalled that more likely she learned of the arguments for and against intelligent design by browsing through such publications as *People* magazine while standing in checkout lines at the grocery store and in Wal-Mart. Then she quickly added that she had also read portions of an issue of *National Geographic* in a doctor's waiting room (the only major piece in this magazine that addressed ID at the time had been "Was Darwin Wrong?" published in November 2004—a month after Yingling voted for the new policy). Then this board member, who regularly voted on the details of Dover High School's science curriculum (as well as its other subjects), added, "I generally don't read books cover to cover unless they are really good. And bio certainly wouldn't be really good in my personal library."

She did, however, try to read *Of Pandas and People* in preparation for her vote, but she couldn't understand it. And when the head of the science faculty, Bertha Spahr, tried to explain to board members her objections to the book and to the inclusion of intelligent design in the curriculum, Yingling recalled—without regret—that she "turned a deaf ear" to Spahr, just as she had done long before, when she came to dislike Spahr as her own science teacher at Dover High. Yingling, still harboring that old grudge, felt as a board member just as she had felt three decades earlier as a student—that Spahr complained too much, though about what, Yingling could not say, as she never listened. "I just see her mouth moving, and I write or read—I do. I don't pay attention at all." The same was true of the other teachers—all they did was "bitch," Yingling said—so she and other members of the board "turned a deaf ear" to them as well.

Here was her final insight into the board members' machinations and their seeming imperviousness to criticism, information, and expertise—they just didn't listen. Instead, Yingling explained, they voted in favor of one scientific idea (intelligent design) and against another (evolution) that none of them really understood, denying publicly that religion had anything to do with their actions even as they privately admitted they were instituting this new policy as a service to God.

And this witness was just the start. There would be so much more that, in the end, Rothschild and his colleagues wouldn't even need to call Angie Yingling to the stand.

Chapter 12

SWORD AND SHIELD,
SHOCK AND AWE

C an you have a creationist on a school board with a secular purpose? Absolutely," the attorney with the hawkish profile and a wrestler's stature is saying. "The problem is, the law is so confused. And because of that, the courts are trying to cleanse America of religion, to remove Christianity from the public square. It's really a cleansing, a kind of genocide. . . . This case is about changing that. Our hope is to get to the Supreme Court from here and, hopefully, change the law. Stop the genocide."

This is classic Richard Thompson—the voice of Christianity, besieged denouncer of secular war criminals, crusader for the day when the Bible, the Ten Commandments, and intelligent design are welcome in any public school classroom. The only part he leaves out of this jeremiad is his usual attack on the ACLU as the "agency that has turned the country into a secular, atheistic society"—an omission made on this day in deference to his opponent from said atheist agency, who happens to be standing a few feet away and with whom Thompson has developed a reasonably cordial relationship.

Thompson is standing on the courthouse steps in Harrisburg, holding forth as he loves to do, talking as long as there is a camera or a reporter's notebook present, or a question in the air; tirelessly theorizing, answering, and pondering; sometimes emphasizing points the reporters remember well from watching trial testimony that day, sometimes creating spin out of whole cloth; perceiving victory where everyone else has observed abject defeat, and defying the reporters to call him on it. They rarely do. He is something of a force of nature, his compact, stocky form unmistakable in the crowd gathered on the weathered granite steps, his shining bald pate and bird-of-prey profile iconic in the gray afternoon drizzle. He does not look or sound his sixty-eight years. He has a habit of chewing his fingers and even his necktie in and outside court when he's absorbed, lost in thought, or agitated. It is a strange sight to behold, weirdly primal, especially given his demeanor at other times, particularly when he is engaged in cross-examination. Then he is at turns graceful and aggressive, though the effect is sometimes shattered by his habit of playing to the jury box, even when he's in a court trial, with only a judge deciding matters and nothing but lowly news reporters sitting in the jurors' seats.

Out on the courthouse steps, he speaks in a rush, sometimes without obvious punctuation. In one short paragraph, he managed to sum up his entire case. Who cares if the Dover school board had a creationist or two on it, and who cares if there was a religious motivation partially behind the intelligent design policy? Doesn't matter. The policy itself is sound, it's secular, and it has a legitimate educational purpose—to teach the controversy. To open minds. To break the grip of the single-minded, dogmatic, fanatic Darwinian cabal that has a stranglehold on American education. And why, exactly, do those Darwinists get so worked up over this tiny little statement? "Is evolution that fragile, is the science that shaky, that a one-minute statement will bring it tumbling down? Says a lot, doesn't it?"

And then there is the endgame. He is frank about this. If he loses the case at the trial level, the setback will simply be the opportunity he has been waiting for, enabling him to take his cause up the judicial line. He hopes, prays, and longs to reach the Supreme Court, in the belief that the establishment-clause case law will become fair game. This is the part of the First Amendment body of cases that governs the separation of church and state—"a national embarrassment," Thompson avers (quoting Scalia in *Edwards*), a mess of opinion and legal tests and contrariness that is notoriously difficult to follow in any consistent fashion, and that is being used to quash important ideas simply because they have a theistic, or Christian, import. "That's wrong, so wrong," Thompson says.

And Dover, if he has anything to say about it, will be his pathway to fixing that wrong.

Even lawyers have a patron saint. His name is St. Thomas More, the sixteenth-century politician, lawyer, philosopher, diplomat, author, and inventor of the word (and concept) "utopia." More was a Christian humanist who was beheaded by the imperious Tudor king Henry VIII for refusing to acknowledge Henry, rather than the pope, as the supreme head of the Church of England. Immortalized as a man of conscience in the play and film *A Man for All Seasons,* More was willing to die rather than betray his faith. That is why Thompson and his benefactor, Domino's Pizza magnate Thomas Monaghan, chose More as the symbol for their Christian law firm. There are aspects of More's background that are less well known and less celebrated, and that make him an even more interesting choice as figurehead for the Thomas More Law Center. He, too, was a zealous culture warrior in his day, vigorously prosecuting heresy. As King Henry's lord chancellor, More ordered six Lutherans burned at the stake as heretics and had dozens of other Protestants interrogated and tortured in his own home. He habitually wore a hair shirt and practiced self-flagellation. Representatives of the Thomas More Law Center and other opponents of evolution frequently criticize Charles Darwin and his "materialistic" theory as the principal inspiration for communism; but ironically, More's book *Utopia*, which described a well-ordered, prosperous, non-Christian society with no private property, greatly resembled Karl Marx's ideal society, and has long been an inspiration for Marxist and communist theorists.

The law center is situated on the rolling acreage of Domino Farms in suburban Ann Arbor, Michigan. Its decor includes a holy water fountain and a large shield and cross emblem on one wall, in which the cross is also a sword. Monaghan started the center with a donation of $500,000 in 1999 and continued providing the bulk of its funds for five years, though now other donors contribute to its $2.3 million annual budget, including more than 50,000 members who each donate $25 dollars annually. Eight lawyers—two based in California and one in Washington—work full time at Thomas More, augmented by more than 400 volunteer attorneys throughout the United States. They have several hundred cases and causes going at any one time: opposing abortion rights, gay rights, assisted suicide, research on human embryos, sex education, pornography, and the removal of Christian symbols from public spaces. These lawyers have helped draft legislation banning partial-birth abortions in Michigan and all abortions in South Dakota, with the goal of forcing the Supreme Court to review *Roe v. Wade*. They advised Governor Jeb Bush of Florida on how best to intervene in Terri Schiavo's case; they sued in California to keep Los Angeles County from removing a Christian cross from its official seal and to stop officials in La Jolla from removing a twenty-nine-foot cross from a public park. The center has sued cash-strapped small cities for refusing to erect nativity scenes at Christmas and a Michigan school district for failing to allow an antihomosexual speaker during a celebration called "Diversity Week." One of its most notorious cases, which it lost on appeal, the "Nuremberg files" case, cast Thomas More lawyers as adamant defenders of free speech on behalf of a virulent antiabortion website run by a dozen activists who, using the format of "Wanted" posters, publicized the names and addresses of doctors who performed abortions. The activists were fined millions of dollars for attempting to threaten and intimidate these doctors, who argued that the Internet site was tantamount to putting out a murder contract on them.

The law center had been looking for a case on intelligent design since the year 2000, when it began actively seeking a school district to adopt *Of Pandas and People*. Lawyers from the center made personal presentations to school district officials around the country, recommending the book and offering free legal representation when the inevitable lawsuit was filed. Dover was the first to say yes. So in this case, it was not the ACLU but Thomas More that went shopping for a client to support its cause.

Thompson cofounded the law center three years after he was voted out of office as prosecuting attorney of Oakland County (Michigan), where he made a name for himself for repeatedly and unsuccessfully prosecuting Dr. Jack Kevorkian—"Dr. Death"—for physician-assisted suicide. Thompson believes that his mission is to use the law—and cases such as Dover—"to return America to the culture established by our Founding Fathers," which he asserts was a Christian culture. His objection to evolution and his support of intelligent design are based on simple reasoning: "Evolution is presented by the humanists, the materialists, as if it were a religion itself—a religion without purpose or direction or meaning. Which means, when that is taught, we are teaching our children that there is no God. . . . And that's a problem. That's beyond science, then, and straying into religion."

Seen in this way, promoting the introduction of intelligent design in public schools isn't injecting religion into the classroom. It's helping to weed out religion—because Thompson considers evolution the most dangerous religious idea of all.

Besides, he says, he just couldn't accept the notion that he could be related to an ape.

After the initial wave of depositions of school board members had passed and the threat of a TRO had been lifted, Thompson's attention turned to the meat of his case, which, he decided, would rest not so much with the actions of the board members as with the quality of the scientific experts he could march into the courtroom. Scientific experts and testimony had been crucial in all the modern cases involving evolution. Creation science had fallen when those experts could not separate themselves from the Book of Genesis. But the Supreme Court had left an opening in the Edwards case, the suggestion that genuinely scientific alternatives to evolution would be welcome. To Thompson, that was not a loophole, it was an invitation, and he intended to accept it. To win, he had to have the leading lights of the ID movement to help him tear apart evolutionary dogma, while at the same time showing intelligent design to be a living, working scientific theory, the next big thing, a scientific alternative to materialism and naturalism that had been unhitched from scripture, from God, and from Genesis.

If he could do this, no judge would throw out that one-minute statement. And then he would have accomplished what his multiple prosecutions of Jack Kevorkian and his bullying of city councils into erecting nativity scenes had never and could never accomplish: He would have struck a fatal blow in the "War on Christians." In Thompson's view—a view shared by many evangelicals and fundamentalists—

the War on Christians was being waged all around them, every day. One case—this case—could change the balance of power and begin to change the culture.

So the A list members of the ID movement were recruited to testify for the Dover school board, and they all agreed—at first. Michael Behe would explain irreducible complexity and the failure of mainstream evolutionary theory to account for it. John Angus Campbell, a professor of communications and rhetoric at the University of Memphis, would explain why science education is best presented as opposing arguments. Darwin structured *The Origin of Species* as a response to the prevailing creationist views of his day, which he clearly outlined in his book; Campbell would explain that, in the same way, ID in the classroom would become a valuable educational tool for helping kids understand evolution. Stephen Meyer, the Discovery Institute's philosopher of science, with his gift for providing digestible explanations of complex ideas, would provide an overview of intelligent design and its history, and explain why it was no less scientific and no more religious than the theory of evolution. Scott Minnich, an associate professor of microbiology at the University of Idaho and a member of the Iraq Survey Group that searched fruitlessly for weapons of mass destruction, would offer the opinion that evolution was a theory in controversy, and that ID provided a valid, scientific alternative.

Each of these experts was affiliated with the Discovery Institute, and each would be paid $100 an hour for his efforts, plus expenses. By contrast, the experts on the other side volunteered to work on the case at no charge.

At the top of the list (and paid $200 an hour) was William Dembski, the mathematician, philosopher, doctor of divinity, and a senior fellow of the Discovery Institute. Dembski, a self-described "design theorist," is often looked on as the intellectual leader of the intelligent design movement. He was one of the original group of scholars who joined forces with Phillip Johnson during the formulation of the wedge strategy. In numerous books and articles and on speaking tours, he has argued that it is possible to infer design from various structures and systems in life and nature for which, mathematically, the probability of having been produced by undirected chance events is vanishingly small. Dembski has developed his own twist on information theory to explain this reasoning and to identify aspects of nature that have been intelligently designed rather than evolved. As one hallmark of ID he adopted the term "specified complexity"—meaning complexity in nature that has a specific pattern, such as the bacterial flagellum Michael Behe has studied. He also uses the term "complex specified information," referring to complicated biochemical structures, such as DNA. (Representatives of the Discovery Institute have implied that other information theorists also use these terms, but they appear to be unique to Dembski and the ID movement.) For evolution to account for the rise of new species and new body structures—man evolving from primitive hominids, or whales (with their vestigial legs) evolving from land animals—Dembski argues that it would have to generate enormous infusions of new complex specified information in the genetic code. He maintains that this new information cannot come about through the Darwinian mechanism of variation and natural selection, any more than letters in a Scrabble crossword game could be arranged properly through undirected, unintelligent forces.

Mainstream scientists and mathematicians have attacked Dembski's ideas and accused him of intellectual dishonesty. His former teacher at the University of Chicago, the mathematician and computer scientist Jeffrey Shallit, now a professor at the University of Waterloo in Ontario, has been a particularly harsh critic of Dembski. "His mathematical work is riddled with errors and inconsistencies that he has not acknowledged," Shallit wrote in an expert report prepared for the Dover trial: "[I]t is not mathematics, but pseudomathematics."

The evolutionary biologist Richard Dawkins produced a simple computer program to challenge ideas similar to Dembski's. It simulates gradual random change and selection to accomplish exactly what Dembski asserts is impossible: creating complex specified information, arranging the Scrabble tiles without a designer. Inspired by the old saw that creationists have used to criticize evolution for a century—that a million monkeys typing for a million years could never produce a Shakespearean sonnet—Dawkins wrote a computer program that generates random strings of twenty-eight characters and spaces, then selects the strings that most closely resemble the phrase from *Hamlet* "Methinks it is like a weasel."[1] This process is supposed to be analogous to the genetic variations that occur in living creatures, from which natural selection favors those that enhance the ability to survive and reproduce. In this case, the program allows to survive then "breeds" the phrases that come closest to the line from Shakespeare, allowing them to pass on their new characteristics to the next "generation." Members of the next generation then "mate" and produce more generations.[2] Since each generation has variations—as in life, there are continued slight mutations—there is always a pool of new letter arrangements to select for survival. The number of possible combinations of characters and spaces for a string of twenty-eight items is huge: 2,042,911,512,229,885,603,274,215,297,897,150, 684,236,521,591,013,376.

Using a variation of this program, the "member" of the first generation with the best "adaptations" might look like this: *UtwKdBkD x, 9MRMAU84VNCMOvid.*

> After 10 generations: UewKdBkD x, 6sRMAktTt 'ZDsɪd.
> After fifty generations: UewKiBkD x, is lAkcIa Feasɪd.
> After 100 generations: UethiBkD it is lAke a Feasɪl.
> After 150 generations: MethinkD it is lAke a Feasel.
> And in 189 generations, the program reaches its target:
>   Methinks it is like a weasel.

Dembski and others, including some evolutionary biologists, criticize this as an overly simplified model that relies on a human designer to make it work, but Dawkins never intended it to be a real model for evolution—he wanted only to show that those monkeys really could type Shakespeare by randomly pecking at the keys. Their efforts simply had to be coupled with a mechanism like natural selection. And in the time scale over which evolution works, 189 generations are barely a heartbeat—in human terms, the number is well within history (as opposed to prehistory), and for fast-breeding bacteria it's practically no time at all. Dawkins's experiment suggests that, contrary to Dembski's findings and calculations, very specific forms of complexity can arise through random, unintelligent means.

Dembski disputes such challenges to his ideas as reflexive opposition from closed-minded evolutionists and biologists with little understanding of the higher math applied in his approach. He casts ID as a rising star in the field of science that, in twenty years, will overthrow Darwinian evolution and be recognized as a new paradigm. His mission in the Dover case would be to defend the school board's new policy—which he found to be "inaptly stated" in parts, but useful and accurate overall. According to the expert's report he prepared for the trial, he would also tell the court about the broader meaning of intelligent design theory, likening detection of design in biology to the mainstream work of cryptographers searching for hidden patterns, archaeologists who separate man-made artifacts from natural forms as they piece together fragments of the past, and astronomers searching the cosmos for signals of advanced alien life, who must differentiate random radio pulses from intelligent ones. Despite having common ground with these other disciplines, Dembski complained in his report, ID gets no respect and has been falsely labeled as creationism. "Not only is this label misleading, but in academic and scientific circles it has become a maneuver to censor ideas before they can be fairly discussed."

The defense attorneys considered Bill Dembski a key part of their presentation, but they also knew he came with a history that the plaintiffs' attorneys would be eager to plumb. Intense and angular, Dembski has a cutting rhetorical style when he is blasting his critics or discussing scientists with whom he disagrees. His style may be satisfying for him at the time, but opposing lawyers could use it to paint an unflattering portrait. He has allied himself with the extremist rightwing pundit Ann Coulter, known for her vehement hate speech and her well-documented history of writing books riddled with factual errors. Her view on the Christian principle that humans should serve as Earth's stewards is typical: "God said, 'Earth is yours. Take it. Rape it. It's yours.'" Still, Dembski proudly noted on his blog, UncommonDescent, that he served as Coulter's consultant for chapters about evolution in her 2006 book, *Godless: The Church of Liberalism*, in which she displays a comic-book understanding of Darwin, asserting that evolution is the left's irrational, godless religion and that it is "less scientifically provable than the story of Noah's Ark."

In an Internet posting mockingly entitled "The Vise Strategy" accompanied by photos of Darwin's tombstone and a Darwin puppet being crushed in a vise, Dembski reduced scientists who accept evolutionary theory to three insulting stereotypes: the Robert Dawkins Darwinists (for the eminent evolutionist and author of *The God Delusion*), atheists who use "evolution as a club to beat religious believers"; the Eugenie Scott Darwinists (for the head of the National Center for Science Education), agnostics who "placate religious believers by assuring them that they can be good followers of their faith as well as good Darwinists"; and the Kenneth Miller Darwinists (for the co-author of the Dragonfly Book), evolutionists who accept God yet maintain that "God's activity in natural history is scientifically undetectable." Dembski then explained the cross-examination method he would like Thomas More's lawyers to use when "squeezing the truth" out of these experts in the Dover case. The Dawkins evolutionists must be goaded "into following the example of Rumpelstiltskin by publicly tearing themselves apart in their rage against religion . . . so that [they] come across as the bigoted extremists that they really are." The Scott evolutionists should be revealed for "the patronizing elitists that they really are." And the Miller Darwinists should "come across as the closet ID theorists that they really are."

Dembski's zeal to turn the vise on these scientists contrasts with his own experience on the receiving end of a similarly pointed interrogation in 2002. The memorable encounter came during a panel discussion at the American Museum of Natural History in New York, where he was left sputtering when questioned by the same Ken Miller accompanied by Robert Pennock, a philosopher of science at Michigan State University. Eugenie Scott served as moderator. The questioners had tried to get Dembski to define "design" and other seemingly simple terms, but he could not or would not do so in any comprehensible fashion. They had also asked him to explain when, in the 3.5 billion years that life has existed on Earth, the "design events" took place. Again, Dembski could not answer the question. He said that design might have occurred at different times, such as the Cambrian Explosion, or that the information necessary for design "could have been put in there right at the big bang." In other words, intelligent design could have occurred at the moment the universe was born, 11 billion years before life appeared on Earth, and everything else unfolded from there—a proposition with a certain majesty, but also vague, unprovable, and unscientific. When the

Discovery Institute placed a description of the exchange on its website, it suggested that Dembski had performed "brilliantly," but it posted a transcript of only his eloquent prepared remarks, with no reference to his halting responses to the questions from Pennock and Miller—or to the laughter of the audience.

Dembski has been a polarizing figure at times, particularly at his former place of employment, Baylor University, whose president recruited him in 1999 to head the first intelligent design think tank at a major research university. When faculty members complained that Dembski's new ID center had little oversight and could threaten Baylor's reputation in the scientific community, a compromise was worked out that satisfied his critics, but that Dembski interpreted as a vindication of his work and of intelligent design as legitimate science. He penned an inflammatory press release stating, "Dogmatic opponents of design who demanded that the Center be shut down have met their Waterloo." Amid outrage from the faculty, the president of the university asked Dembski to retract the press release, and when Dembski refused, he was relieved of his duties. He spent the rest of his five-year contract at Baylor with no courses to teach, working from home, giving speeches on ID, and writing. He called it a "five-year sabbatical." Then he left to take a new position at the Southern Baptist Theological Seminary in Louisville, Kentucky.

Although he argues forcefully that ID is a scientific proposition completely separate from creationism, Dembksi has also cast it as part of a wider campaign to defeat materialism in society and establish a more Christian-focused culture—the basic premise of the wedge strategy. He has written that intelligent design "should be viewed as a ground-clearing operation that gets rid of the intellectual rubbish that for generations has kept Christianity from receiving serious consideration"—evidence, his critics say, of ID's creationist core.

Nick Matzke's colleague at NCSE, Wes Elsberry, began assembling a large file on Dembski for the plaintiffs' attorneys to use during the forthcoming depositions of the scientific experts—his writings, his comments, his papers, his blog; every available comment, from the most profound to the most intemperate. Of all the experts, Dembski's cross-examination file was the thickest.

The process of using scientific experts in court is a laborious one. Unlike "regular" witnesses, experts must demonstrate their qualifications in their field, then submit a written report outlining their official opinions relevant to the case, and finally sit for depositions, allowing the other side in the case to question them under oath. This process was well under way with other witnesses when Dembski's turn came and trouble began.

In the portion of his expert report about *Of Pandas and People*, the lawyers for the plaintiffs noticed two things. First, *Pandas* held a more exalted position within the ID movement than they had previously realized. Dembski wrote: "*Of Pandas and People* was and remains the only intelligent design textbook. In fact, it was the first place where the phrase 'intelligent design' appeared in its present use. Since the second edition of this book, intelligent design has gone from a small and marginalized challenge confronting neo-Darwinian evolution to a comprehensive scientific research program for reconceptualizing biology." In other words, *Pandas* wasn't just a high school textbook that extolled ID and showed its creationist roots, as the plaintiffs' attorneys had intitally thought. It was a pillar of ID, a foundational document—the first major publication in a movement.

The second thing Dembski mentioned in his expert's report about *Pandas* violated a cardinal rule of the scientific expert game: Don't identify anything in your report as a basis for your expertise or opinions that you don't want to be questioned about, and that you don't want subpoenaed. Dembski, however, volunteered that he had become academic editor for the publisher of *Pandas*, the Foundation for Thought and Ethics, in Texas; and he had edited and was the principal author of a new, not yet published, edition of *Pandas*, to be renamed *The Design of Life: Discovering Signs of Intelligence in Biological Systems*. This work, he suggested, made his opinions particularly relevant in *Kitzmiller*.

Because Dembski included this information in his expert's report, the lawyers for the plaintiffs argued it was fair game and that they could subpoena the new manuscript. Any changes Dembski had made in the new edition—or failed to make—would, in the lawyers' view, be very telling. The publisher, which maintains a close relationship with the Discovery Institute and employs a number of its fellows as writers, was not pleased by this turn of events. The president of the company, Jon A. Buell, said he'd rather go to jail than reveal his unpublished book to courtroom scrutiny and possible competitors.

Just before the scheduled depositions of three of the experts from the Discovery Institute—Dembski, Meyer, and Campbell—they all decided that they wanted their own attorneys present to watch out for their legal interests. (The other witnesses from Discovery, Minnich and Behe, had already been deposed by that point, without their own lawyers.) The attorney retained by Dembski, Meyer, and Campbell happened to be the attorney who represented the publisher of *Pandas*, the Foundation for Thought and Ethics, and it was clear from comments made by Bill Dembski on his blog that the push for legal representation was coming more from the publisher, and perhaps the Discovery Institute, than from him. The publisher wanted its own lawyers on the case, fearing that *Pandas* was going to be painted during the trial as a creationist text— which could hurt sales—and that lawyers who were focused on representing the Dover school board would not necessarily care about protecting the publisher's reputation.

Thomas More's attorneys refused to go along with this. They had hired the experts and were paying them handsomely on behalf of the Dover school board, and they certainly did not want other lawyers representing different interests getting involved—lawyers who would be there to protect a publisher's reputation, not sustain a school board's policies. It was unheard of, they said, for paid expert witnesses to arrive with their own counsel for depositions. None of the experts on the other side had lawyers, Richard Thompson later observed. What, he wanted to know, was the Discovery Institute so afraid of?

But the experts would not relent, and the Thomas More Law Center fired both Dembski and Campbell, canceling their depositions and leaving them no role in the case. The leading scholar in intelligent design would not be part of the first intelligent design trial. The Thomas More lawyers then relented and said they would allow Meyer, as an officer of the Discovery Institute, to bring his own attorney. But by then, Meyer would later say, he had lost confidence in Thomas More's attorneys because of their "capricious handling of this matter." He came to the conclusion that the Dover school board and the Discovery Institute were not really on the same side in the trial—that championing Dover's policy was conflicting with his primary role of promoting the scientific case for intelligent design.[3] Meyer decided that even with his own lawyer involved, he no longer wished to testify in the case, and he quit as an expert. This happened so late in the case—more than six months after the lawsuit was filed, and less than four months from trial—that the deadline had passed for adding more experts. A public rift had opened between the Discovery Institute and the lawyers for the Dover school board, which had just lost half its force of expert witnesses, even as one of them suggested, very pointedly, that Dover's policy was not a good thing for the cause of intelligent design.

In a strange twist, the publisher of *Pandas* immediately tried to join the case, asking to intervene just as the group of pro-ID parents had done months earlier. Lawyers for the publisher said they would bring Bill Dembski and at least one of the other fired experts back as their own witnesses—witnesses whose departure from the case they had helped engineer.

Neither the parents' attorneys nor Thomas More wanted any part of this. "They should not be rewarded for meddling with defendant's case by being allowed to intervene and then bringing those same experts back," Eric Rothschild griped.

Then, in a preview of the devastating cross-examination skills he would bring to the trial in a few months, Rothschild showed no mercy for Jon Buell, the *Pandas* publisher, who had arrived from Texas to testify on behalf of his plea to join the case. Buell testified that he was trying to intervene so late in the case because he had only just learned that *Pandas* was going to be portrayed as a creationist text, and he wanted to be sure that contention was rebutted to protect the sales of his book.

Rothschild handed Buell a copy of a letter the publisher had written about the book after its publication—a promotional letter Nick Matzke had unearthed from the NCES archives. It stated: "Our commitment is to see the monopoly of naturalistic curriculum in the schools broken. Presently school curriculum reflects a deep hostility to traditional Christian views and values, and indoctrinates students to this mind-set through subtle but persuasive arguments." The letter also discussed how important it was to stop schools from denying the notion that man was created in God's image. How, Rothschild wanted to know, could Buell write such a letter yet have no idea that *Pandas* might be viewed as a creationist text? Buell had no answer.

The judge rejected the publisher's request to join the case. And Meyer, Campbell, and Dembski—the latter had to threaten legal action against Thomas More to get paid for more than 100 hours he had spent on the case—were out for good.

As the pace of trial preparations picked up during the spring and summer of 2005, so did the broader culture war nationwide. In the hearings on evolution in Kansas; in the matter of Terri Schiavo's right to live or die; in the issue of public displays of the Ten Commandments; and in a controversial conference, "Confronting the Judicial War on Faith," attended by some of the nation's top political leaders, tension between the secular and the sacred continued to rise during the months leading up to *Kitzmiller v. Dover*.

In March 2005, a new survey by the National Science Teachers Association found that nearly a third of those who responded felt pressure by students and parents to include creationism, intelligent design, or other alternatives to evolution in their science classrooms. Thirty percent of science teachers felt they were being pushed to soft-pedal and steer away from teaching evolution.

Bruce Alberts, head of the National Academy of Sciences, wrote an open letter to his colleagues about the "growing threat to the teaching of science." He called on scientists nationwide "to confront the increasing challenges to the teaching of evolution in public schools." In a story in *USA Today* about the poll and the letter, Stephen Meyer of the Discovery Institute commented: "My first reaction is we're seeing evidence of some panic among the official spokesmen for science."

For its April Fools' Day issue, *Scientific American* magazine drew attention to these concerns by publishing an editorial headed, "OK, We Give Up," in which it apologized for promoting evolutionary theory, saying, "As editors, we had no business being persuaded by mountains of evidence . . . [or] thinking that scientists understand their fields better than, say, U.S. senators or best-selling novelists do."

That same month, three Republicans in the U.S. Senate—Richard Shelby of Alabama, Richard Burr of North Carolina, and Sam Brownback of Kansas—proposed a Constitutional Restoration Act. If passed, it would bar federal courts from getting involved in any cases in which local, state, or federal officials made "acknowledgment of God as the sovereign source of law, liberty, or government." In other words, if the governor of California decided to post the Ten Commandments in every classroom, or the Texas state legislature decided that every school day should open with a Christian prayer, or the Dover school board instituted a biology curriculum that some parents believed was religion disguised as science, the federal courts could not get involved. Under this legislation, there would be no *Kitzmiller v. Dover*; nor would it be possible to sue a school district to stop it from teaching full-blown creationism. And if a judge went ahead anyway and heard such a case, the act would call for immediate impeachment.

Legislatures in Missouri and Idaho enthusiastically endorsed the bill and vowed to post the Ten Commandments in every public space if it became law. Lawmakers in New York, meanwhile, introduced legislation mandating the teaching of intelligent design in public schools, but it failed to pass.

Also in the same month—April 2005—it was revealed that U.S. Park Service stores at Grand Canyon National Park had begun stocking and selling a creationist history of the Grand Canyon. Scientists protested, but the books remained on the shelves.

In Bristol, Virginia, meanwhile, an open secret was revealed: A veteran biology teacher, Larry Booher, had for fifteen years been teaching creationism along with evolution at John S. Battle High School, using a 500-page tome he had personally assembled for his students. The book, *Creation Battles Evolution,* a compendium of articles and other material, included the statement, "Evolution is, in reality, an unreasonable and unfounded hypothesis that is riddled with countless scientific fallacies. Biblical creationism, on the other hand, does correlate with the known facts of science." Elsewhere, his book stated, "The widespread influence of evolution is largely responsible for our moral decline of recent years. . . . Herein lies the awesome danger of this Satanic delusion."

The book and Booher's unofficial lesson plan were common knowledge in the Washington County, Virginia, School District—even the former president of the PTA had known about it for years. But school officials maintained that they had been unaware of Booher's lessons on creationism throughout the fifteen years he offered them, until an anonymous tip was phoned in. The lessons and Booher had been extremely popular in this conservative community and were assigned as "extra credit," which almost every student elected to do. Once the situation was publicized, he agreed to stop the lessons—but he received an outpouring of public support from parents and disappointed prospective students, who demanded that he be allowed to resume his teachings on creationism. Numerous letters from graduates posted on the Internet in support of Booher, however, showed that the writers' understanding of standard evolutionary theory was poor or nonexistent, perhaps as the state of Virginia intended. Virginia's statewide standards did not even contain the word "evolution."

While that controversy simmered, the Smithsonian Museum of Natural History in nearby Washington, D.C., agreed that the Discovery Institute could hold a reception and a private screening of a new film on intelligent design and the universe, *Privileged Planet*. This film is based on a book of the same title cowritten by Guillermo Gonzalez, who is an astronomer at Iowa State University and a fellow of the Discovery Institute. In it, Gonzalez argues that the universe shows evidence of being "fine-tuned" to allow life to exist and to facilitate scientific observation, and that the probability that such a universe could evolve on its own through random processes is infinitesimal. This, he argues, provides evidence of design. *Privileged Planet* is ID on a cosmic scale and, like its biological counterparts, has been criticized by mainstream scientists as slickly updated creationism. When science bloggers learned of the event and accused the Smithsonian of violating its own policy against providing screen space for religious or political films, museum officials expressed shock, withdrew their cosponsorship of the event, and then refunded the $16,000 screening fee. The screening occurred as promised, however, and Discovery's officials got a free event and were still able to complain of censorship and persecution by "ideologues" of mainstream science.

The Discovery Institute's scholars became a constant presence in the news media during the time leading up to the Dover trial. Michael Behe had an op-ed piece in *The New York Times*; Stephen Meyer appeared on Fox News, in *The Big Story with John Gibson*, and other shows; and David Berlinski—a mathematician, author, and one of Discovery's fellows—wrote a column criticizing evolution that was syndicated in newspapers around the country.

Their greatest triumph came less than two months before the Kitzmiller trial was supposed to start. In August 2005, President Bush told a press gathering that he believed evolution and intelligent design should both be taught to public school students (an update of sorts of his previous statement that creationism and evolution should be taught together). "Both sides ought to be properly taught . . . so people can understand what the debate is about," Bush said—implying the existence of a genuine scientific debate that mainstream scientists adamantly denied.

Critics dismissed this as part of the Bush administration's general hostility to mainstream science, of a kind with its denial and inaction on global warning; its appointment of a bioethics adviser who based her opposition to stem cell research on an episode of *Star Trek*; its denial of an impending energy and oil crisis; and its suppression or rewriting of scientific findings of the Food and Drug Administration, NASA, and other government agencies for political, rather than scientific, reasons. But officials at the Discovery Institute were ecstatic, commending the president and using his remarks to solicit donations on the Internet to support a new campaign that would help scientists, teachers, and students "under attack for questioning evolution." *The New York Times* published a major three-part series on the evolution-ID "controversy" with headlines such as "Politicized Scholars Put Evolution on the Defensive" and "In Explaining Life's Complexity, Darwinists and Doubters Clash." Evolutionary scientists complained that the coverage seemed to put the tiny minority of ID proponents on an equal footing with thousands of mainstream scientists worldwide.

The loose association of pro-evolution bloggers scattered around the country stewed over this coverage and furiously sought to counter the wave of pretrial pro-ID publicity. The popular science blogger P. Z. Myers, a biology professor at the University of Minnesota, was singled out for criticism on the Discovery Institute's website as an enemy of free

speech because he called for "the public firing and humiliation" of teachers who present the "pseudoscience" of ID to impressionable students.
Myers became a one-man wrecking crew, blogging furiously against
Bush, Berlinkski, Behe, *The New York Times*, and Discovery, and posting links to research papers and other information that refuted every
one of their claims.

Myers is an interesting, dynamic science writer, and his excitement at
the revolution going on in his discipline—developmental biology, one of
the hottest fields in science—is palpable, as is his anger at the claims of
ID proponents. His is one of the most read personal science blogs on the
Internet—Pharyngula.org, which routinely draws 25,000 readers a
day—and he uses the platform to talk science, post weekly exotic photos
of cephalopods (octopuses), and attack intelligent design incessantly. "It
is infuriating," he explains when asked why he spends several hours a
day attempting to rebut ID claims. "There is all this excitement in our
field, new discoveries all the time, and yet I have to push back against
these idiots who want to bring us back to the Middle Ages. You have
places like Discovery Institute that claim there's a rising tide of scientists
speaking out against evolution. It's just an absolute lie. . . . And then you
have the president stepping into it."

On his blog, Myers called Bush "the Moron-in-Chief" for endorsing
ID, but he agreed with the president's idea that both sides should be presented in classrooms, so long as this was done P. Z. Myers's way: "We
need to stand up and plainly state that creationism is a lie and any attempt to incorporate faith and the supernatural into science is as destructive to the enterprise as would be requiring religion to provide concrete,
repeatable tests of their beliefs. That's the only rational version of 'equal
time' that will work."

Even as he got in his shots, however, Myers lamented that the weight
of the press coverage, the kibitzing by the president, and even Myers's
own public engagement of the other side (Discovery began using one of
his tirades in its fund-raising appeals) had created an environment that
helped the proponents of intelligent design make their case that a scientific controversy existed. Meanwhile, the message of design and creation
continued to resonate with the public far more than the message of evolution. In Bristol, Virginia, there was almost no condemnation in the

many letters written about Larry Booher—no concern that religious ideas were being passed on, without permission and against the law, to impressionable high school sophomores. There was only praise.

A letter by one of Booher's former students displayed her views on the Constitution, biology, and evolution, revealing exactly what was learned in that classroom, and what a "teach the controversy" approach can truly achieve:

> If there is no place for creationism in schools, there is no room for biology. . . . Larry Booher didn't use poor judgment; the system did when it chose to separate church and state. In school, we have to learn in every classroom in different ways because people are afraid to talk of religion even though evolution is a religion as well that requires plenty of faith.

With the Dover trial fast approaching—it was set to begin on September 26, 2005—a daunting prospect loomed for the proponents of evolutionary theory. They might win the battle against the Dover school board, but continue to lose the public in the process. The issue could not be simply about winning the verdict; the scientists who had joined the case to fight for evolution wanted to win hearts and minds, too, and they recognized this for what it was—an uphill battle.

For the advocates of intelligent design, the prospect of losing the Dover case was not so troubling. On the eve of trial, Bill Dembski, who was no longer a witness, predicted that the specific policy of the Dover school board would almost certainly be overturned by the court, thanks to the overtly religious statements of board members. He assigned a 70 percent probability to that outcome, acting more as an oddsmaker than a mathematician, for this prediction was admittedly subjective. He felt there was only a 20 percent chance of victory for the school board. The least likely outcome, in his opinion, was that the science of intelligent design would go down in flames at trial as stealth creationism. The chance of this outcome was only 10 percent, Dembski said.

But even a ruling that intelligent design was not truly scientific, Dembski confidently predicted, would not be its Waterloo. In fact, such a setback would have the beneficial effect of forestalling the complacency that inevitably would accompany victory. In the end, a big loss would spur even more interest and research in intelligent design, he suggested: "It will continue to grow regardless of the outcome. . . . Whether favor or adversity is, at least for now, the best tonic for ID's intellectual vitality remains to be seen."

In Dover, the conflict over the intelligent design policy, which had been mostly quiet in the winter months, heated up again in the spring of 2005. The "monkey dance" man stepped up his appearances as the Rehm family passed by. Other plaintiffs had similar experiences. Tammy Kitzmiller, whose name was now the most prominent of the group, regularly received hate mail. The Rehms eventually stopped going to restaurants in town, because of several incidents in which strangers approached and insulted them, in one case shouting profanities at them with their children present.

"I've never been called an atheist before," Christy Rehm says.

It got even harder when the campaign for the school board began in earnest, and a coalition of newcomers to Dover politics, including Rehm, formed an organization called Dover Citizens Actively Reviewing Educational/Economic Strategies (Dover C.A.R.E.S). Its members ran as a united group to defeat the incumbent board. For the first time in Dover's history, candidates for the school board advertised on billboards in town, held fund-raising barbecues, and walked through the neighborhoods, talking, persuading, and passing out literature. Bryan Rehm's photograph appeared on the billboards, and though he soon had a number of well-wishers on his side, the harassment increased, too.

Seven the of the nine board members faced challenges in the election—and then an eighth seat opened up as well, when Bill Buckingham, who was still ailing, announced by a letter to the board that he was resigning and moving to North Carolina. Buckingham had been mostly silent since his deposition, missing meetings, and when he did come, appearing in terrible temper. He looked ill and seemed shaky. He said he had decided to leave because of his bad back and his ongoing recovery from addiction to OxyContin—he no longer used the medication, but the craving and the related physical problems were not so easily conquered. He apologized in his letter if, because of his illness, he had offended anyone. And then the main mover behind the intelligent design policy, the man who had brought in the Thomas More Law Center to wage a momentous battle of his own creation, quietly left town before his cause could reach the courtroom.

His departure meant that almost the entire board—and certainly majority control—was up for grabs. Unlike most elections, in which only a few seats at a time would be open, because of staggered terms, all the resignations meant that this time the community faced a stark choice: stay the course or throw out the incumbents. And so, though both sides tried to talk about taxes, academics, teachers' contracts, and other mundane though vital issues, the election quickly became a referendum on the intelligent design policy. The incumbents, led by Alan Bonsell, stood behind the ID policy, to no one's surprise. Dover C.A.R.E.S. vowed to revoke it, although the group expressed support for creating a course in the social studies department or some other appropriate place within the curriculum to explore cultural issues such as origins and belief. Anywhere, they said, but science.

The candidates hoped that the primaries in May would indicate which way the community was leaning, but the division remained even. The two factions split the vote in the primary. All the candidates had filed to run in both the Democratic and the Republican primaries, as was permitted in Pennsylvania's open-primary system. The incumbents won on the Republican side, and the challengers took the Democratic vote. This seemed to bode well for the incumbent board, as York County as a whole was heavily Republican. The two slates of candidates would have to face off again in November—and so they would be campaigning during and after the big trial, which would begin in September and continue right up to Election Day. Most if not all of the testimony would take place before the vote, and would be likely to prove pivotal in influencing voters. Either the school board or the parents suing the board—who were allied with the Dover C.A.R.E.S. slate—would take a fall in the trial, quite possibly deciding the election.

The incumbents did all they could to generate support in the community before that time came. First they encouraged a group of parents to gather signatures supporting their ID policy. A petition praising the district policy was eventually presented at a school board meeting with 280 signatures and a statement that "intelligent causes, rather than random chance, provide a better explanation for the complex biological systems of life." The two women who handed over the petition also congratulated the board for exposing the gaps and problems in evolution.

Steve Stough, who had begun attending board meetings conscientiously after joining the lawsuit, approached the two women who presented the petition and demanded, "Name three gaps and problems in evolutionary theory."

The women just stared at him. "If you can't do that, then you're just repeating what someone else told you to say," the middle school science teacher told them.

One of the petitioners then suggested that the sudden appearance of biological chemistry out of nothingness was her biggest problem with evolution. But Stough pointed out that this objection made no sense. She was talking about the origins of life, which Dover does not teach—and which is not part of evolutionary theory. The argument went nowhere, for it was not really about evidence or logic anymore. It was about belief—and, more specifically, about what people choose to believe, because it is a choice. As Stough would later realize, the women bearing those petitions would not be convinced by anything he could possibly say. They chose to believe something that they did not understand and that sounded scientific—ID—and to reject something else that they did not understand and that sounded scientific: evolution. They did this because one appealed to their worldview and their religious convictions, while the other scared and challenged them. They did not really want to know more than this, and though Stough's convictions that he was right to join the lawsuit never wavered, he began to wonder what it would really take to open such minds.

Tammy Kitzmiller was also there, listening, fuming, and biting her tongue. She felt far less patient and reasonable than her new friend Stough. She hated it that people in her community would reflexively support intelligent design out of what she considered to be ignorance, bias, or the idea that it would fit better with their religious beliefs—without attempting to inform themselves about what they were embracing. It particularly bothered her that the members of the school board were campaigning on the claim that their ID policy was designed to promote critical thinking. Kitzmiller bluntly called this claim "a crock."

"The superintendent reads a statement saying ID exists and leaves the classroom," she told a reporter covering the meeting: "The kids aren't even allowed to discuss the theory or ask questions. How does that inspire critical thinking if you can't even talk about it in class?"

Tension heightened when the board brought the biochemist Michael Behe to town to lecture community members in the auditorium at Dover Area High School. Behe gave his stock lecture about irreducible complexity and the inadequacies of evolution to an audience of about 100. The Dover C.A.R.E.S. candidates attacked the incumbent school board for engineering such a one-sided, politicized presentation at the high school; but in truth, Behe's lecture, like Behe himself, was low-key and nonconfrontational. Several people rose to challenge his ideas during a question-and-answer session, suggesting that a high school biology classroom might not be the appropriate place for introducing such complex and controversial ideas. Behe genially disagreed, saying that the controversy would stimulate interesting discussions. He seemed to be unaware that Dover's policy specifically forbade students and teachers to discuss the controversy once it was introduced into the classroom.

With the help of the Thomas More Law Center, the school district also produced a slick newsletter that was sent to every resident in the school district—the school board's attempt to win hearts and minds. The newsletter included an article praising the board, written by Senator Rick Santorum of Pennsylvania, who sat on the board of directors of the Thomas More Law Center. The newsletter also included an extensive and flattering explanation of intelligent design, and a list of criticisms of evolutionary theory. It described ID as a scientific theory—which not even the official Dover curriculum policy and statement had done, choosing the word "explanation" instead—and it asserted that ID was "endorsed by a growing number of credible scientists," without mentioning that this assertion was open to debate. The newsletter added, "In simple terms, on a molecular level, scientists have discovered a purposeful arrangement of parts, which cannot be explained by Darwin's theory. In fact, since the 1950s, advances in molecular biology and chemistry have shown us that living cells, the fundamental units of life processes, cannot be explained by chance." This is the Discovery Institute's standard line, which the Dover school board presented without any qualification, and without any acknowledgment that the vast majority of biologists and all major scientific organizations in the country consider this statement to be fundamentally dishonest.

In short, the newsletter told the citizens of Dover, unequivocally, that evolution had been falsified and that scientists had discovered that life and the universe were products of a supernatural designer.

The newsletter also took a shot at the parents who had filed suit, calling them "a small minority" who were not telling the truth. The newsletter asserted that there was no religious agenda behind the ID policy, and that ID was not being taught to any students—the students were "only being made aware of ID." Evolution was specifically linked to atheism in the newsletter, which asserted that ID was no more religious than "Darwinism"—a term that few evolutionary scientists use and most avoid. Creationists and ID proponents prefer to use the term "Darwinism" because it encourages a view of evolutionary theory as the tired nineteenth-century dogma of a long-dead naturalist, rather than the continually updated and greatly expanded body of science that modern evolutionary biologists embrace. "Darwinism" is a pejorative term to many, like "Marxism" or "Maoism," and the newsletter used it liberally to describe scientists in a way they would never describe themselves. But the term "creationist," which many mainstream scientists use to describe proponents of ID (who hate that term as much as their opponents hate "Darwinist"), was not used anywhere in the newsletter in conjunction with intelligent design.

Kitzmiller, the Rehms, Stough, and the other parents in the lawsuit were outraged that the school district would use public funds and resources to send what Steve Stough called "blatant propaganda," designed to influence opinions and votes in the community. "It is full of misinformation," Stough complained. "Lies."

But the parents' lawyers were delighted by the newsletter, and told their clients not to worry about it. One constant refrain of the board and of Thomas More was that the board's ID policy resulted in nothing more than a simple one-minute statement, that no details of intelligent design were being given to students, and that ID was not being "taught" in any real sense—in short, no harm, no foul. That ID wasn't being "taught" had become a huge part of their legal defense.

And yet now, the lawyers observed, the school board had sent out to every family, parent, and student in Dover a detailed two-page newsletter extolling intelligent design as a valid scientific theory—and delivering standard creationist arguments against evolution. In the short term, the newsletter might be a slight political gain for the incumbent board members. But it also looked as if the school board members were doing something else with the newsletter, something that was not going to help the arguments of the Thomas More Law Center one bit.

They were teaching the community about ID.

Chapter 13

PALEOZOIC ROADKILL, KENTUCKY FRIED CHICKEN, AND BAD FROG BEER

When the trial finally played out in the fall of 2005, the greatest revelation in the case of *Kitzmiller et al.* vs. *Dover Area School District et al.* ended up having nothing to do with the candor (or lack of it) displayed by the school board members, the subtle machinations of the wedge strategy, or the rope-a-dope tactics that lured a star of the intelligent design movement to his rhetorical downfall behind a pile of research papers he had claimed didn't exist. These would, undeniably, turn out to be eyepoppers during the course of an epic trial, but there would be something even more sensational on view in the courtroom in downtown Harrisburg.

The real star in this forty-day trial would be science.

Scattered throughout the trial were moments, sometimes whole hours, and on a few occasions entire days, when cutting-edge science truly stole the show. The testimony, abundantly illustrated and explained with a clarity many university classes would envy, revealed just what evolutionary biology can say about the natural world. The presentations were at times so riveting that the judge presiding over the case later confessed he wished he had granted a request he had reflexively denied (as federal courts have always done): to allow portions of the proceedings to be televised. There were images of fossilized dinosaurs frozen in time over their nests, protecting their eggs just like modern brooding birds. There was a visual unspinning of the human and chimpanzee genomes to reveal the very footprints of evolution and the place where man and ape diverged as species. And there was slide after slide of ancient, extinct, fossilized life-forms that appeared to refute the claim of creationism (and ID) that there are no "transitional" fossils linking animals to their common ancestors. These links did not exist, creationists had long argued, yet the courtroom was treated to vivid pieces of evidence pulled from the earth—images of birdlike dinosaurs with feathers preserved in sandstone with almost photographic clarity; extinct ancient fish with primitive limbs and heads the shape of amphibious frogs; the four-legged, hoofed, ponderous *ambulocetus*—the walking whale—from the Eocene era 60 million years ago; and a line of animal skulls each slightly different from the next, beginning with reptiles and ending with mammals, transforming across the millions of years in just the gradual manner that Charles Darwin predicted paleontologists one day would discover.

Kevin Padian, the Berkeley paleontologist and curator of his university's Museum of Paleontology, entertainingly brought the bone hunter's perspective to the courtroom, the sort of character on whom the fossil-hunting hero of the film *Jurassic Park* was modeled. Padian happily showed slides of his "critters," as he tended to call the ancient fossils and bones he used as a window on the past. He is one of the world's leading authorities on the evolutionary transition from dinosaurs to birds ("Birds *are* dinosaurs," he says impatiently) and on the gradual evolution of flight and wings, painstakingly re-created from the fossil record and not, he would testify indignantly, the fully formed product of an abrupt creation, as *Of Pandas and People* and leading intelligent design advocates assert. He produced a photo of a remarkably preserved fossil of an oviraptor, a type of small feathered dinosaur found in Mongolia, frozen in time, its three clawed fingers splayed out protectively over its clutch of eggs just as a modern hen would splay her wings. Such sharing of characteristics and behaviors across time and related species, Padian explained, is predicted by the theory of evolution, and in turn confirms Darwin's venerable idea—bone after bone of hard, real evidence of gradual adaptations, of walking dinosaurs to flying birds.

Padian, with evident fierce joy, debunked the often repeated claim that the absence of "transitional fossils" was a problem for evolution and an argument for creation or intelligent design. He displayed a raft of slides of just such transitional creatures, beginning with the half-eaten bony remains of a fishlike creature called Eusthenopteron, whose fins showed the development of rudimentary finger bones. "You'll have to excuse me; I'm showing you some Paleozoic roadkill," he told the court enthusiastically. In documenting the gradual fusing over time of finger bones in dinosaurs (five digits), to the primitive dinosaur-like bird archaeopteryx (three clawed fingers on the end of each wing), to modern birds (all finger bones fused into one), he would explain, in his characteristic rush of words, that the traces of this evolutionary path were available to anyone with a penchant for greasy fast food.

> You would know them better as the pointy part of the wing in the Kentucky Fried Chicken. So if you were to dissect your Kentucky Fried Chicken, which I don't recommend, but I can tell you about turkeys and Thanksgiving, which is lot of fun, you will find that you can get to the individual hand bones, we can watch the bird develop, and these are individual bones that later become fused. And this is because the bird is no longer using its hand for anything except flight. It's not using its fingers to pick up things or claw or scratch anymore. And early in the evolution of birds, when they dedicated themselves to flying with four limbs and very little else, there was no further need to use these fingers for anything.

So there it was, with the presiding judge suppressing a smile at the idea of it: evolution for all to see, tucked neatly inside a bucket of wings, hold the hot sauce, please.

This is not to say there were no moments of boring scientific monotony. The term "bacterial flagellum" was repeated 353 times, as its structures were reviewed and discussed and displayed on video screens in endless and mind-numbing detail. Late in the case, the thirty-third and last witness pulled up on the big screen an image of the now familiar flagellum diagram with its gears and pistons drawn in lieu of protoplasm. "We've seen that," the judge groaned.

The witness, University of Idaho microbiologist Scott Minnich, smiled ruefully before doggedly continuing with his point. "I kind of feel like Zsa Zsa's fifth husband, you know?" Minnich told the judge.

"As the old adage goes.... I know what to do but I just can't make it exciting."

It is often said, with good cause, that a trial is a poor place to have a reasonable, informative debate about science, because the goal of a trial and the goal of science are so often at odds. Science, at least as it has been practiced for the last century or two, begins by assembling facts—the data—and then seeks an overarching theory to unify and explain those facts. Whether it's the big bang theory or plate tectonics or germ theory, from the cosmic to the microscopic, the approach is the same. Ignoring or denying inconvenient facts is not permitted. Trying to uncover facts that *disprove* a treasured theory is encouraged. This is part of the modern scientific method, which holds that theories should be subjected to rigorous attempts to prove them false before they become widely accepted (or discarded as incorrect).

In the law, however, the process works in exactly the opposite direction. Each side of a legal dispute starts with a theory: "The accused is guilty" versus "My client is not guilty" (or "Intelligent design is religion" versus "No, it's science"). Once the opponents have settled on their mutually exclusive theories of the case, each side lines up the facts that support its own preferred version of reality. And each side studiously ignores, minimizes, or attempts to disprove the facts that belie its own theory of the case.

One consequence of this time-honored approach is that, as a general rule, the courts don't do science very well. Judges aren't trained for it, and jurors are often overwhelmed by it. (Watch the eyes glaze over in criminal trials in which the technology of DNA fingerprinting is explained or challenged—such proceedings are torturous.) As a result, the tendency is to avoid or deemphasize science wherever possible, and to focus on other issues to resolve a case. In *Kitzmiller*, the simplest course would have been to address only the intentions of the Dover school board, and what effect its new biology curriculum had on students and the community, perhaps with an additional, limited foray into whether *Of Pandas and People* advocated a nonreligious or a creationist perspective.

But this did not happen in *Kitzmiller*. The trial did, of course, thor-

oughly explore those narrow issues, but it also ranged across vast philosophical ground, attempting to answer such sweeping questions as these:
What is science? Where is the line drawn between faith and reason?
Does evolution leave room for God? Can a scientific idea also be a religious one, and if so, can it be taught in public school? Charles Darwin's
Origin of Species frequently discusses God and creation—should it, too,
be banned from public school classrooms? At the same time, the trial
dived into the nitty-gritty questions of science with serious scholars who
were also skilled teachers and lecturers, true believers in their own disciplines and in the power of science to explain at least some small piece of
the universe. During the sometimes grueling cross-examinations, the
experts coached the lawyers to ask the questions scientists ask, and so
this one time, the witness stand became a place to serve, rather than foil,
science. The power and weaknesses of evolutionary theory and intelligent design were plumbed in remarkable and unprecedented ways.

Kitzmiller became at root everything the original Scopes trial had
started out to be but was not. Back then, the leading scientists had been
ready to testify, only to be ruled irrelevant by the creationist judge who
presided in Dayton. Eighty years later in Harrisburg, a very different
tone was set with the appointment—through random and undirected
selection—of the presiding judge in the case, John E. Jones III.

There was nothing in his background to prepare Jones for this case
or to suggest the approach he would end up taking. On the basis of his
résumé, his long career as a Republican politician in a particularly conservative region of the state, and his appointment to the bench in 2002 by
President George W. Bush, the early handicapping in the trial suggested
a judge who would take a narrow approach to the case, and who would
be inclined at least philosophically to defer to the Dover school board.

The handicappers didn't know John Jones very well.

Before taking the bench, he had happily lived the life of a country
lawyer for twenty-two years in his hometown of Pottsville, in the heart
of Pennsylvania's coal country. Jones's grandfather, a descendant of
Welsh coal miners, found a new calling for the family when he gambled
his life savings to build a public golf course on converted farmland, offering an alternative to expensive, exclusive country clubs. The business
thrived, and the Jones family now operates five golf courses. They put

Jones through college and law school. As a lawyer, he did a bit of every-thing, representing small towns and municipal agencies, many of them quite like the Dover school board—citizens who needed legal help in their public service. He handled juvenile cases, family law cases, contract disputes, and criminal trials. Over time, he also became increasingly in-volved in Republican politics, first at the local level, then statewide.

Jones is of the old school of Pennsylvania Republicans, moderate and pragmatic, what used to be called a "John Heinz Republican" after the late senator, an environmentalist who was heir to the ketchup fortune. Jones's friend and political mentor was another Republican moderate, Tom Ridge, the former Pennsylvania governor appointed by President Bush to be the first director of the Department of Homeland Security, established after 9/11. In 1995, early in Ridge's first term as governor, he appointed Jones to head the state Liquor Control Board. There, Jones stirred up controversy, leading the governor's unsuccessful campaign to privatize Pennsylvania's anachronistic system of state-owned liquor stores (the only legal source of hard liquor in the state). He gained addi-tional notoriety for banning the sale of a brand of ale, "Bad Frog Beer," because the label featured a drawing of the namesake amphibian flip-ping the bird to prospective buyers.

After toying with the idea of running for governor, then deciding to step out of the way of the juggernaut campaign of former Philadelphia mayor Ed Rendell, Jones pursued something he had wanted for some time, a federal judgeship, and secured the recommendation of both of Pennsylvania's senators. He had been on the bench for just over two years when the random selection process landed *Kitzmiller* on his docket. He had, as the lawyers predicted, no knowledge of intelligent design, but at the time he was quite pleased to have such an interesting case before him. His most celebrated case so far had been another First Amendment suit involving several students who were banned by Shippensburg Uni-versity from posting anti–Osama bin Laden posters on campus because of a school speech code. Jones ruled for the students as a matter of free speech.

Throughout the course of *Kitzmiller*, Jones kept his personal views on religion and evolution out of sight. It seems clear, however, that he personally senses no conflict between scientific principles and articles of faith, though neither is he inclined to reject out of hand the possibility that there might be a scientific argument for purpose or plan in the universe. Raised a Presbyterian, Jones later joined the Lutheran church of his schoolteacher wife, Beth Ann, and he has attended it for the past two decades, joined by his son and daughter. Lutherans, he points out, have no issues with evolution, so his participation in the case did not create any conflicts for him in that regard.

Margaret Talbot, the *New Yorker* magazine writer who covered the trial, described the judge in the most glowing terms: "Jones has the rugged charm of a nineteen-forties movie star; he sounded and looked like a cross between Robert Mitchum and William Holden." Jones admits that he was flattered by this portrait, but to avoid looking like a braggart, he invariably qualifies this recollection: "It caused my wife and children to go screaming [with laughter] through the house." Still, he had an extra spring in his step when he went to work the next day, Jones recalls, until his twenty-something law clerks looked at him quizzically and asked, "Judge, who exactly are Robert Mitchum and William Holden?"

As the pretrial hearings for *Kitzmiller* progressed and it became clear that both sides felt there was too much at stake to even contemplate settling out of court, Jones began to realize that his new case would not be simply an interesting legal battle. He realized that he would be living through—and ruling on—a piece of history, and that the choices he made could have an impact far beyond Dover and the limited jurisdiction of his trial court. This is what led him to fashion a playing field for the lawyers that would be as broad as possible: Intelligent design would get its full day in court, and so would the scientific heirs of Charles Darwin, and may the best science win. "I had no dog in this hunt," he would later say, "but I was a very interested observer."

The model for his approach to the case became the late Judge William R. Overton, who in 1981 had presided over *McLean v. Arkansas*—the first test of creation science. Like Jones's federal district court in Pennsylvania, Overton's jurisdiction had no reach beyond his district in Arkansas, and so his ruling did not have the force of precedent. Federal trial judges decide cases, not law, except in two special circumstances: the rare instances when their decisions become the basis of a Supreme Court holding (with the High Court either embracing or rejecting the trial court's reasoning); or the even rarer instance when the opinion of a trial court is so well reasoned, so groundbreaking, and so clearly correct that it becomes by general acclamation the standard, and even without the force of appeal, it is cited, relied on, and incorporated into other courts' rulings in similar matters. Overton's opinion achieved the second status, without ever reaching an appeals court. He brought in the best experts on both sides of the creationism question, and for the next quarter century his eloquent opinion that creation science was actually religion and therefore inappropriate for public school classes became the basis of case law and school practice, though it was the Supreme Court's opinion in *Edwards* six years later that set the actual nationwide precedent.

With luck, with attention to detail, and with a thorough airing of the scientific arguments, Jones hoped his ruling could have the same impact as Overton's—whether he decided in favor of evolution or intelligent design. For better or worse, his courtroom would be the setting for the latest incarnation of the age-old debate about man's origins. He felt it was his duty to do what he could to settle the question, so that some other school district and some other court wouldn't have to traipse—to use one of the judge's signature words—over the same ground again.

Opening arguments in a trial are the rough equivalent of the first round of betting in a hand of poker. They prove nothing, but as a general rule, if you've got the cards, you come out blazing. If not, you hedge your bets and hope for an opening later. Because *Kitzmiller*, like all First Amendment cases, was being tried before a judge, the opening statements were not nearly as critical as they would have been with a jury deciding the case, but they were no less telling. It was clear to everyone in the courtroom which set of lawyers came to the table confident about the cards they held.

Before a gallery packed with representatives of more than fifty news organizations from around the country, as well as from Germany, Italy, and Britain, the plaintiffs' lead attorney, Eric Rothschild, had no hesitation about revealing some of his best (though certainly not all) cards. He confidently predicted for the court what the evidence would show: that the Dover school board had a religious intent in formulating its ID policy; that some of its members subsequently lied about their motives; and that their chosen vehicle, intelligent design, had been devised specifically to sneak creationism back into public schools, an attempt to circumvent court rulings by crafting a form of creationism that did not mention God or the Bible.

"Board members did everything you would do if you wanted to incorporate a religious topic in science class and cared nothing about its scientific validity," Rothschild said, delivering his parting comments with the force of a karate chop. "Intelligent design is . . . the twenty-first-century version of creationism."

Next came the far more muted Pat Gillen, a young attorney from the Thomas More Law Center with a soft, hoarse voice, a pained expression on his face, and a poker hand he didn't seem to relish as he gamely outlined a three-part defense strategy. First, the defense would paint the "one-minute statement" as a modest curriculum change with little significance or impact, which left the teaching of evolution intact at Dover High, just as before. Gillen used the phrase "modest change," or something very close to it, seven times during his five-minute opening statement.

Next, the defense would paint Bill Buckingham—the leading force behind the curriculum change—as a rogue element who had little influence over the board or its policy. This was news to anyone in the courtroom who had also attended meetings of the Dover school board. Bryan Rehm gaped in astonishment when Gillen added, "The board listened to the science faculty more than it listened to Bill Buckingham." Rehm would later mutter, "I wonder what school board he was talking about, because it sure isn't Dover's." Still, this argument was the only spin on the evidence that had even a remote chance of allowing the school board's policy to survive one of the legal analyses Judge Jones was required to make: whether or not the board had a religious intent behind its change in policy. If Buckingham and the creationism connection could be disregarded as a sideshow, perhaps the judge could be persuaded that the policy was, as Gillen described it, "for the purpose of enhancing science education," not advancing Christianity.

Finally, intelligent design would be shown to be science, not religion,

Gillen promised, and there would be plenty of expert testimony to prove it. The fact that ID was a minority position did not make it unscientific—the big bang theory, the heliocentric solar system, and even the theory that lightning is caused by natural forces rather than an angry divinity all had been minority positions when they were first proposed, and had been bitterly opposed by mainstream science. "Intelligent design theory is really science in its purest form, the refusal to foreclose possible explanations," Gillen concluded, closing on what sounded like a winning note. But it would soon be shown that this was one of the ID movement's weakest points.

For Gillen neglected to mention what he meant by "refusal to foreclose possible explanations." He was referring to the refusal of intelligent design advocates to rule out the supernatural—the one limitation that modern science has imposed for the last four centuries, and that became the cornerstone of a period of unrivaled expansion of human knowledge known as the Enlightenment.

Out in the hallway, waiting for his turn to take the witness stand, Ken Miller, the Dragonfly Book coauthor and Brown University biology professor, paced about, pleasant to those around him but distracted, a boxer preparing for his match. Slim and tall, he wore as a tie clasp a partially dismantled mousetrap—an effective taunt for ID proponent Mike Behe, whom he has debated numerous time and criticized regularly. When in the same room, the two men were painfully cordial to each other, but only because they were both nice guys and gentlemen, even if there was a certain amount of tooth gritting involved in the effort.

The mousetrap had been Behe's favorite metaphor for his signature issue, irreducible complexity—until Miller had the bright idea of dismantling one of the homely little devices and using various pieces for a key chain, a paperweight, a money clip, and, most annoyingly from Behe's point of view, a ridiculous-looking tie clasp. For Behe, the mousetrap illustrated the same quality he saw in the bacterial flagellum—take away one mechanical part of the trap, or one sequence in the fifty-protein structure of the flagellum, and neither "machine" would work. That constitutes irreducible complexity, Behe maintained, something evolution could not explain or construct. And Miller said, no way—just see how the parts of a mousetrap can have other functions or their own, just as many of the proteins inside the flagellum have other uses in the body. The beauty of evolution (and mousetraps), Miller says, is that "used parts" can be co-opted to make new structures—there was nothing irreducible about it. Behe, in turn, says that Miller's reasoning is nonsense, because it still did not explain how evolution constructed the whole irreducibly complex flagellum. The two men had been debating this for years, neither one convincing the other, and Miller couldn't resist wearing the trap on the first day of the big case. Miller had hoped that Behe, who wasn't scheduled to testify until later in the trial, might show up to watch the start of testimony, which Behe did. But the trial's first witness eventually removed his makeshift tie clasp in the interest of courtroom decorum. "The judge might not take it the right way," Miller confided. "I'm a bit concerned about that."

On the witness stand, Miller's job was to set up the parameters of the case, with Vic Walzcak asking questions that, the science-phobic lawyer cautioned, should be answered so that his mother would understand. ("And me," Judge Jones chimed in at one point.) The questions were at times extraordinarily basic, and therefore fascinating, because most Americans really don't remember or know the answers. Who was Charles Darwin? What is evolution? What is the scientific method? And then there was an even more basic question, the one that lies at the heart of the case: "What is science?"

"Science," said Miller, "is the systematic attempt to provide natural explanations for natural phenomena." The explanation was so simple, so clear—and so deadly to design, which, at root, involves unnatural explanations. That was what Gillen was getting at when he spoke of ID as not precluding any explanations. Gillen asserts that the exclusion of the supernatural is unreasonable, a kind of scientific discrimination with ID as its victim. Miller says that the exclusion of the supernatural is unavoidable and correct, a kind of methodological guardrail keeping science from driving off a cliff.

With his definition, sometimes called the scientific method, sometimes called "methodological naturalism," Miller could explain why evolution is a widely accepted science, a well-tested theory that makes robust predictions that can be tested and verified, and have been tested and verified, time and again. No biology course would be complete without a thorough explanation of evolution, Miller added, and he ought to know; he teaches the large freshman "intro to bio" course at Brown most years, in addition to advanced courses in molecular and cellular biology.

Intelligent design, on the other hand, is neither scientific nor testable, because it looks outside natural causes for its answers. Therefore, it is fundamentally religious, Miller told Judge Jones, a new brand of creationism.

The power of this simple statement, so early in a trial that was to take months to complete, was almost lost on the gathering, where the media (at least during the first week of trial) outnumbered ordinary spectators ten to one. But if the judge accepted what Miller said as accurate—this statement that on its face seemed so eminently sensible, almost axiomatic—then the case was already over.

Miller is a skilled and practiced lecturer, and he offered his comments with jovial authority. He has experience with this sort of thing. When he was a young professor in 1981, his students at Brown cajoled him into debating Henry Morris, the author of *The Genesis Flood* and, at the time, the most prominent creationist in the country, who had been invited to Brown by a Christian student association. Miller finally agreed to the debate, but only after his students rounded up enough materials on Morris's theories for him to study and prepare in advance, enabling him to refute one creationist argument after another that in the past had left less-prepared scientists flummoxed. He soon rose to national prominence as a crusader for evolution and as the author of a best-selling book, *Finding Darwin's God*, which is divided into two parts: a very effective attempt to demolish creationist arguments, and a somewhat more difficult reconciliation of science and faith. Miller is a devout Roman Catholic who finds religion and science complementary (another reason why the legal team for *Kitzmiller* cannily chose him as its first witness), but this is a position with which many scientists struggle. Miller also made a name for himself—and became a permanent fixture in the national media's Rolodex files under "evolution expert"—when he stole the show during a debate in 1997 on the political commentator William F. Buckley's public television show, *Firing Line*. Buckley and three anti-evolutionists were pitted against Miller and three prominent defenders of evolution. Afterward, Buckley surprised Miller by pulling him aside and commenting, "Young man, that was the most astonishing performance I've ever seen." Miller found Buckley brilliant as well—but lacking in scientific knowledge. "Like many brilliant people, he is also capable of profound self-delusion," Miller observed—a comment he would also apply to some proponents of ID involved in *Kitzmiller*.[1] It was around this time that high school science teachers began chasing Miller down at conferences to get him to autograph copies of his Dragonfly Book, which he says 35 percent of American high school students lug around in their backpacks and may even read.

Walczak led Miller through a series of questions to explain why supernatural causes should not be part of science, and the teacher who knows how to keep his student's—or a judge's—attention leaped at the opportunity.

> I hesitate to beg the patience of the Court with this, but being a Boston Red Sox fan, I can't resist it. One might say, for example, that the reason the Boston Red Sox were able to come back from three games down against the New York Yankees was because God was tired of George Steinbrenner and wanted to see the Red Sox win. In my part of the country, you'd be surprised how many people think that's a perfectly reasonable explanation for what happened last year. And you know what, it might be true, but it certainly is not science.

As the judge smiled, Walczak nailed the point home. "Does science consider issues of meaning and purpose in the universe?" Miller answered:

> To be perfectly honest, no. Scientists think all the time about the meaning of their work, about the purpose of life, about the purpose of their own lives. I certainly do. But these questions, as important as they are, are not scientific questions. If I could solve the question of the meaning of my life by doing

an experiment in the laboratory, I assure you I would rush off and do it right now. But these questions simply lie outside the purview of science. It doesn't say they're not important, it doesn't say that any answer to these is necessarily wrong, but it does say that science cannot address it. It's a reflection of the limitation of science.

And there it was. Science doesn't rule out the supernatural—it doesn't rule out God as a cause—because scientists are small-minded or conspiring to cover up evidence of divine design, as creationism and ID often allege. Science rules out the supernatural because it is *science* that is limited, whereas God is not.

Excluding God from science does not deny God, in Miller's worldview; it glorifies God. By contrast, containing God within science, as ID attempts, would shrink and limit God, reducing a supposedly almighty force into a sort of tinkerer who has to step in and design a bacterial flagellum, rather than create a grand universe and a natural world that can handle such small matters on its own.

Then the slide show began, and Miller attempted to illustrate how evolutionary theory has been confirmed by modern genetics, which was unknown in Darwin's day. One of the serendipitous developments in preparing for this case, Miller explained, was that, three weeks before the trial began, an article appeared in the premier science journal *Nature* on the decoding of the chimpanzee genome, a breakthrough in the analysis of primate genetics. This decoding allows researchers to compare the DNA sequences of the human genome with what evolutionary theory asserts is man's closet relative, the chimp (their common ancestry dating back as little as 5 million years). Miller explained that this genomic comparison, never before possible, posed a major test for evolutionary theory: Certain genetic features should be the same between creatures so closely and recently related. If not, then the theory of evolution and common descent would be proved false, and intelligent design would be busily elbowing its way to a place at the table.

The test was simple in concept. Miller explained that one relatively recent discovery about DNA sequences is that they are filled with harmless mistakes—bad copies called pseudogenes. These mistakes are stacked next to the "real" genes, which contain the blueprints for cells, organs, and everything else that makes a living creature tick. The pseudogenes, identical in almost every respect to a real gene, are missing a bit of code and so are rendered inert, mere placeholders. Pseudogenes appear in every organism, and they occur in a completely random way. Imagine a box of fortune cookies, all apparently uniform, from a factory that, every now and then, unpredictably produces a cookie without a fortune inside.

The thing about pseudogenes is that they don't evolve; natural selection doesn't touch them, because most perform no obvious function and therefore cannot be selected or weeded out. They just stay. So there is no way for two different species to have the same random mistakes in the same place in their genome unless the species are related—unless they have a common ancestor that passed on the same pseudogene to each lineage. If the lineage of *Pan troglodytes* and the lineage of *Homo sapiens* truly are connected through an extinct apelike relative, then man and chimp should possess many of the same pseudogenes. "Evolution makes a prediction," Miller explained to Jones. And it's a make-or-break test.

The result of analyzing pseudogenes in the blood cells of both species: Evolution passes the test. Man and chimp share exactly the same genetic mistakes.

But there's a problem elsewhere in the genome, Miller added, turning to the next slide. Humans have twenty-three pairs of chromosomes. Together they form the blueprint of human life. Chimpanzees, however, have twenty-four pairs in their blueprint. Chromosome pairs are not insignificant: An organism or species cannot just lose one. Each pair contains so much information that such a loss would be fatal. There's no way to evolve that. "And so the question is," Miller said, "if evolution is right about this common ancestry idea, where did the chromosome go?"

Evolution makes yet another testable prediction here, according to Miller. Sometimes, chromosomes fuse—two pairs combine end to end to make one pair. Fusions are easy to spot in the genes, because chromosomes are distinct in their structure, with specific ends and middles: telomeres and centromeres. And if evolution is true, this fusion would have occurred after the lineage of man and the lineage of chimp went their separate ways, leaving evidence behind in the human genes—a forensic clue, like fingerprints at a crime scene. Evolution predicts that there should be a pair of chromosomes in humans that have double sets of ends and middles—seams where the genes were spliced together— and they should correspond to two unfused pairs of chromosomes in the chimpanzee genome. If they are not there, Darwin goes down.

A click and the next slide pops up—evolution has passed the test again. The chromosomes match. The fusion of two pairs of human chromosomes is exactly where it should be for evolution and common descent to be true. Evolutionary theory made a very risky prediction: that someday genetic codes would be found that fit in with the theory rather than falsifying it. This fusion may mark the point where humans and chimps became distinct species, or this particular change in the human genome may have come later. But either way, it is powerful support for evolution, and offers nothing to support intelligent design.

"To me, as a scientist," Miller said, "the most remarkable thing about evolutionary theory is that, as the science of biochemistry has developed, as the science of cell biology, genetics, molecular biology, and other elements of science have developed, all of these have fit beautifully into the general framework described by Darwin almost one hundred fifty years ago."

As the trial progressed, the scientific battle became increasingly one-sided. At the start, the plaintiffs had six volunteer experts lined up, with one more waiting in the wings to undo any damage inflicted by the defense. That extra expert would never be called.

The defense, however, suffered multiple casualties. The lawyers from Thomas More had started with eight paid experts, including the leading lights of the intelligent design movement. But before the trial began, three of them had departed in the flap over their insistence on personal legal representation. That left five experts, all of whom were mentioned in the defense's opening statement as important to the case, but during the course of the trial two more vanished from the witness list. That left only three experts, and only one of them came close to being a "star" witness. Michael Behe would have to shoulder most of the burden of defending intelligent design.

The departure of the defense experts in the middle of the trial was an embarrassment to the defense, and it could not be chalked up to a perception that the plaintiffs' case was weak, requiring little response. On the contrary, the presentations that followed Ken Miller's began to seem like a scientific juggernaut to most observers of the trial. Nick Matzke began referring to the departed ID experts as jumping off a sinking ship. Even Richard Thompson, who appeared on C-Span while the case was still being tried, complained bitterly about the Discovery Institute's witnesses fleeing the case, "victimizing" the Dover school board, and ruining his own ability to counter the plaintiffs' experts.

One day, there would be Kevin Padian, not only tending his "critters" but delivering a scathing summary of ID proponents as untrained in the disciplines they criticize, relying instead on "a smoke-and-mirrors pantomime of scientific logic." Another day brought Brian Alters, an expert in science education from McGill University in Montreal, who dissected the "one-minute statement" as an enemy of critical thinking and of good science education. He asserted that the statement misused the word "theory," fostered doubt about evolution where none legitimately exists, and imposed a gag rule on discussion—inviting students to conclude that ID is a special, seductive "secret science" that must not be questioned.

Through Robert Pennock, a philosopher and computer scientist at Michigan State University, the plaintiffs made a unique two-barrel attack on ID. As a philosopher, Pennock traced the origins of "methodological naturalism"—the scientific method—from Hippocrates's rejection in 400 BCE of divine possession as a cause for epilepsy, through Ben Franklin's investigation of lightning as a natural phenomenon rather than a manifestation of an angry God, leading directly to his lifesaving invention of the lightning rod. Ruling out the supernatural in science was not simply the province of evolution, Pennock explained, but a basic rule of the Enlightenment—and of every significant scientific advance that followed. Yet, he said, proponents of ID want to turn back the clock and create a theistic science, and to prove that they meant business, he quoted Bill Dembski saying so in the most unambiguous way: "The ground rules of science have to be changed. . . . The scientific picture of the world championed since the Enlightenment is not wrong, but massively wrong."

Wearing his computer geek hat, Pennock then described a program he had helped design, "The Ancestor," which allows "digital life" to reproduce, compete, go extinct, and "evolve" in a simulation of the rules of nature—self-replicating, independent mini-programs that rewrite themselves across generations, compressing millions of years of evolution into moments. Pennock and his colleagues just sit back and watch, and one thing they look for has been the evolution of supposedly "irreducibly complex" systems—the digital version of Behe's flagellum. They found these, Pennock said, following precisely the rules of evolution that Behe said could never produce such complexity. "It's a direct refutation of that challenge to evolution," Pennock told the court.

Finally, the plaintiffs brought in a Catholic theologian from Georgetown University to paint ID as a purely religious proposition, attempting to disguise its special brand of creationism by refusing to identify the designer. John Haught said the ID movement made the mistake of asserting that religion needed to fit inside the world of science, an idea commonly associated with creationists and biblical literalism. Most theologians resolve this by treating the Bible as metaphor (if a video camera had been present during the Resurrection, it would have recorded nothing, Haught says, because seeing that moment required faith) and by understanding that science addresses "how questions," while religion tackles "why questions." By mingling the two, ID proponents create an "appalling theology" that "belittles God," reducing him to a meddler and a tinkerer, Haught says, adding that ID itself could be considered a sin and a violation of the Ten Commandments: "That's known as idolatry."

For ten days this barrage continued unabated. Almost half of the time was taken up by lengthy but largely unproductive cross-examination that failed to shake the witnesses or bring down the theory of evolution. "Just wait until it's our turn," Dick Thompson would say at the end of each day when intelligent design was used as punching bag, spinning the situation as best he could out on the courthouse steps. "This is the plaintiffs' case; they get to do all the talking for now. Just wait for our guys."

Thomas More's inability to inflict damage during cross-examination was not a matter of bad lawyering—the witnesses were very good, leaving the opposition little room to maneuver—but there was an increasingly obvious problem with the defense lawyers' scientific knowledge. The plaintiffs' lawyers had a full-time science adviser, Nick Matzke, sitting in court ready to come up with new lines of questions, to dig reports out of the NCSE archives on his laptop, to respond to an unexpected jab from the defense, or to recruit help from the international group of anti-ID science bloggers—who were following the trial's every word almost obsessively and then posting daily dissections on the web. The crew from Thomas Moore had no such daily backup, and it showed, time and again. Thompson's cross-examination of the theologian Haught proved a painful example.

Thompson had scored a few rare points early on, surprising Haught by producing a line from one of his books criticizing other experts, including Pennock, for indiscriminately lumping advocates of intelligent design in with creationists. This forced Haught to concede that there were legitimate differences between the two, making his previous testimony—that ID was a kind of "special creationism"—sound contradictory. But instead of consolidating this small victory, Thompson launched an ill-conceived attack on the whole idea of common descent, which even many ID proponents accept as at least possible, but which he personally disbelieves.

"Let me give you an analogy," he said. "I have some nuts and bolts. I take some nuts and bolts and make a car."

Haught listened politely. Thompson continued, "Then I take some other nuts and bolts and make an airplane. They have the same parts, but does that mean that the airplane came out of the car?"

The reporters in the jury box gaped at this, exchanging looks. What was Thompson talking about? The judge maintained a poker face, but Haught wasn't sure what to say, other than to agree. Thompson was correct: an airplane built with some of the same parts as a car did not "come out of the car."

Thompson looked triumphant. He had just demonstrated, in his own mind at least, that the similarities in the ape and human genomes that Ken Miller had illustrated did not prove common descent. Couldn't God, he asked Haught, have just used similar raw materials to do the job of creating both man and ape—the same nuts and bolts—without their being related? Well, sure, Haught answered, God can do anything. He could also have let evolution do the job. So what?

Thompson's inquiry, near the end of the day, was a disaster on two fronts, though he appeared to be oblivious of both. First, he butchered the basic principles of both genetics and evolution—humans don't come out of apes in evolutionists' view; they have a common ancestor that passed on the same *unique* "bolts" to each species, and no other. That's what made the genome evidence so compelling: The matching bolts that mattered in this case were the accumulated mistakes in the pseudogenes, not the original "design." The fact that the whole genomes were constructed of the same bolts meant nothing—this was as true for man and alligators as it was for man and apes. The point was that, once those bolts were in place, they had been, figuratively speaking, overtorqued and stripped in exactly the same way in both species. And this powerful evidence of common descent seemed to have eluded Thompson.

Second, and worse still for Thompson, he had inadvertently hammered home Haught's point about why Hippocrates and Franklin and all of modern science attached such importance to sticking with natural explanations, all of which can be tested. Thompson's assertion that "God did it" could never be tested, proved, or disproved. Thompson had unwittingly demonstrated the problem with insisting that intelligent design is science—it could explain any finding, because, ultimately, the "designer" can do anything. That might well be true. But if a proposition cannot be falsified, it is not scientific.

Thompson, in essence, had disproved his own case.

Afterward, Eugenie Scott, the director of the National Center for Science and Education, stood outside with Nick Matzke and exulted with reporters, keenly aware of what Thomas More's chief counsel had done. "Thompson was not even wrong," she said. "He missed the boat completely."

But at the Harrisburg Hilton, where the crew from Thomas More was staying during the six-week trial, Thompson visited in the lobby with the easygoing and erudite science writer Gordy Slack. Slack was covering the trial for the Salon.com web magazine, and Thompson granted an interview. As Slack would later report, Thompson said excitedly, "Did you see me show that there's no scientific evidence for man coming from an ape? I shouldn't have. But I couldn't help myself. I just wanted to do that."

Slack tactfully changed the subject.[2]

———

The rest of the plaintiffs' scientific case came down to shooting holes in intelligent design. Ken Miller assumed the role of Michael Behe's nemesis, attempting to debunk in advance his rival's anticipated testimony on biological systems that are supposed to be irreducibly complex—the blood-clotting cascade, various vital protein sequences inside cells, and the familiar flagellum. None are irreducible, Miller maintained. A subset of the proteins in the flagellum, for instance, turns out to serve quite nicely as the part of the bubonic plague organism that injects toxins into its victims' cells. Some species—dolphins, for one—have blood-clotting factors that are missing several of the components that Behe asserts must be present for the whole system to work, yet dolphin blood clots just fine. If irreducible complexity were true, dolphins would bleed to death from the slightest cut, Miller said. But they don't.

Miller also condemned *Of Pandas and People* as riddled with errors, bad science, and "fundamental dishonesty." It makes boneheaded mistakes about evolution on nearly every page, he said, such as the flat assertion that evolution cannot create novel biological "information," meaning that it cannot cause new species or body structures to evolve. Miller, sounding almost angry at this notion, said it is simply false, that the scientific literature is full of evidence and experiments to the contrary, and he suggested that *Pandas* made this and many similar errors for a very conscious purpose: They lead automatically to the argument that a supernatural designer was needed to create life on Earth.[3]

Miller heaped scorn on the statement the Dover school board had mandated be read to students. Short it might be, Miller said, but modest it was not. He said it was packed with lies and designed to poison the well for kids who might be interested in science. Before kids learned word one about evolution, they would hear that the school would teach them about the theory only because the state mandated it—not because anyone really wanted to teach it or because it was important, but because some bureaucrats insisted. Further, students learned from the "one-minute statement" that evolution, singled out from the many theories they would learn about in school, was suspect. It had gaps. Unlike gravity or electricity, it had alternatives. Evolution was special in this regard, and not in a good way. He called the statement a "science stopper," a perfect vehicle for leading kids to dismiss scientists as untrustworthy and their profession as dishonorable, at precisely a time when America needed more kids interested in science, not fewer. "I think this is a tremendously dangerous statement."

Miller was then cross-examined by the third member of the legal team from Thomas More, Robert Muise, whose otherwise pleasant face now seemed locked in a furrow-inducing frown. He spoke in a monotone and adamantly refused to participate in any of the light moments that Miller injected into the testimony and that Judge Jones seemed to appreciate. Perhaps this humorlessness had something to do with the fact that Muise scored very few, if any, points against the practiced debater Miller. Muise couldn't even coax Miller into admitting that when Michael Behe testified, the biochemist would probably cite scientific evidence and studies of his own to refute Miller point by point. Miller just shook his head and said he had no way of knowing what Behe would say.

"Perhaps he'll get up here in a couple days and say, you know, I listened to everything Dr. Miller said and, by God, he's got it exactly right."

Judge Jones chuckled and said, "We'd have a real story then, wouldn't we?"

"Exactly," Miller said with a broad smile.

Muise was not remotely amused and responded as if he weren't quite sure that Jones and Miller were kidding. "I doubt that will happen," he said. "Do *you* think that will happen, Dr. Miller?"

"I'd much rather make a bet on the outcome of the World Series this year than to make that kind of bet."

That was what Muise was up against: a cheery, energetic witness who knew his stuff, would not be bullied, and who was having fun. All Muise managed to extract from Miller was the admission that evolutionary theory was a theory, not "absolute truth," though it's unclear how this helped the defense. "No theory is ever regarded as absolute truth," Miller explained in a helpful tone, a smile still visible beneath his trim salt-and-pepper beard. "We don't regard atomic theory as truth. We don't regard the germ theory of disease as truth. We don't regard the theory of friction as truth."

Okay, Muise said, that means all theories continue to be tested? Yes,

Miller said, all theories are subject to testing and revision. And he further agreed with Muise that it's reasonable to say there are many unanswered questions about scientific theories, including evolution. But he balked when Muise got to the meat of this line of inquiry—when the lawyer tried to substitute the word "gaps" for "unanswered questions." This was the little trap Muise tried to spring. He wanted to suggest that the one-minute statement—which told students that there were "gaps in the theory of evolution"—was just referring to these unanswered questions, and so did not contradict mainstream science in the least. This was part of the defense strategy to render the statement noncontroversial. If Miller said, yes, sure, there are gaps, it would be a blow to the plaintiffs' case.

But Miller said that the statement was completely wrong, that there were no gaps in the *theory*. The theory was solid and unified and amazingly complete after 150 years. There may be gaps in the *evidence,* he said—there were gaps in his family tree, for that matter, because he didn't know who his great-great-great-uncle was. But the underlying theory behind a family tree, like the theory of evolution, had no gaps in it.

Muise wouldn't let it go, though. Isn't the theory of evolution tentative? Isn't it incomplete and subject to revision?

"All science is necessarily incomplete," Miller answered placidly. "On the day that physics becomes complete, for example, it will be time to close every department of physics in the United States because we'll know everything. I don't expect to see that happen."

Nor could Muise get Miller to bite on a key point Behe often makes: that there is no step-by-step explanation of how a cell could evolve through Darwinian processes. This is part of Behe's fundamental argument for irreducible complexity and intelligent design—the missing "proof" Behe feels is so damning. But Miller responded that this lack of explanation for all the steps of cellular evolution was not the fault of evolutionary theory—it was the fault of cell biologists like Miller and biochemists like Behe, who had yet to figure out fully how cells and their many complex structures work. And until you know how something works in the present, Miller said, you can't even begin to say how it evolved in the past.

Muise did get Miller to concede that he was a creationist "in the ordinary sense of the word"—that is, he believed that God created the universe. Or as Miller put it, "I believe God is the author of all things seen and unseen." But this was not news—Miller wrote a whole book about his spiritual fulfillment as a believer in theistic evolution, the idea that evolution is just one of the tools God created to shape the universe.

A clearly frustrated Muise was finally reduced to carping about a small error in the definition of evolution in a six-year-old edition of Miller's Dragonfly Book ("In bold print!" Muise announced triumphantly), and to bringing up a fringe idea of abiogenesis advanced many years ago by Nobel laureate Francis Crick. The idea, called "panspermia," hypothesized that life on Earth could have been seeded by space-faring aliens. Muise introduced this idea to suggest that an intelligent designer need not be supernatural, as Miller maintained, because alien visitors, though likely to be more advanced than humans, would still be part of the natural universe. Somehow, the idea of relying on UFOs to account for life on Earth in order to avoid the implication that the intelligent designer was God seemed more than a bit ridiculous—particularly coming from a representative of the "sword and shield for people of faith." Miller seemed amused. Unstated, but rather obvious, was a fundamental problem with the panspermia hypothesis: It doesn't really remove the need for a supernatural designer in ID, because the question immediately arises of who or what designed the aliens.

Later, at the daily press gaggle on the courthouse steps, Thompson inexplicably pronounced portions of Miller's testimony helpful to the defense, and disparaged the rest of what the professor had to say. "We don't agree with him. We don't believe ID is a science stopper," he said. "The big bang theory was developed by a Belgian priest. He was accused of advancing a religious idea as well. Einstein called him a buffoon. Now it is accepted science." Then Thompson added somewhat cryptically, "The designer doesn't have to be supernatural. It might be some element we haven't discovered yet." Apparently it was decided that an unknown element sounded better than spacemen; there was little further discussion of panspermia after that.

The Reverend Jim Grove offered a more pointed and briefer commentary after Miller stepped down from the witness stand: "How can he believe in God," Grove asked incredulously, "if he doesn't believe in creationism?" Nearby, a woman stood reading softly from a Bible, holding a one-person prayer vigil to ask God to help Judge Jones make the right decision, a Christian decision—a decision in favor of intelligent design.

Chapter 14

OF PANDERS AND PEOPLE

On the fourth day of the Kitzmiller trial, the Reverend Jim Grove passed out small green squares of paper to the journalists and onlookers—handbills advertising a presentation called "Why Evolution Is Stupid," featuring Kent "Dr. Dino" Hovind, the creationist scourge of evolution. The flyer gave the impression that Hovind would be present at the Dover Fire Hall that Thursday night, unless you read the fine print and noticed that the event was "a public service showing" of "Why Evolution Is Stupid." This meant Grove was going to roll out a few TV screens and play one of Hovind's endless series of DVDs on the evils of Darwin and the vacuity of every scientist in the world who disagrees with Dr. Dino's pronouncements, which would be just about every scientist in the world. Nevertheless, a group of slightly bored news reporters covering the trial during the first week decided to head over to the Fire Hall that night, hoping to find something juicy to write about that did not involve the word "flagellum."

There is a certain dynamic to long trials that draw large contingents of the national and international press. On the first day, the courtroom is packed. During breaks, reporters frantically try to find the corners of the courthouse that have decent cell phone reception so they can call in their stories to Chicago or New York or Milan. Outside on the steps, four or five video crews will be shooting stand-ups for the evening news. *Nightline* will be booking guests. Everyone wants a front-page story on the big opening day of the Scopes Monkey Trial II. In the mornings and when returning from lunch, the plaintiffs self-consciously file past the rows of cameras—"the perp walk," Tammy Kitzmiller called it—providing the setup shots each news report will broadcast that night.

By the next day, all the TV people, except the locals, will be gone, and so will half of the out-of-town print correspondents. By Thursday, all but a few writers for magazines and science journals will be gone, on to the next story. The major news organizations simply can't or won't staff a trial for six weeks. Better to parachute back in for some pivotal moment—the verdict, or perhaps the much anticipated testimony of the expert on the bacterial flagellum, Michael Behe—and let the Associated Press wire service coverage fill in the gaps. In this regard, the Kitzmiller case was not at all like the Scopes trial, which was covered breathlessly every day in the national press. In Harrisburg, by Thursday, only the local reporters remained in force, the ones who have covered the story from the beginning and know it better than anyone, whose coverage—particularly in the two local papers—has been exemplary, and who will see it through to the very end. They tend to rush at high speed to the neighborhood Starbucks during breaks, hoping to fortify themselves with enough caffeine to remain focused on the next interminable round of cross-examination on the topic of the blood-clotting cascade, which, as Vic Walczak likes to joke, is not nearly as interesting as it sounds.

The Dover Fire Hall is a sparse meeting room with hard plastic seats and many long tables suitable for bingo, and an unfortunate echo that makes it hard to hear the welcoming speaker at the front, who is the Reverend Jim Grove. The event begins with about seventy-five people in attendance, though the audience will double before the evening is out.

"I'm not opposed to teaching evolution in the public schools. Teach it!" Grove tells the gathering. "But I don't think you want it taught, and I don't want it taught, with a bunch of lies."

He then gives a very interesting recap of the week's trial in Harrisburg, which seems to have little to do with what was actually said in the courtroom, though he makes it clear that bunches of lies are pretty much all that is being tossed around in there. He shakes his head in disgust and says he saw Ken Miller and Robert Pennock testify. Doctors, they call themselves, Grove says mockingly. P-H-Ds. He draws out the letters.

"Far as I'm concerned, one's Dr. Flapdoodle, the other one's Dr. Flapping Lips." Grove has special scorn for Pennock—Dr. Flapping Lips. He complains that Dr. Flapping Lips claims to be an expert in theology, but Grove knows this cannot be true. "He doesn't even believe the Bible!" Grove howls, in reference to Pennock's nonliteral interpretation of scripture. "He'll tell you that!"

A short time later, Grove dims the lights and cranks up the DVD. Hovind appears on-screen to deliver his usual mix of cornpone humor and utterly incorrect science. Virtually nothing the man says for the next two hours—a very long, mind-numbing two hours—is scientifically accurate, and his arguments against evolutionary theory consist principally of "That's stupid," "Kids, that's just stupid," and, when he's really incensed, "Quite frankly, that's stupid!"

He seems to think the big bang has something to do with evolution, so he brands that as stupid, too. A giant explosion was supposed to create an ordered universe, he says of the big bang. But adding energy to something doesn't organize it, Hovind says; it causes chaos. This is wrong scientifically in so many ways that it would take a book to catalog them, but for starters, the big bang did not add any energy to the universe (or subtract any energy from it), because, as most schoolchildren learn (except, apparently, the former science teacher Hovind and his unfortunate students), energy can be neither created nor destroyed. But never mind this first law of thermodynamics—Hovind knows how to make his point about how adding energy leads to chaos: "Just ask the folks who live here," he says. On-screen, a picture of the nuclear devastation of Hiroshima flashes into view. Hovind finds this hilarious.

"All the evidence for evolution has been proved wrong," he says. "There is no scientific reason to reject the Bible."

Wait, he says, you're wondering, what about the fossil record? That proves evolution, right? Wrong! There is no fossil record. It's just bones in dirt. His way of refuting paleontologists like Kevin Padian and the late Stephen Jay Gould is to invoke this enticing image: "If I got buried on top of a hamster, does that prove he's my grandpa?" Hovind also mentions that he sent some of his Dr. Dino tapes to Gould—America's leading evolutionary scientist—before Gould died, in an effort to "save" him. "I hope he watched them. I hope he got saved. Maybe he did."

The smugness of Hovind is almost unbearable to watch. He recalls a chat with some poor fellow who told him he believed in evolution and didn't believe in hell. Hovind's response, "You can go to hell without believing in it. That won't bother God a bit. . . . Creation is the proof of a creator. You don't believe, you go to hell."

And there's the simple alternatives Hovind likes to throw at people: You can believe me, believe in God, believe in a 6,000-year-old Earth, and believe in a conspiracy of the world's scientists to promote Satan's lie (i.e., evolutionary theory): or you go to hell.

At last the DVD ended, followed by a burst of applause. Grove then opened up a question-and-answer session by denouncing the ACLU, a "humanist organization" that had brought in outside experts to make sure Dover's children were fed lies about evolution. And this was being done with the help of some of Dover's own, Grove added. There were mutters in the crowd about this treachery, but no one volunteered any commentary. This was not really the sort of event staged for the curious to learn about evolution and its critics, or to even pose questions; this was an occasion for the true believers, an affirmation of opinions about science from someone who puts the title "Dr." in front of his name (Hovind received a doctorate in "Christian education" from an unaccredited correspondence school, Patriot University). So it was somewhat astonishing when a young man in a white shirt stood up, straightened his glasses, set up a portable mini-lectern before him on his bingo table, and, reading from notes, launched into a defense of evolution and the parents who had sued the school district—who, he pointed out, were all Christians.

"You can believe in God and also believe he evolution," he said, and immediately an angry buzz began to build in the crowded room. "You can get a lot of people afraid by saying Darwin's going to take your souls away. . . . But I don't want your religious misgivings to govern what's taught in any kid's science class."

The shouting and catcalls began then.

"There is no evidence for evolution," one man yells.

"Are you a science teacher?" a woman calls out. This is meant as an insult—to her, the science teachers who refused to read the board's statement on ID are the enemy.

"You were brainwashed in college!" another woman shouts with a manic glee. She is standing up in the middle of the hall, screeching at him. The man tries to continue, but he is drowned out and finally gives up.

His name is Burt Humburg, and he is a medical resident at Penn State University. He has been following the Kitzmiller case and imagined that he might try to turn a few minds around by standing up alone in the Fire Hall. He is a native of Kansas and was active in the anticreationism movement there, although he grew up as a creationist. His passion for science led him to learn and to examine his beliefs critically—and this turned out to be the "cure" for his creationist beliefs. It is no coincidence, he says, that areas of the country that have low rates of college education (Dover is one of them) have the highest rates of belief in creationism and the lowest rates of acceptance of evolution. It's simply a matter of education, Humburg says, and that's why the fight in Dover is so important.

"Brainwashed," the screeching woman says. "That's what college does to you. Takes your soul."

In a calmer moment later, the woman, who says her name is Joan, explains that she wholeheartedly supports the school board's ID policy, and wishes that the schools would teach intelligent design outright instead of the "lies of evolution." Then she shows, once again, how the refrain from ID leaders that they are not advancing a religious proposition is belied by their most enthusiastic supporters, who have no doubt what the core proposition of intelligent design really is: "This country needs to get God back into our public schools," Joan says. "That's all the school board wants to do, to make us a good Christian nation again. What's wrong with that?"

Back in Harrisburg, the attorneys for the plaintiffs began the second week of the trial by calling their star witness, a soft-spoken professor from Louisiana whom Kevin Padian nicknamed "Creationism Kryptonite." She was the only witness in the trial that the Thomas More Law Center fought to keep out of the case, attacking her in legal motions in unusually personal and vituperative terms, calling her "little more than a conspiracy theorist and a web-surfing 'cyber-stalker' of the Discovery

Institute." She is also the only expert the Discovery Institute continually criticized, called names, and belittled on its website, also in highly personal terms, while confidently opining—incorrectly—that Judge Jones had "skewered" her in pretrial hearings and would severely limit, perhaps exclude entirely, her testimony. As it happened, Thomas More and Discovery could not have been more mistaken: Jones would find this witness's testimony among the most compelling, dramatic, and thorough presentations in the entire case. Its effect was devastating not only to the arguments in favor of the Dover school board's policy, but to the entire intelligent design movement's claim of scientific legitimacy.

Barbara Forrest, a professor of philosophy at Southeastern Louisiana University, first became concerned about the influence of creationism in public schools when the controversy erupted years ago in her own community. Her unpardonable sin, in the eyes of Discovery and Thomas More, was that she, alone among opponents of intelligent design, decided to compile the speeches, articles, web postings, and public utterances of the leaders of the ID movement. And then she used their own words against them.

When she was through, it became quite clear why Richard Thompson had fought so hard to keep her from testifying, and why the Discovery Institute tried to make her look foolish, nicknaming her "Barking Forrest" on its website. Her information was devastating. If the scales had not already been leaning in the plaintiffs' favor before her time on the witness stand, Forrest's testimony all but tipped them over, leaving the school board in a very precarious position when it was their turn to present a defense.

First, there was the revelation Forrest provided the court about the book *Of Pandas and People,* based on the old manuscript drafts that Nick Matzke had helped uncover, then pored through with Forrest. Side by side, the pages presented to Jones showed that *Pandas* had been written originally not as a work of intelligent design, as it ended up, but as work of creation science. One of its two principal coauthors, Dean Kenyon, was a creationist who had submitted a sworn affidavit in the Edwards case in 1987—attesting to the scientific validity of creation science, using precisely the same arguments that were now being used to promote intelligent design. Kenyon wrote that affidavit at the same time he was working on *Pandas.* And when the Supreme Court rejected creation science as religion a year later, suddenly *Pandas* stopped being about creation science and started being a book about intelligent design.

This in itself might not have been a problem—there is nothing sinister about correcting a book to make it consistent with a Supreme Court ruling. Indeed, that would have been a good thing, Forrest said, except the book was not changed in any substantive way. Instead, every reference to creation and creationism in the book was simply changed to "intelligent design," and "creator" was changed to "designer" or "intelligent agency." There were more than 250 such substitutions in all—but there were virtually no other changes in the text. It was, she suggested, just a cut-and-paste operation. Even the definition of creation science was retained from the early drafts—the book was orginally called *Creation Biology*—right through to the final version, retitled *Of Pandas and People*.

Then Forrest put up a slide showing the definition of "creation" in the drafts before the *Edwards* decision outlawed creation science in the schools, and displayed it next to the definition in the post-*Edwards* draft. Jones's eyebrows shot up at this stark evidence; he would later call it "astonishing." From the draft of 1986:

> *Creation* means that the various forms of life began abruptly through the agency of an intelligent *creator* with their distinctive features already intact—fish with fins and scales, birds with feathers, beaks, and wings, etc.

From the version published in 1993:

> *Intelligent design* means that various forms of life began abruptly through an intelligent *agency*, with their distinctive features already intact—fish with fins and scales, birds with feathers, beaks, and wings, etc.

What did Forrest conclude from her analysis of *Pandas* and her other research on ID? "It is my view that at its core intelligent design is a religious belief," she said. ". . . Right out of the creationists' playbook. It's not new at all."

"You just saw the smoking gun," Matzke told a reporter during a break, clearly proud of his role in uncovering the old drafts of *Pandas*. Nor was he exaggerating. The testimony not only showed that the Dover school board's new policy referred students to a starkly creationist text; it also revealed that the foundational textbook of the intelligent design movement, one that its leaders endorsed and were in the process of updating even as Forrest spoke, was indistinguishable from a textbook on the very same creation science that had been banned from public school classrooms as unconstitutional in 1987. Forrest, her slight drawl growing stronger when she strove to hammer home a point, suggested her analysis showed that intelligent design had begun life as a trick, a bit of word substitution designed to get around that court decision, just as the movement's critics had suspected all along.[1]

If that was not enough, Forrest, who is the coauthor of the book *Creationism's Trojan Horse: The Wedge of Intelligent Design*, provided Jones with a complete primer on the wedge strategy and then began a long recitation of statements by ID leaders that seemed to suggest that their movement was religious at its roots, not scientific.

A barrage of objections interrupted Forrest as the lawyers from Thomas More angrily complained that she was trying to use an ad hominem argument (attack the messenger, not the message) against proponents of ID, and that it was improper to impugn the leaders of the movement for having religious beliefs. They argued that a scientist can be religious, even a creationist, and talk about the religious implications of a scientific theory such as ID, without invalidating the theory itself. Their example of this proposition was the atheistic star of evolutionary biology, Richard Dawkins; they incessantly quoted his line from *The Blind Watchmaker*, "Darwin made it possible to be an intellectually fulfilled atheist," as an instance of a scientist whose statements about his personal religious take on evolution had nothing to do with the *science* of evolution. Proponents of ID should be granted the same leave to comment on religious matters without having their scientific statements impugned, the defense lawyers said.

No one in the courtroom disagreed with this reasoning. The problem, Judge Jones said, was that it didn't apply. Dawkins never equated Darwinism with atheism (although that's what many creationists and advocates of ID do all the time, engaging in a true ad hominem argument). Dawkins was just saying that evolutionary theory made him feel good about being an atheist. Fine. That's got nothing to do with science. But this was very different from Barbara Forrest's statements, which did not show intelligent design leaders talking about ID making them feel fulfilled; nor did they discuss any other religious implications of ID. Instead, the statements revealed the stars of the ID movement talking about the movement *as* religion. The judge eventually determined that Forrest was not making an ad hominem argument. She was simply making a factual presentation—and using these leaders' own words to do it. Jones overruled every significant objection to Forrest's testimony.

When that strategy failed, Richard Thompson began a seemingly interminable attempt to undermine Forrest's credibility by establishing that she was a "card-carrying member of the ACLU" and a secular humanist. When she readily admitted this to be true, Thompson demanded that she address a list of increasingly bizarre topics, which he asserted were the consequences of believing in humanism and evolution, including this declaration by an obscure philosopher: "Evolution teaches us that we are animals, so that sex across the species barrier ceases to be an offense to our status and dignity as human beings." Thompson read this quotation aloud, then looked at Forrest accusingly and demanded, "Have you ever heard him say that?" Forrest hadn't.

Thompson went on so long and so tediously in this vein that the courtroom began to empty. The one area Thompson showed little gusto for addressing was Forrest's analysis of the book *Of Pandas and People*. His principal complaint about this was, inexplicably, the fact that Forrest did not know off the top of her head the total word count of the book, which had nothing to do with her analysis. And that was it. On the most devastating testimony Forrest—or any other witness—had offered in the plaintiffs' case, Thompson spent all of five minutes. He spent much more time establishing that Forrest was a person who believed "nature is all there is" and who rejected the supernatural. Thompson clearly savored that admission as a crowning victory—to those who support what the Thomas More Law Center stands for, Forrest had just shown bias of the worst kind: bias against God. Unfortunately for Thompson, as a matter of legal principle, it did nothing to undermine her testimony. Atheism is not grounds for disbelieving a witness or ruling her testimony inadmissible, though it seemed very clear that Thompson would not mind in the least having such a rule.

Finally, Eric Rothschild was able to lead Forrest through the roster of intelligent design proponents, as she quoted statement after statement that Thomas More and the Discovery Institute very much wanted kept out of open court:

> Phillip Johnson, ID's godfather: "My colleagues and I speak of theistic realism, or sometimes mere creation, as the defining concept of our movement. This means that we affirm that God is objectively real as creator, and that the reality of God is tangibly recorded in evidence accessible to science, particularly in biology."
>
> William Dembski, ID leader and Discovery Institute fellow: "Intelligent design should be viewed as a ground clearing operation that gets rid of the intellectual rubbish that for generations has kept Christianity from receiving serious consideration."
>
> The "wedge strategy" document, on the goal of the ID movement: "To defeat scientific materialism and its destructive moral, cultural, and political legacies; to replace materialistic explanations with the theistic understanding that nature and human beings are created by God."

Thomas More's lawyers dismissed the wedge document as a mere "fund-raising proposal" of little significance, but this hardly seemed to help their case. Either the wedge strategy was the foundational document of the ID movement, clearly making it a religious movement; or else the document was a disingenuous attempt to court the religious right by pretending that ID was religious in order to rake in monetary donations. Neither interpretation served ID or the Discovery Institute very well.

> Paul Nelson, philosopher, Discovery fellow, young-Earth creationist, and founding member of the wedge: "Without an ability to turn to the supernatural, science will be left with

hopelessly false naturalistic speculations about the reason physical objects exist."

At the end, Rothschild led Forrest back to Dembski, the expert who had abruptly left the case and who was the ID movement's most prolific, harshest critic of the "mechanistic" philosophy behind evolutionary theory. Rothschild turned to an article Dembski wrote in 1999, "Signs of Intelligence: A Primer on the Discernment of Intelligent Design," and asked Forrest to read the last paragraph to the judge. Forrest adjusted her glasses and read:

> The world is a mirror representing the divine life. The mechanical philosophy was ever blind to this fact. Intelligent Design, on the other hand, readily embraces the sacramental nature of physical reality. Indeed, Intelligent Design is just the Logos theology of John's gospel restated in the idiom of information theory.

Rothshild asked, "So . . . William Dembski locates intelligent design in the Bible in the Book of John?"

"He specifically locates it. He defines it as beginning with the Book of John."

There were no more objections now. There was just the back-and-forth between a well-prepared lawyer and a professor with a story to tell. Thompson was rocking in his swivel chair and staring at the ceiling, a disconcerting posture he assumed regularly in the courtroom, when not chewing his necktie. "And can you tell us," Rothschild asked, "how the Book of John begins?"

Forrest knew the lines by heart, the New Testament take on Genesis: "In the beginning was the word. And the word was with God. And the word was God."

The loss of Bill Dembski as an expert for the defense was felt most keenly in the aftermath of Barbara Forrest's dramatic testimony. Without him, Thompson had no one with the stature, background, and rhetorical skills to take her on, to reply to her analysis of *Pandas* and of the religious statements of ID's leaders. Forrest's critique provided an overarching narrative that, unrebutted, could easily frame the whole case.

The Discovery Institute made a last-ditch effort to file pleadings in the case and to get Dembski's and Steven Meyer's expert opinions before the judge, but Jones rejected the attempt. It would be unfair to allow such opinions—and their excoriating criticisms of Forrest and the other experts—into the case "through the back door" without putting them on the witness stand to be cross-examined like everyone else, Judge Jones ruled. "I'm not going to have some rogue cavalry come riding in here at the last instant," he griped to the attorneys during a sidebar.

Had he testified, Dembski's expert reports suggest, he would have forcefully argued that Forrest was little more than a name-caller, in need of some education in critical-thinking skills, and given to using guilt by association as her method of attack.

He argued that her list of statements made by members of the ID movement proved nothing about design theory itself. He denied that his own observation about ID and John's gospel were a statement about the science of intelligent design; they were merely a discussion of the theological implications of ID. And he accused the plaintiffs of rank hypocrisy in claiming otherwise, citing a televised statement by Barry Lynn, head of Americans United for Separation of Church and State, which provided legal assistance for the plaintiffs in *Kitzmiller*. During the same debate on *Firing Line* in which Ken Miller appeared in 1997, Lynn proclaimed, " 'In the beginning was the word. . . .' Indeed that word just might turn out literally to have been a command: 'Evolve!' "

How is this any different, Dembski demanded, from the sorts of statements made by proponents of ID that are supposedly so damning? By Forrest's type of analysis, Dembski suggested, evolution should be declared a religion and subjected to First Amendment attacks, too. He assembled a long list of quotations, in mimicry of Forrest, but his selections were from evolutionists, all pointedly suggesting that Darwin's theory could be used to support genocide, rape, atheism, and bestiality (quoting the same philosopher Thompson threw at Forrest). Should this cause evolution to be declared invalid? No, Dembski wrote: "The religious mileage associated with ID is as separable from the actual science of ID as the anti-religious mileage associated with evolution is separable from the actual science of evolution."

Dembski wrote passionately in his report that evolutionists had adopted a kind of zero-tolerance approach to ID—they would concede nothing, no matter what. If papers were published and peer-reviewed, they would be attacked or, worse, their editors would be ostracized. Nothing the ID movement did would be viewed by mainstream science as useful, scientifically interesting, or theoretically valid. To put this approach into perspective, Dembski invoked Machiavelli, who observed that there was nothing more dangerous than attempting to establish a new order of things, because those who support the status quo will reflexively attempt to stamp out any new ways of thinking. Naturally, ID proponents were the new order in this analogy, and evolutionists stood in as Machiavelli's antagonists.

Dembski also quoted the Nobel laureate William Lawrence Bragg on the nature of science: "The important thing in science is not so much to obtain new facts as to discover new ways of thinking about them." That is what Einstein did working as a lowly patent clerk in Switzerland, Dembski observed. "Intelligent design is attempting to do that for biology—discovering fruitful ways of thinking about and interpreting well-established facts of science that pertain to biological complexity and diversity."

Whether Dembski could have successfully rebutted Forrest or the other experts in the case, and whether he could have really shown that the statements by proponents of ID were about religious implications rather than the essence of ID itself, remains an open question. His report is well written and thought-provoking, but his opponents take issue with most of its assertions. And even if he succeeded in minimizing the damage done by the ID leaders' statements, this does nothing to address the more fundamental claim of all the plaintiffs' experts that, simply by invoking design, the ID movement is by definition unscientific and religious because it is necessarily also invoking a supernatural designer. Another way of stating Dembski's argument that design can be detected or inferred in nature is to say that the work of a designer can be inferred or detected. To deny this is to play a semantic game, critics of ID say. This is one of the main reasons why Dembski has to redefine the meaning of "science" to accommodate ID, while simultaneously calling mainstream scientists unreasonable and rigid. Even the conservative commentator Rush Limbaugh discerned this same problem and therefore called the ID movement—which he generally supports—"disingenuous":

> Intelligent design is a way, I think, to sneak it into the curriculum and make it less offensive to the liberals because it ostensibly does not involve religious overtones. . . . But I think that they're sort of pussyfooting around when they call it intelligent design. Call it what it is. You believe God created the world, and you think that it's warranted that this kind of theory . . . be taught.

In the end, there would be no Bill Dembski in the courtroom to voice his ideas and objections; there was just a report that would not be considered, containing ideas safely advanced in writing but never subjected to the whirling knives of expert cross-examination. There is an underlying tone of bitterness in Dembski's report, a suggestion that he and intelligent design were not being given their due in *Kitzmiller*. If this is true, it cannot be blamed on the "old order." The responsibility for Dembski's absence from the trial is ultimately his own. Though he has repeatedly said he was eager to testify and had simply been caught in legal cross fire, it was Dembski's highly unusual demand that his own lawyer take part in the case that led to his removal from the witness list—and from the chance to defend his movement in the biggest public test ID has ever faced.

As the plaintiffs' scientific case unfolded, a very different but no less important and powerful parallel case evolved—almost a separate trial within a trial. This was the portion of the case that pitted the parents suing the school district against the members of the school board, neighbor against neighbor, friends against (former) friends.

And so, amid analyses of the chimpanzee genome, and signs of common descent in Kentucky Fried Chicken, and sweeping questions probing the definitions of science, religion, and the bright line between, there was the more earthly, poignant, and heartbreaking testimony of Julie Smith to remind the court just what this case was really about.

"What kind of Christian are you, anyway?" her sixteen-year-old daughter, Katherine, asked her one day.

Julie was stunned. Katherine had just come home from class at Dover High School. It was late 2004, after the new policy on intelligent design had been adopted and the community was abuzz with the controversy. Katherine had been talking with her friends at school and in her Bible club, and had decided that evolution was a lie, that nobody was going to call her a monkey girl, and she just couldn't understand why her mother accepted this theory that said there was no God. "You can't be a good Christian and believe in that lie," her distressed daughter said.

Julie is a medical technologist. She works in a blood bank, and she sees the consequences of evolution every day: resistant bacteria, new strains of diseases, the AIDS virus. Evolution is not a lie, she told her daughter, it's real and it's part of nature and it is not any more evil than the law of gravity. Nor does it conflict with Julie's belief in God.

Teenage girls can be headstrong, and Katherine was no pushover when she made up her mind—it took patience and effort to even get her to consider her mother's point of view. But they talked quite a bit about it. Katherine told her mom that the teachers at Dover High barely spoke about evolution, that they would hardly answer questions. Julie supposed they were scared—and her instinct would turn out to be on target—but this would never occur to Katherine or the other students, who saw teachers as the ultimate authority figures. No, the kids figured that the teachers didn't talk about evolution because something was wrong with it. And then when the school board started talking about Christ dying on the cross and how evolution denied God—at least, that's what the kids at school had heard—they knew evolution had to be a lie. Katherine wanted to be a good Christian, she didn't want to go to hell, and she was worried and angry.

This is what happens when a school board injects religion into the classroom, Julie Smith told Judge Jones. This is how my family has been damaged. It has shaken our house, shaken our faith, turned my child off science. Just as Ken Miller predicted, intelligent design was a "science stopper." It didn't matter that Katherine had already been through biology, before any statement had been read. The effect went far beyond just the ninth grade—it hit everyone in the school and, thanks to the newsletter about intelligent design that the board had sent out, the entire community, Julie Smith said.

The Smiths are Catholic, and this topic had come up at Julie's Bible study group, so she had some ideas how to respond. She told Katherine about her discussions at church, how the deacon had talked to them about evolution, how even the pope had declared that it did not conflict with faith. And Katherine, after a very long time and in spite of peer pressure at school, finally came around: Maybe she could be a good Christian and still believe in science, too.

This testimony was short and sweet compared with the marathon sessions with Forrest and Miller and the rest, but that made what Julie Smith had to say all the more memorable. One by one, all the parents came to the stand and recalled their shock and hurt over the school board's injection of intelligent design and of doubts about evolution into the curriculum and the community.

For Tammy Kitzmiller, it was all about her children—she was outraged that her elected school board would dare to inject itself into the personal and intimate matter of a family's faith. "The biggest thing for me as a parent, my fourteen-year-old daughter had to make the choice whether to stay in the classroom and listen to the statement, be confused, not be able to ask any questions, hear any answer, or she had to be singled out, go out of the classroom and face the possible ridicule of her friends and classmates."

Barrie Callahan told the court about her time as a board member and Alan Bonsell's repeated statements at the board's retreats that he wanted creationism to balance evolution. As she spoke, on the big screen flashed those official minutes and notes from the meeting corroborating Callahan's recollections. Later, after she left the board but continued attending meetings, she witnessed Bill Buckingham's crusade. She confirmed as accurate the news reports about his talk of creationism, and his remarks about America being a Christian nation where "students should be taught as such."

Christy and Bryan Rehm heard the same things and swore to Judge Jones that they had witnessed the board members advocate creationism for the classroom. As a result, they said they feared for the future of their children and the integrity of education in Dover. Christy, nine months' pregnant at the meeting of June 2004, when Buckingham and his wife had turned a government meeting into a virtual tent revival, had gotten so upset that she feared she'd go into labor then and there.

The science teachers had their day in court, too, revealing the behind-the-scenes discussion of creationism; the board members' refusal to compromise on anything; the board's ignorance of science; and the insistence by Bonsell that if they taught evolution without "balance," then the kids would come home and say the teachers were lying to them. Bertha Spahr, the indomitable head of the science faculty, explained that it had been her job to unpack the boxes containing *Of Pandas and People* after the donated books arrived at Dover High School. At the bottom of one box, she had found an invoice from the publisher that identified *Pandas* as a creationist textbook. This was one of the documents she had secreted in a safe until the time of trial.

Bit by bit, a wall of evidence was assembled for Judge Jones to consider: the testimony of more than a dozen men and women, backed by documents and other hard evidence, that the Dover school board had a religious agenda, and that it had injured people.

And then there was Fred Callahan, Barrie's husband, a businessman, a father, not an educator or a politician—just a quiet man who had no desire to get involved in the Dover intelligent design fight or in any lawsuits about it. But in the end, he felt he had to get involved. He had read about ID in advance of the board meeting at which the new policy was adopted, had concluded that it was a religious proposition, and had gone to the meeting to tell the board what he had learned. He waited his turn to speak, then urged the board members to step back from this "slippery slope." But when he suggested that they seemed intent on violating their oath of office and their fiduciary duties to the township, the gavel banged and he was pronounced out of order, his concerns dismissed.

As he sat on the stand, looking ill at ease and not altogether happy to be in court, he was asked to explain how the board's policy had harmed him—proving harm being one of the legal requirements of any lawsuit.

Steve Harvey was the attorney examining him, and the question was routine, the last one he planned to ask, and neither he nor the judge expected the burst of eloquence and clarity of thought that Fred Callahan summoned at this moment and that riveted everyone in the room:

I'm a taxpayer in Dover. I'm a citizen of Dover. I'm a citizen of this country. I think the heart of my complaint, my wife's complaint, is that this is just thinly veiled religion. There's no question about that in our minds. If you were to substitute where it says "intelligent design" the word "creationism," which, in my mind, it is, there would be no question that this would be a violation of the First Amendment. I've come to accept the fact that we're in the minority view on this. You know, I've read the polls. A lot of people feel that this should be in [the classroom], that it doesn't cross the line. There are a lot of people that don't care.

But I *do* care. It crosses *my* line. And, you know, I've been— there have been letters written about the plaintiffs. We've been called atheists, which we're not. I don't think that matters to the court, but we're not. We're said to be intolerant of other views.

Well, what am I supposed to tolerate? A small encroachment on my First Amendment rights? Well, I'm not going to. I think this is clear what these people have done. And it outrages me.

Chapter 15

UNDER THE MICROSCOPE, DEER IN THE HEADLIGHTS

On the tenth day of the trial, the plaintiffs rested their case and the defense took over, leading off with its biggest remaining gun: the biochemist Michael Behe, champion of irreducible complexity.

Suddenly, a trial that had appeared to have exhausted its energy reserves (the long yet fruitless cross-examination of Barbara Forrest being the low point) found a second wind. The empty seats in the stuffy courtroom filled again, as new spectators came, and many of the out-of-town journalists returned to Harrisburg to hear the first word from the intelligent design side of the second Monkey Trial.

They would listen to Behe's direct testimony, of course, but his views were well known and no surprises were expected there—indeed, his ideas had been discussed at such length as a target of the other experts' criticisms that it seemed he had already testified before he began. But the cross-examination was another matter. The duel between the canny debater Behe and Eric Rothschild, who had proved himself to be a master of the jovial, polite form of impaling witnesses, was highly anticipated. Behe's testimony in Kansas five months earlier had been largely unchallenged. But in Harrisburg, the plaintiffs had made no secret of the fact that they planned to go after Behe hard, while the defense had set high expectations that the biochemist would refute the scientific case so painstakingly assembled against ID in the first half of the trial. The plaintiffs were ahead on points and had drawn blood. Now the question was: Could Michael Behe get the defense back into the fight?

Although the first part of his courtroom presentation—under the friendly questioning of Thomas More's attorneys—seemed to go on rather long (a day and a half), Behe bore the burden of supporting the entire defense quite well. He cheerfully contradicted almost everything Ken Miller had said about him and his bacterial flagellum, using the rather polite word "mischaracterization" almost as many times as the word "flagellum" had come up in the trial.

He challenged even the most rudimentary points, including Miller's definition of intelligent design, which had been: "the proposition that some aspects of living things are too complex to have been evolved and, therefore, must have been produced by an outside creative force acting outside the laws of nature."

Mischaracterization, Behe said. It was an understandable one, Behe added generously, because Miller was trapped by his own evolutionary perspective, and could see ID only as an attack on Darwinian theory. "And it is not that. It is a positive explanation."

Behe's definition of ID was subtly different: "Intelligent design is a scientific theory that proposes that some aspects of life are best explained as the result of design, and that the strong appearance of design in life is real and not just apparent." He later added that the "positive argument" for design lies in the detection of a "purposeful arrangement of parts" in an organism through observation and logical inference.

That did sound scientific, and Behe helped this impression along by glossing over the "logical inference" part of his definition, with good reason: It is basically the same as William Paley's venerable watchmaker's analogy as proof of God's existence. In other words, ID proponents believe that they can see what appears to be design everywhere in biology. In human experience, design (such as words on a page) requires intelligence, so they argue that it is reasonable to infer that design in nature (such as the genetic code) requires intelligence, too. This does not sound so scientific, which is probably why Behe largely avoided it.

His definition was telling in another way, too, because unlike Miller's, it made no mention of a designer. The phrase Behe used to finesse this—"intelligent design proposes that some aspects of life are best explained as the result of design"—is the stock recommended phrase used virtually word for word by most, if not all, of the Discovery Institute's speakers and affiliates. They object vehemently to Miller's version, insist-

ing that the supernatural intervention in nature he includes in his definition as a matter of common sense is in reality an *implication* of the theory,
not part of the theory itself. This distinction allowed Behe to insist to
Judge Jones that ID was not creationism in disguise, was not religious,
did not require God to be the designer, and was in very much the same
position as the big bang theory when it was first proposed in the 1930s,
only to be harshly criticized as a religious idea. Eventually, the big bang
was supported by a large body of evidence and was accepted. The same,
Behe said, is happening with intelligent design.

Then he spent the next day and half explaining his familiar examples
of biological systems that he asserted could not have evolved. Miller's re-
buttals of Behe's opinions on these systems, the witness said, were just
plain wrong.[1] There still was no scientific evidence, no detailed, step-by-
step explanation, of how these systems could evolve naturally, he claimed.

Behe was careful to say that he did not intend to indict all of evolutionary theory. Some systems did evolve—there was hard evidence for
that. And he had no trouble with common descent as a principle. But
there was no evolutionary explanation for some parts of biology. And
that, he said, left design as the only alternative.

Michael Behe is a good witness, and a persuasive one. Though sometimes his jargon was overwhelming, more than a few spectators re-
marked later that they had walked into the courtroom thinking ID was
a crock but had walked away at the lunch break thinking maybe there
was something to it after all.

When it was finally time for cross-examination, Behe's mood—chip-
per, even ebullient, under the friendly questioning of Robert Muise—
shifted. As Eric Rothschild posed questions, Behe took on a distracted,
seemingly evasive air at times, and his usual smile faded. He continually
asked Rothschild to repeat questions, as if he had developed a hearing
impairment.

Trouble started immediately. Rothschild read him an excerpt from
Pandas, the part asserting that life began abruptly on Earth through an
intelligent agency—birds with feathers, fish with scales. In the original
draft, this has been the definition of creationism, and so Rothschild
asked the obvious question: Couldn't you substitute the word "creationism" for "intelligent design" in this definition?

Judge Jones already knew that the answer was yes—Barbara Forrest had shown him the contrasting drafts of the book—yet Behe inexplicably answered, "I don't think so, no." Then he explained that what the book really meant was that life appears abruptly *in the fossil record*.

"But it talks about life *beginning* abruptly, not just appearing abruptly, correct? Beginning is different than appearances in the fossil record, correct, Professor Behe?"

Behe was trapped, in the first five minutes of cross-examination, by a book that he had endorsed as excellent during his direct testimony and that he had critically reviewed for the publisher. He would not give in, however, even though his recalcitrance reinforced the image Rothschild and the other attorneys for the plaintiffs had been constructing throughout the trial—of ID advocates as dissemblers and purveyors of double-talk. "I don't take it to mean that way, no," Behe insisted.

But Rothschild was already moving on to his next target, keeping Behe off balance. He asked if Behe was a coauthor of the forthcoming revision of *Pandas* along with Bill Dembski and several other leaders of the ID movement. Behe said no, whereupon Rothschild smiled brilliantly and handed him copies of a deposition by the publisher of *Pandas* and Dembski's expert report, in which both men identified Behe as a coauthor.

"That's false?" Rothschild asked.

Behe then made the perplexing claim that, no, it wasn't false, but neither was it true. He might be a coauthor later, he said. "The way I read that is that he's seeing in the future."

"Seeing into the future is one of the powers of the intelligent design movement?" Rothschild asked. Then he shrugged off the inevitable objection and moved on to yet another point of attack: Behe's claim that ID was a scientific theory.

"When you call it a scientific theory," Rothschild asked, "you're not defining that term the same way that the National Academy of Sciences does?"

Behe nodded. "Yes, that's correct."

"You don't always see eye to eye with the National Academy?"

"Sometimes not."

Rothschild then reviewed the National Academy's definition of a scientific theory, which did not include ID: a well-substantiated explanation of some aspect of the natural world that can incorporate facts, laws, inferences, and tested hypotheses. Then Rothschild asked Behe: "Your definition is a lot broader?"

Behe preferred a definition that excluded "well-substantiated." "That's right, intentionally broader. . . ."

"Sweeps in a lot more propositions?" Rothschild asked.

"It recognizes that the word is used a lot more broadly than the National Academy of Sciences defined it."

"And using your definition, intelligent design is a scientific theory, correct?"

"Yes."

And then the next trap was sprung. "Under that same definition, astrology is a scientific theory, correct?"

Astrology. Horoscopes. Magic. Behe hesitated and at first did not answer directly, but when Rothschild pressed him, he admitted, "Yes, that's correct." The definition he used to qualify ID as a scientific theory also encompassed astrology.

And so it went, point after point. Next Rothschild ran through statements issued by the nation's leading scientific organizations denouncing ID as unscientific, and asked Behe if he was familiar with each one. Yes, the National Academy. Yes, the American Association for the Advancement of Science. Yes, the National Center for Science Education. Yes, yes, yes. Worse still, Rothschild forced Behe to acknowledge a statement that his own school, Lehigh University, had issued. This statement expressed support for Behe's academic freedom, but added that the entire biology department at Lehigh took the collective position "that intelligent design has no basis in science, has not been tested experimentally, and should not be regarded as scientific." Behe's own school and colleagues considered ID no more scientific than astrology.

"So you've not even been able to convince your colleagues, any of them, Professor Behe?" Rothschild asked.

Behe shot back that the statement also has his colleagues expressing unequivocal support for evolutionary theory. "What does that mean?" Behe said. "To commit oneself to a theory, to swear allegiance to a theory? That's not scientific. . . . It doesn't carry the weight of a single journal paper."

No, Rothschild responded, they're just referring to 140 years' worth of accumulated research. He then forced Behe to admit that the number of peer-reviewed articles in scientific journals arguing for the irreducible complexity of molecular machines was zero. Behe admitted he had never published such an article, and neither had anyone else.

However, Behe added, his book *Darwin's Black Box* was peer-reviewed. And that had laid out his entire case for ID.

Rothschild nodded. "You would agree that peer review for a book published in the trade press is not as rigorous as the peer review process for the leading scientific journals, wouldn't you?"

No, Behe said, he would not agree with that at all. His book was sent through the same sort of careful process. "I think, in fact, my book received much more scrutiny and much more review before publication than the great majority of scientific journal articles."

Now, these are the moments that attorneys live for. Cross-examination is an art. It's not just a debate, a back-and-forth; it's a maneuvering of a witness into an impossible position, into making an assertion that the attorney already knows he can demolish. Eric Rothschild had just entered that zone, thanks to his courtroom science geek, Nick Matzke, whose memory was jogged by a comment Behe had made a day earlier about his book being peer-reviewed. Matzke had hurriedly searched online for an article he thought might help Rothschild, and a few hours before court was to begin he handed Rothschild a piece of dynamite.

Rothschild serenely asked Behe if one of the peer reviewers who had performed this intensive examination of *Darwin's Black Box* was named Michael Atchison. Behe said yes, and Rothschild handed him a magazine article titled "Mustard Seeds," written in the first person by Atchison, a department chair at a veterinary school. Rothschild read the article to Behe, who sat stone-faced, listening:

While I was identifying myself as a Christian in Philadelphia, a biochemist named Michael Behe at Lehigh University was writing a book on evolution. As a biochemist, Behe found the evidence for Darwinian evolution to be very thin. In fact, when he looked at the cell from a biochemical perspective, he believed there was evidence of Intelligent Design. Behe sent his completed manuscript to The Free Press publishers for consideration. The editor was not certain that this manuscript was commercially viable or worth stirring up the inevitable controversy it would create. There were clearly theological issues at hand, and he was under the impression that these issues would be poorly received by the scientific community. If the tenets of Darwinian evolution were completely accepted by science, who would be interested in buying the book?

The article went on to explain that the editor told his wife about this dilemma. The wife happened to be a student in Atchison's class at the veterinary school, and she suggested that the editor give him a call.

We spent approximately 10 minutes on the phone. After hearing a description of the work, I suggested that the editor should seriously consider publishing the manuscript. I told him that the origin of life issue was still up in the air. It sounded like this Behe fellow might have some good ideas, although I could not be certain since I had never seen the manuscript. We hung up and I never thought about it again.

The article concluded with an anecdote about Atchison's finally meeting Behe years later, when the book was a best seller. Atchison wrote of his meeting with Behe: "He said my comments were the deciding factor in convincing the publisher to go ahead with the book."

In other words, according to this article, the critical peer reviewer of *Darwin's Black Box* hadn't even read the book.

"And is this your understanding," Rothschild asked Behe, "of the kind of peer review Dr. Atchison did of your book?"

"No, it wasn't. . . . I was under a different impression."

Rothschild then turned to a passage in *Darwin's Black Box,* in which Behe wrote, "Intelligent Design theory focuses exclusively on proposed mechanisms of how complex biological structures arose." So, Rothschild asked, can you please identify those mechanisms for us?

But Behe couldn't. No proponent of ID has ever identified these mechanisms. Instead, he asserted that the real focus of intelligent design theory was to detect design, not explain its mechanisms, and he said this was done by "inferring" design whenever biochemical structures showed a "purposeful arrangement of parts." But that definition had its own problems, and Rothschild soon pointed them out: Behe had conceded in various articles that evolutionary theory can account for systems that have a purposeful arrangement of parts. One of them—the antifreeze proteins that keep the blood of cold water fish from freezing—had been well documented by evolutionary biologists. Rothschild had undermined yet another element of Behe's proposal: the design test that Behe proposed didn't work. It was a "god of the gaps" argument, Rothschild suggested: Behe inferred design in cases where mainstream science hasn't yet figured out the evolutionary pathways, while he conveniently ignored well-understood systems that show the same evidence of "design."

Rothschild wasn't through yet. Earlier in his testimony, Behe had spoken of a peer-reviewed research paper he wrote with another scientist, David Snoke, that examined a computer simulation of the evolution of certain proteins. This was the paper that Behe had described so glowingly and without being challenged months earlier in Kansas, and that had been touted by the Discovery Institute as a nail in the coffin of Darwinian evolution. The paper did not discuss ID or irreducible complexity directly, but Behe said that it provided strong evidence to support his theory that evolution cannot produce complex protein structures. The study took a simulation of a simple, extremely common type of microorganism called a prokaryote (organisms with no cell nucleus, mainly bacteria), and purported to show that complex proteins would take far too long and far too large a population to evolve in nature. In Behe's view, that left design as the most likely explanation.

But then Rothschild did something no one had thought to do during the one-sided hearings in Kansas: He asked Behe to walk him through the numbers. Behe happily agreed. The simulation concluded that a population of 10^9 (1 billion) prokaryotes would take 10^8 (100 million) generations to produce a novel protein feature through Darwinian evolution.

"That's correct," Behe said, pleased.

"And yesterday, you explained about bacteria, that ten thousand generations would take about two years in the laboratory, correct?"

"Yes."

Then Rothschild asked the question Behe did not expect. "So one hundred million generations would take about twenty thousand years?"

Behe blinked. "I'm sorry?"

"One hundred million generations, which is what you calculated here, that would take about twenty thousand years?"

"Okay," Behe conceded, "yes."

Rothschild then handed Behe a different journal article, which counted the number of prokaryotes in a normal ton of soil. There are 10^{16} prokaryotes in every ton of soil. That would be 10 quadrillion (10,000,000,000,000,000). "And we have a lot more than one ton of soil on Earth, correct?"

"Yes, we do."

It takes a moment for this to sink in, but what Rothschild had just accomplished was to show that the paper Behe claimed as support for ID, which had been celebrated in Kansas and by the Discovery Institute as strong evidence of irreducible complexity, actually proved the opposite. Behe's simulation assumed a population of bacteria on Earth far less than the number found in a single ton of soil, then set up rules for evolution that were far more conservative than those in nature (he considered only one type of mutation, whereas there are many other sources of variation in the evolutionary process). Yet the simulation still found that it would take only 20,000 years for these creatures to evolve a complex protein structure—exactly the sort of structure Behe's theory of irreducible complexity claimed would almost never evolve.

The one research paper Behe produced that he claimed supported ID had just been spectacularly dismissed.

The courtroom was riveted by the exchange between Rothschild and Behe. Word had gotten out, and lawyers in the courthouse came in to watch what Judge Jones would later call the most effective cross-examination he had ever witnessed in a quarter century of legal practice. "It was rather painful at times to be that close to it," Jones said. "When all the hoopla dies down and you take apart what will endure, what will be featured in the law books, it's his cross-examination."

Behe also claimed that his theory of irreducible complexity was testable in a laboratory. All that had to be done would be to take a species of bacterium with no flagellum, place it in an environment that gives survival advantages to highly mobile microorganisms, and then grow it for 10,000 generations to see if a flagellum begins to evolve. This would take about two years, according to Behe, and if something evolved, his entire theory would be falsified. Proof or disproof, in just two years—and yet Behe has not performed that test of his own theory.

When Rothschild asked Behe why he didn't perform such experiments, the response seemed to shock the courtroom: "It would not be fruitful."

The lasting image of Behe, however, came near the end of the cross-examination. The biochemist had testified that there were no published papers to explain the evolution of the immune system, which he considered irreducibly complex. Rothschild proceeded to pile a stack of books and fifty-eight peer-reviewed articles on the witness stand, all about the evolution of the immune system.

"So these are not good enough?" Rothschild asked.

Sitting there surrounded by the scientific literature, Behe said, "They don't address the question I'm posing."

The closing drama of the trial brought the case full circle, shifting from the microscopic back to the political, from the science of evolution and the challenge of design to the nitty-gritty reality of what was and wasn't done by the Dover school board.

There were two more scientific experts for the defense to dispense with first, but they added little to the case and seemed to do as much damage as good to the cause of intelligent design. Scott Minnich, the microbiologist from the University of Idaho, reiterated Behe's testimony about the flagellum, but also admitted that in order for ID to be considered scientific, science would have to be expanded to include the supernatural. Coming at the very end of the case, and after a mind-numbing return engagement by the bacterial flagellum, this surprising agreement with the critics of ID was barely noticed among the exhausted spectators; but as the plaintiffs' attorney Steve Harvey later noted, "We could win the case on that admission alone." And Steve Fuller, a British sociology professor from the University of Warwick, testified that Dover's policy was good because ID needed an "affirmative action" plan to "recruit" young people to the "revolution" in order to overcome the scientific establishment's refusal to accept intelligent design. Fuller also admitted that ID was a form of creationism—something the other experts had denied and that Thomas More's lawyers had tried throughout the trial to rebut.

With the science out of the way, the Dover School District took center stage, and William Buckingham returned to Pennsylvania from his new home in North Carolina, ready to defend his actions, his honor, and the school's policy. This did not go well.

Although he testified during the defense portion of the case, Buckingham had been called by the plaintiffs; the conflicting schedule of out-of-town experts and his own travel needs delayed his testimony until the defense half of the trial was well under way. But because Buckingham was the plaintiffs' witness, Steve Harvey rose to set the tone by questioning him first, with Jones's permission to treat him as a hostile witness, which is a legal definition as well as a plain fact.

Buckingham, who was in poor health, with his knee and back killing him, did not want to be there. On the witness stand, he was clearly no longer the booming-voiced ideologue who had spoken his mind at school board meetings with no thought of being politically correct. Nor did there seem to be much left of the cocky cop who strode into biker bars with six bullets and a stare. It would be wrong to say that Bill Buckingham seemed humbled, contrite, or fearful on the stand; "surly" would probably better describe his demeanor. But he was also diminished, no longer larger than life. He had the look of a boxer on the bottom of the fight card and near the end of his career, resigned to the fact that he would be taking a beating before he'd be allowed to trudge from the ring.

Harvey, like everyone else on the plaintiffs' team, appeared meticulously prepared and seemed to know what Buckingham would say before the witness did. The lawyer began, with ominous pleasantness, by finding some things they could agree on. Buckingham believed in a literal Genesis, correct? He felt that an alternative to evolution should be taught. He understood little about intelligent design and less about evolution. He objected to teaching that man came from monkeys. During one contentious board meeting, he had challenged the audience members who supported evolution "to trace your roots to the monkey you came from." He had held up the purchase of Ken Miller's Dragonfly Book because it was "laced with Darwinism," and he had said as much at the now infamous meeting of the board on June 7, 2004, when the controversy began. At that same meeting he had said that the separation of church and state was mythical. He had expressed concern that if the theory of evolution were taught over and over, students would begin to accept it as fact. And he had announced that an alternative to Darwin should be offered, in class and in any textbook the district selected for biology.

True, Buckingham said. All true. Harvey beamed. But both men well knew that the agreeing was about to end.

For next came the critical point: Buckingham swore he had never used the term "creationism" at that meeting or at any other. Just intelligent design.

When asked if he had read the news coverage of the fateful school meetings that summer, Buckingham said he had wanted nothing to do with the two newspapers from York that thumped on his doorstep each day. "They came, but I didn't read them," he said. "It wouldn't make sense to do that because I don't believe a darn thing they print. . . . I would be told by people there are things in there, but my experience with the reporters were the articles almost got to be laughable. They'd come to the meetings and we talked intelligent design, and you could almost bet your house they were going to say creationism the next day. And it just got disgusting and I just wouldn't pay for it or read it anymore."

Harvey showed him one article after another from the two York newspapers that reported his creationist and religious statements during board meetings. "You actually said that?" Harvey would ask after showing each article.

"No, I didn't."

How about board members talking among themselves, Harvey wanted to know; did creation come up then? Never, Buckingham said. His memory was hazy at times, because of his addiction to OxyContin, Buckingham said, but he remained certain that no member had ever used the word "creationism" at a board meeting. "It's just something we didn't do."

How about after a board meeting? Did you ever talk about creationism with a news reporter after any board meeting?

"No."

Harvey used the same sort of relentless, sharklike cross-examination style that Rothschild had used with Michael Behe, but with less smiling and no good-natured bits of humor. This was business, and Harvey's purpose was to discredit this man who had started the whole controversy. Harvey would later say that, personally, he did not enjoy destroying Bill Buckingham. The lawyering was exhilarating in its own way, and it was necessary, without doubt—like performing surgery. But it wasn't fun to do, and it wasn't fun to watch. He had brought his wife and kids to sit in on parts of the trial, as had the other attorneys, for they were proud of what they were doing and felt this would be one for the history books. But he did not bring his family in for this episode, though it was arguably one of the most critical moments of the case. The issue of whether the school board's policy had a religious basis came down to a matter of credibility, of whom the judge would believe. Judge Jones could choose to believe the board members, who vigorously denied harboring a religious purpose; or he could side with the teachers, reporters, and plaintiffs who swore that they had heard creationism and God regularly invoked, and that the controversy over the curriculum had begun as—and remained throughout—a religious crusade. If one side seemed dishonest on the witness stand, then the credibility scale would tip the other way. That was how the law worked, and that is how the game was played. And Harvey had Buckingham cold.

"You're going to need to look at the monitor," the lawyer told him. "The television screen."

Buckingham, the judge, and the rest of the courtroom then watched a news video from June 2004, in which a Fox 43 reporter from York conducted an impromptu interview in the parking lot after the school board meeting. And there was Buckingham on the screen, confidently and clearly saying what he had just sworn he had never said: "The book that was presented to me was laced with Darwinism from beginning to end. My opinion, it's okay to teach Darwin, but you have to balance it with something else, such as creationism."

Such as creationism.

The tape stopped. Every eye in the courtroom was on Buckingham. "That was you speaking, wasn't it?"

"It certainly was."

"Need to see it again?" Harvey asked.

Buckingham shook his head slightly. "No."

Bill Buckingham, before he ever walked into the courtroom, must have known that he had been set up as the villain in this drama. He believes that this was manifestly unfair. But there can be no doubt that he brought on many of his own problems. He is a man of strong faith and conviction, which are admirable qualities, but he is also a man of strong opinions that are not always based on fact, and this is a problematic quality in civic leaders in America today, from school boards to the White House. Buckingham believes that the constitutional separation of church and state is a myth, but he is wrong. He believes this because of a book which purports to be scholarly but which is, in fact, propaganda, and discredited propaganda at that. But he believes it still, because he wishes to believe it, and he is not alone.

He also believes that America was founded as a Christian nation, but he is wrong on that, too—the Founders were adamant about this. In 1796, the Treaty of Tripoli was read aloud in the U.S. Senate, approved unanimously without a single word of dissent, and signed by President John Adams, a signer and coauthor of the Declaration of Independence and one of the greatest patriots in the nation's history. The treaty, intended to forge good relations with the Muslim world, included the following passage:

> As the Government of the United States of America is not, in any sense, founded on the Christian religion; as it has in itself no character of enmity against the laws, religion, or tranquility of Mussulmen; and, as the said States never entered into any war, or act of hostility against any Mahometan nation, it is declared by the parties, that no pretext arising from religious opinions, shall ever produce an interruption of the harmony existing between the two countries.

Buckingham is not persuaded by such evidence; his beliefs are too much a part of who he is, and mere facts do not easily dent them. And so he also disbelieves evolution and embraces intelligent design, without much understanding of either. This would be no one's business or concern but his own, except that those beliefs found their way into the business of the Dover Area School District. And then they became everyone's business.

Yet even as he was being set up to shoulder the blame for the mess that followed—not so much by the plaintiffs, who wanted to spread the fault, but by the school board's own attorneys, who needed a lone scapegoat—there was more to Buckingham than the two-dimensional, cardboard villain he has been made out to be.

Months after the trial ended, he revealed in an interview that he attributes many of his woes, including his abrasive conduct at board meetings, to his OxContin addiction, and to the fact that his first round of detoxification and counseling was cut short. This had been his choice—a bad one, he says now. He mistakenly believed that once the most severe physical symptoms of withdrawal had passed, he would be okay. But he had been wrong. The drug hadn't just protected him from the physical pain of his damaged back; it had insulated him from personal loss and grief, and those later came roaring back. He poignantly described what happened next, his own words revealing a side of him that few have seen since the school board controversies began—the side that was in turmoil throughout the time he championed the new biology curriculum at Dover High:

> While I was taking the OxyContin, my mother Betty died, my father Walter died, my dog died, two of my close uncles died, my closest aunt died and my neighbor's daughter committed suicide.
>
> The OxyContin kind of protected me while I was taking it. It was like circled wagons during an Indian attack.
>
> When the OxyContin left, everything came rushing in all at once. It got to the point that I became extremely depressed. When my neighbor's daughter committed suicide, I was the first one there. It was a bitch. I spoke to her the day before and I now realize she was saying good-bye. I wish I could have helped her in some way. While she was killing herself, I was

sitting on my back porch, just 50 feet away. She was a very nice person.

I felt like I was fighting a war with the emotions of these deaths. Coupled with the kids that died in front of me while [I was] working with the police department, I now realize that I came very close to cracking. At the school board meetings, I just wasn't up for any "crap" from anyone. My Marine and Police Training contributed to my, at times, harsh responses to the verbal attacks that kept coming from the same people, over and over again. I just couldn't sit there and take it. I became very defensive.

As time went on, I kept having a harder and harder time. I tried to find out what the effects of OxyContin, long and short term, would be. No one knows. I am like a guinea pig. I still don't know.

One day I just couldn't take it anymore. I went fishing and this time I took my gun. I didn't plan to come home again.

As God would have it, I found my third grade teacher at home. Hadn't been by her house in at least 40 years. I had a nice talk with her and her husband, who was a friend of my dad's.

She doesn't know it to this day, but she saved my life. She was so nice and glad to see me that it picked me up.[2]

Bill Buckingham is a flawed man who pursued a flawed policy. He did so at a time when his life was in crisis and his memory and judgment were in question, and when he was in no mood to countenance criticisms or to consider the views of others. He became the catalyst for events that eventually divided his community and cost it dearly, and then he left. Yet the most haunting aspect of the whole sad story may well be the image of this wounded man attempting to construct a legacy out of a bureaucrat's one-minute statement, to bring a small piece of salvation to the young people of Dover—and to himself.

No one in Judge Jones's courtroom would see the vulnerable and pained side of Bill Buckingham. Instead, everyone saw a man on videotape uttering exactly what he swore he had never said. As the television monitor went dark, it seemed almost inconceivable to the spectators in the courtroom that Buckingham had so blindly walked into this trap. The tape had been no secret. Eric Rothschild had played it during his opening argument. It had been disclosed to the defense. It had been in the newspapers. Alan Bonsell had sat in the courtroom as the trial began and watched it, seeming relaxed, confident, and unconcerned—as he seemed throughout the trial. But apparently no one told poor Bill Buckingham about it. Either that, or he knew about it and forgot.

Harvey pointed out that the statement Buckingham made on the television news video after the meeting and the statements the newspaper attributed to him *during* the board meeting were very similar.

"That doesn't make it accurate," Buckingham snapped.

But what about the phrase "laced with Darwinism"? Harvey asked. You already agreed the newspapers were correct in reporting you said that during the meeting. Then you said it again on the video. Did the newspaper reporters get that portion right, but the creationism part wrong?

"That's not what I said," Buckingham insisted, and then he offered an explanation. "What happened was, when I was walking from my car to the building, here's this lady and here's a cameraman, and I had on my mind all the newspaper articles saying we were talking about creationism. And I had it on my mind to make sure, make double sure nobody talks about creationism, we're talking intelligent design. I had it on my mind, I was like a deer in the headlights of a car. And I misspoke. Pure and simple, I made a human mistake."

Harvey said, "Freudian slip, right, Mr. Buckingham?"

"I won't say a Freudian slip. I'll say a human mistake."

Unfortunately for Buckingham, he had just made another mistake: He told Harvey he recalled having in mind during that television news interview "all the newspaper articles saying we were talking about creationism"—articles that he had, just moments before, sworn he had never read.

Buckingham explained this, too, by saying he had known what the articles had said "through talking with the board members . . . not through reading any of their papers." Then he reiterated again the remarkable idea that he had "concentrated so hard on not saying creationism, I made a human mistake and I said it." He chalked this up to being afraid of being interviewed, which did not sound very much like Bill Buckingham.

"Let's take a look at the tape again," Harvey suggested brightly. The courtroom watched again, transfixed at the image of a calm, collected Buckingham on the screen. Then Harvey said, "You didn't look like you were very pressured to me."

"I can't help how it looks. I'm telling you I felt pressured at the time."

Things did not get better after that for Bill Buckingham. Harvey forced out an admission that the school board had consulted with no one—not science organizations, not science education experts, not local universities—in formulating its new policy on biology. Only the Thomas More Law Center and the Discovery Institute had been consulted, Buckingham conceded—and the school board was blocking the court from learning the details of those consultations on the grounds of lawyer-client privilege.

Then there was the donation of the sixty copies of *Of Pandas and People*. In his deposition taken in January, Buckingham had sworn he didn't know who had provided the money for those books, and though he had some ideas, he couldn't say for sure. He had added, "I didn't want to know." But the story he told at the trial turned out to be quite different. Buckingham had appealed to his church congregation about the need for charitable donations for purchasing the books so that tax dollars would not have to be used, and $850 was raised in that way. Buckingham then wrote a check for that amount from his own bank account, filled it out to Mr. Donald Bonsell, and then gave the check to Bonsell's son, Alan—who was then the school board president.

"Mr. Buckingham, you lied to me at your deposition," Harvey said bluntly. "Isn't that true?"

"How so?"

"By not telling me . . . that you knew that a collection had been taken at your church."

"I did not take a collection," Buckingham replied. "I didn't consider it a collection. I didn't ask for it. They just did it because there was a need there. I didn't ask them for it."

Buckingham's definition of "not" asking for a collection was this: standing in front of his church congregation on a Sunday morning and saying, "There's a need, we don't want to use taxpayer dollars, and if you feel led to donate, fine. I'm not asking for money, I'm just letting you know there's a need."

He denied trying to hide that information in his deposition testimony.

When Harvey continued to press the point, Judge Jones said he had heard enough. Then, ominously for Buckingham, the judge told the attorney, "I get the point, and you've made the point very effectively."

There wasn't much more to say after that. Pat Gillen, who had been left alone at the lawyers' table by his colleagues with increasing frequency as the trial wore on, spent a very short time questioning Buckingham. The shadow of the videotape hung over everything.

Buckingham's parting words were plaintive, a last grasp for a legacy that seemed to be eluding him: "We were doing it for the students, to give them an alternative scientific theory to go along with their biology class. We thought we were doing something good for them."

Steve Harvey, bristling at the idea of leaving that remark unchallenged, considered having another go at Buckingham and started to say he had just a few more questions. But then, looking at the man on the stand, he shook his head and said, "In fact, it's over."

Chapter 16

FORTY DAYS AND FORTY NIGHTS

The situation never improved for the Dover School Board after Bill Buckingham's unnerving time on the witness stand. He was followed by the two reporters he had denounced, and the contrast could not have been greater.

Heidi Bernhard-Bubb, a stay-at-home mother of two who freelanced for the *York Dispatch*, and Joseph Maldonado, an Air Force veteran who split his time between running PBJ's sandwich shop and freelancing for the *York Daily Record*, confirmed for the court that their news stories were accurate. Buckingham and other board members had, indeed, used the word "creationism" and expressed a desire to have it taught before switching to the topic of intelligent design at subsequent meetings, the reporters said.

This was critical, because—given the board's insistence on erasing the audiotapes of meetings—the articles were the best (or only) record of what actually was said at the meetings. And they had almost been excluded from the case—a decision that would have been disastrous for the plaintiffs.

Both reporters had faced contempt of court and jail time for initially refusing, on First Amendment grounds, to testify. Thomas More's lawyers, insisting that their clients had never said the things the newspapers had reported, wanted to question Bernhard-Bubb and Maldonado in depth about their biases, motives, and sources and, if they resisted, wanted them locked up. The plaintiffs, on the other hand, were content to accept sworn affidavits from the reporters attesting to the accuracy of the news articles.

The impasse continued right up until the day the reporters were to either testify or go to jail. A resolution came at the last minute through a deal brokered by Judge Jones, who insisted that the reporters appear in person to testify, but agreed to limit the attorneys' questions to what the reporters had seen and heard at school board meetings. This meant they could come into court and "authenticate" their articles, qualifying the stories as evidence in the case—and they did, quotation by damning quotation.

The reporters also pointed out that, despite all the claims that their stories were inaccurate, no board member or school district official had ever directly challenged the specifics of their reporting or asked for a retraction. Left unsaid, though it was widely known in Dover, was that board members regularly made insulting comments about the reporters during public meetings; the journalists would have to sit there red-faced and take it. Some board members took special delight in trashing the two freelancers when they knew that reporters from national publications were present in the audience. (A glimpse of this sort of venomous sentiment emerged at the trial when the former board member Jane Cleaver took the stand and said, "Joe [Maldonado] doesn't know how to tell the truth. He only knows how to tell a lie.")

The defense theorized that the two reporters made up stories about creationism because they had an economic incentive to cook up a controversy in Dover where none existed: Hot stories would make the front page, and front-page stories would earn them more pay as freelancers. Judge Jones appeared far less impressed by this line of argument than he was by the reporters' calmly dignified testimony and principled stand when faced with jail. In any case, it was hard to imagine that advantages to be gained from such a conspiracy would be worth the risk: Bernhard-Bubb received $40 or $50 per story and wrote six to ten articles a week to help make ends meet; Maldonado earned $50 for a basic story, but if a story made a section cover he received $62.50, and a front-page item bumped it up another five dollars.

Eric Rothschild appeared upset by these insinuations: "Two hard-working freelance reporters had their integrity impugned and were dragged into a legal case," he complained, "solely because the board members would not own up to what they had said."

The reporters were followed on the stand by the board member Heather Geesey, who smiled broadly throughout her testimony regardless of the seriousness of the topic being discussed, described her occupation chirpily as "full-time mommy," and referred to Bonsell as "President Alan."

Vic Walczak drew the job of cross-examining her, and his questioning about why Geesey voted for the new ID policy in October 2004 was revealing—not because she provided any real answers but because of the sheer spectacle of an elected member of a school board smilingly admitting to near-total incuriosity and ignorance regarding the curriculum changes she had mandated for her constituents' children:

Walczak: You supported the change?
Geesey: Yes.
Q. Because it gave a balanced view of evolution?
A. Yes, I mean . . .
Q. It presented an alternative theory?
A. Yes.
Q. And the policy talks about gaps and problems with evolution?
A. Yes.
Q. Yes. You don't know what those gaps and problems refer to, do you?
A. No.
Q. But it's good to teach about those gaps and problems?
A. That—yes, that's our mission statement, yes.
Q. But you have no idea what they are?
A. It's not my job, no.
Q. Is it fair to say that you didn't know much about intelligent design in October of 2004?
A. Yes.
Q. And you didn't know much about the book *Of Pandas and People* either, did you?

A. Correct.

Q. So you had never participated in any discussions of the book?

A. No.

Q. And you made no effort independently to find out about the book?

A. No.

Q. And the administration had made copies of the book available to board members?

A. Yes.

Q. But you never read the book?

A. No.

Q. And no one ever explained to you what intelligent design was about?

A. No.

Q. And you never got any instructional materials or tapes about intelligent design?

A. No.

Q. And you never viewed any or read any books about intelligent design?

A. No.

Q. And you didn't study it independently?

A. No.

Q. You didn't go on the Internet and look it up?

A. No.

Q. So you didn't really think too much about intelligent design?

A. No.

Q You just knew it was something else that the kids were going to learn?

A. Yes.

Q. And it was a theory that was different from Darwin's view?

A. Yes. . . .

Q. You were relying on the recommendation of the curriculum committee?

A. Yes.

Q. So the two people you were really listening to and talking
to about this were Bill Buckingham and Alan Bonsell?

A. Yes.

The real blockbuster, however, came when Geesey contradicted the
sworn testimony she had given in her deposition six months earlier, when
she said the specific term "intelligent design" never came up during those
board meetings of June 2004 at which the biology textbooks were first
discussed. In the deposition, she had recalled that Buckingham expressed
a desire to find a textbook offering an alternative to evolution, but nothing
specific was mentioned. Neither creationism nor intelligent design was
suggested as an alternative at the time, she had sworn. That came later.

At the trial, however, her story evolved. Suddenly, she remembered
that intelligent design had been mentioned all along, beginning with
those first discussions in June. Her testimony now agreed completely
with Buckingham's, and with the overall defense narrative that creationism had never been mentioned, that the issue was always about intelligent design, and that the reporters (and other witnesses) were lying if
they said otherwise.

When Walczak challenged her, she said she had genuinely forgotten
this point during the deposition—all the meetings "run together," she
explained. But for the trial, she had prepared herself by reading a letter
to the editor that she had written in June about the curriculum controversy, and the letter refreshed her memory. That's when she remembered
the alternative theory Buckingham wanted to balance with evolution
had been discussed at those June meetings, and that it had been intelligent design. "It must have come up," she said, "because I wrote that."

When the questioning by the lawyers was done, Judge Jones did
something he hadn't done during the entire trial: He questioned the witness directly. "I have a question before you step down, Mrs. Geesey, because I'm confused."

"So am I," Geesey said happily.

"Well," Jones said, "it's more important that *I'm* not confused than
you're not confused." Then he asked her what part of the letter to the
editor reminded her that intelligent design had been discussed at the
meeting.

Geesey looked blank, then stammered. Jones rephrased his question. "Point me to what in the letter, not generally, but specifically . . . refreshes your recollection."

Geesey continued to stammer. Jones took a breath and said, "I asked you that question because I don't see the words 'intelligent design.' "

The judge had in front of him the letter to the editor, in which Geesey had written:

> I do not believe in teaching revisionist history. Our country was founded on Christian beliefs and principles. We are not looking for a book that is teaching students that this is a wrong thing or a right thing. It is just a fact.
>
> All we are trying to accomplish with this task is to choose a biology book that teaches the most prevalent theories. The definition of theory is merely a speculative or an ideal circumstance. To present only one theory or to give one option would be directly contradicting our mission statement.
>
> You can teach creationism without its being Christianity. It can be presented as a higher power. That is where another part of Dover's mission statement comes into play. That part would be in partnership with family and community. You as a parent can teach your child your family's ideology.

This is the letter she had just testified that she wrote because the name of the alternative theory "must have come up" during the meetings in June. But the only alternative to evolution identified in this letter is "creationism."

Her response to the judge's question was gibberish:

> Right. The part where it says, "what we are doing." I— since all the meetings run together, I didn't realize back then that I knew everything that was going on because it's not my committee. But by me saying that what we were doing was to choose a book that teaches the most prevalent theories, I mean that—that's what I was talking about. I mean, I already knew that they were doing something, and before I couldn't tell the

meetings apart. So I kept saying, no, I don't remember, because
I couldn't tell them apart when they would say did it happen
then, and I wasn't sure of when it happened.

In other words, there was nothing in the letter that could have reminded her of intelligent design. Yet somehow she had managed to make
her sworn testimony match the stories of her fellow board members.

The judge tried a few more times to cast some light on the subject,
then gave up and asked the attorneys if they wanted to give it a go. Pat
Gillen said he had one more question, to which the judge replied, "I
don't know what you could possibly hope to achieve."

Lightning struck twice when Alan Bonsell testified a short time
later, taking the stand with a confident air and a stick of gum in his
mouth.

He freely admitted that he had spoken of creationism at two school
board retreats. He didn't remember doing it, but he trusted the accuracy
of the school superintendent, who had taken notes at the retreats. But he
was quite certain that he wouldn't have been advocating teaching creationism in school, and he certainly never suggested it at any board
meetings.

Though he admitted that he personally is a creationist and biblical
literalist who disbelieves the theory of evolution and common descent,
he feels that creationism is something to be taught and discussed at home
or church, not in public school. The newspapers had it wrong when they
suggested otherwise, he said.

Far from being in Bill Buckingham's camp, Bonsell said, he had been
"taken aback" when his fellow board member started using such terms
as "laced with Darwinism." Bonsell allowed that creationism did come
up once in a board meeting: Buckingham's wife, Charlotte, came to the
microphone during a public comment phase and gave a very long, very
religious speech that included her belief that the district should be teaching creationism. But that wasn't the board talking; indeed, Bonsell had
thought her comments went on too long and were inappropriate, but he
did not feel that, as president of the board, he should gavel a fellow board
member's spouse into silence.

Confident, convincing, relaxed, and articulate, Bonsell, the college-educated businessman, was a very good witness. His testimony was in sharp contrast to Buckingham's surly contradictions and Geesey's smiling incoherence. But that was under his own attorney's questioning.

When it was Steve Harvey's turn, he called up on the big screen an image of the $850 check donated for *Of Pandas and People*, the check Buckingham had handed to Bonsell.

Wasn't it true, Harvey asked, that at a board meeting at the time, Bonsell had been asked where the money came from and wouldn't say? Yes. Then Harvey reviewed what Bonsell had said in January, when asked similar questions at his sworn deposition. Under oath, Bonsell had offered a variety of answers. First he said he didn't know. Then he said, "They wanted to remain anonymous." Then, "I don't know all the people that donated them, no." Finally, he said he knew one person who donated: "Donald Bonsell."

At his deposition, he never mentioned Buckingham's giving him the check or raising funds at church, Bonsell admitted. But he denied being deceptive, claiming he had answered the questions as they were asked.

When Harvey finished, Judge Jones again stepped in, telling Gillen, who was ready for his round of redirect questions, that he'd have to get in line. "I want to exercise my prerogative," he said sternly, "and I have some questions before we break today."

He took a few moments to read the relevant portion of Bonsell's deposition, then asked the witness why he never mentioned during his deposition that Buckingham was involved in supplying the money for buying *Of Pandas and People*.

Bonsell started to say that he had been asked who donated the books, and he didn't know the actual identity of the donors, but Jones angrily cut him off. Look at line nine of your deposition, the judge commanded. No one asked who donated the books. You were asked: *Who was involved?* "Now, you tell me why you didn't say Mr. Buckingham's name."

Bonsell hesitated, his confidence evaporating and something resembling fear clouding his expression. "Then I misspoke. . . . Yeah, that's my fault, Your Honor. . . . I should have said, 'Mr. Buckingham.'"

This did not satisfy the judge. "Now you were under oath on January third of 2005, is that correct? And your reason that you didn't mention Mr. Buckingham's name is because you said you misspoke?"

Bonsell stammered a lengthy reply that did not directly address the question, and the judge finally cut him off. Jones was deeply disturbed by what he was hearing, and it led him to slip into his old style as a trial lawyer, twenty-two years of knocking heads with dissembling witnesses instinctively coming to the fore. "Why did you utilize your father," the judge demanded, "as a conduit for this money?"

More stammering followed, and more pressing of the issue by the judge, before Bonsell finally said of his father, "He agreed to—he said that he would take it, I guess, off the table or whatever."

The judge's eyes narrowed. *Off the table.* And did Bonsell have any idea why Buckingham, at his own deposition, also professed ignorance about who was involved in the donation of *Pandas*?

"I don't have any explanation for that," Bonsell said.

The judge turned away, saying he had no more questions and that the trial was done for the day. Then he swept from the room, red in the face.

A day after Bonsell finished testifying, the assistant superintendent Mike Baksa—caught in the middle yet again, his hands shaking visibly as he took the stand—swore to tell the truth about the Dover school district that employed him. And he swore that at those June meetings, Bill Buckingham really had talked about creationism, just as the newspapers reported.

"So if anyone that attended the June meeting says that Mr. Buckingham didn't talk about creationism," Eric Rothschild asked, "you know that's not correct?"

"Well, I remember him saying that."

On November 4, 2005, after twenty-one days of trial spread across six weeks, the attorneys stood to make their closing arguments. Each side argued passionately and, fittingly for a case that had divided a community, each outlined a take on the case that shared not a shred of common ground with the other.

As expected, Pat Gillen pointed the finger of blame at William Buckingham while exonerating the rest of the Dover school board.

"Bill Buckingham is not the board," Gillen reminded Judge Jones.

"And . . . it is simply not the law that the mere mention of the word 'creationism' is illegal in these United States. . . . The main support for the plaintiffs' claim is a mountain of press clippings built on a molehill of statements allegedly made by one board member who, troubled and wrestling with the addiction of OxyContin, occasionally allowed people to put words in his mouth."

The rest of the board, Gillen said, pursued a valid policy to improve science education.

As for the rest of the case presented by the plaintiffs, Gillen urged the judge to disregard it. He argued that the information on the wedge strategy and Phillip Johnson and the Discovery Institute was interesting, but the school board—the actual defendants in the case—knew nothing about any of it. Nor did the school board have anything to do with the writing and revisions of *Pandas*. They had never seen all those old drafts. *Pandas* meant nothing in this case, as Gillen saw it.

What mattered was the "modest" four-paragraph, one-minute statement—that's what the case was about. And Gillen defied anyone to detect a religious purpose anywhere in its words or execution.

As for intelligent design itself, Gillen described it as legitimate and thriving, and not at all beyond the bounds of science, even though it may consider the possibility of the supernatural. "Quite the contrary, intelligent design theory's refusal to rule out this possibility represents the essence of scientific inquiry, precisely because the hypothesis is advanced by means of reasoned argument, based not on the Bible, but on empirical evidence and existing knowledge."

Eric Rothschild presented a starkly different view, and he sought to voice the outrage of his clients, and his own:

> What I am about to say is not easy to say, and there's no way to say it subtly. Many of the witnesses for the defendants did not tell the truth. They did not tell the truth at their depositions, and they have not told the truth in this courtroom. . . .
>
> There are consequences for not telling the truth. The board members and administrators who testified untruthfully for the defendants are entitled to no credibility. None. And furthermore, and perhaps more importantly, this court should

infer from their false statements that defendants are trying to conceal an improper purpose for the policy they approved and implemented, namely an explicitly religious purpose.

The board's behavior mimics the intelligent design movement at large. The Dover board discussed teaching creationism, switched to the term "intelligent design" to carry out the same objective, and then pretended they had never talked about creationism. As we learned from Dr. Forrest's testimony, the intelligent design movement used the same sleight of hand in creating the *Pandas* textbook. They wrote it as a creationist book and then, after the *Edwards* decision outlawed teaching creationism, simply inserted the term "intelligent design" where "creationism" had been before.

This subterfuge, Rothschild argued, extended to the "modest" one-minute statement, which is not modest at all, having an impact that extends far beyond the brief time in which it is read aloud. Part of the policy behind the statement, Rothschild reminded the judge, was that the origins of life would not be taught. And the members of the school board defined "origins of life" as the "origin of *species*"—the principle of common descent—and demanded that it not be taught. This exclusion is a classic creationist argument, he said, and the one-minute statement came attached to a much more momentous policy to undermine the teaching of evolution. The effect, argued Rothschild, is to try to block "real science," such as the presentation by Ken Miller on the human and chimpanzee genome, from making its way into the classrooms of Dover High. In its place, the school board suggested a bogus substitute science that invokes the supernatural and tells would-be scientists to give up because there's nothing more to learn, because a mysterious designer did it all:

How dare they. How dare they stifle these children's education, how dare they restrict their opportunities, how dare they place a ceiling on their aspirations and on their dreams. This board imposed their religious views on the students in Dover High School and the Dover community. You have met the parents

who have brought this lawsuit. The love and respect they have
for their children spilled out of that witness stand and filled
this courtroom.

They don't need Alan Bonsell, William Buckingham,
Heather Geesey, Jane Cleaver, and Sheila Harkins to teach
their children right from wrong.

The lawyers sat down then, the long case done. All that remained of
the second Scopes Monkey Trial would be the decision, which the judge
promised by the end of the year. For now, he congratulated the lawyers
on both sides for their fine work, and for reminding him why he had
become a lawyer and a judge.

Pat Gillen, soon to return to Ann Arbor and the offices of Thomas
More, had one more question to ask: "By my reckoning, this is the forti-
eth day since the trial began, and tonight will be the fortieth night, and
I would like to know if you did that on purpose?"

"Mr. Gillen," Judge Jones responded with a smile, as the courtroom
erupted with laughter and applause, "that is an interesting coincidence,
but it was not by design."

Chapter 17

BREATHTAKING INANITY

Four days after the trial ended, the voters of Dover went to the polls in the election of November 8, 2005, to choose their school board. The election pitted eight incumbents against eight challengers. And it pitted evolution against intelligent design.

The general election campaign had been the ugliest and most expensive in memory, with the trial as a constant backdrop. Even as the current school board members were being cross-examined and accused of lying, they released a mass campaign mailer denouncing the group of challengers, the Dover C.A.R.E.S. coalition, for its ties to the lawsuit and the ACLU. The letter detailed what it claimed was the ACLU's record of defending special rights for terrorists and of protecting the right of the North American Man/Boy Love Association "to put out information on how adults can lure young children into having sex with them."

"We were setting the facts straight with this letter," Alan Bonsell told reporters when the letter came to light. "There's nothing in this letter that isn't true."

Added pro-ID incumbent Ron Short, who had been appointed to the board, "I fear the ACLU more than I fear al-Qaeda."

Vic Walczak had heard it all before, though it galled him that a guy like Bonsell, who had been excoriated on the witness stand for his contradictory accounts, could turn around and impugn the integrity of candidates who had nothing to do with the ACLU. "Freedom of speech is a wonderful thing," he said. "But one of the consequences is irresponsible and misleading speech."

The school board's challengers saw desperation and an attempt to deflect responsibility for the lawsuit. There would have been no suit, Bryan Rehm said, if the board hadn't mixed religion and the public schools in the first place.

"They brought the ACLU to Dover," Bonsell shot back. "Now everyone knows it."

As it turns out, a majority of voters in Dover saw the matter differently. The headline in the *York Dispatch* the next day was, "Dover Dumps Designers," as the citizens sent a clear message by rejecting the Republican slate of candidates in an overwhelmingly Republican district. In a surprising upset, every one of the eight incumbents who favored intelligent design was ousted in the election. Perhaps because of his less than stellar performance on the witness stand, or perhaps because he was reported to have done the research for the letter that mentioned the Man/Boy Love Association, the incumbent who received the fewest votes was Bonsell. Next to last in votes was the current board president, Sheila Harkins.

Overall, it was a close election. The candidate from Dover C.A.R.E.S. who finished in eighth place was only twenty-six votes ahead of the highest vote-getter among the incumbents. But it was still a clean sweep for the challengers.

The incumbents seemed stunned. "The public has spoken, that's about all I have to say," Ed Rowand told reporters. Bonsell didn't return calls for comment from the local press.

The winners were largely noncommittal about the fate of the ID policy, saying that the community was still closely divided and that they would wait to hear what Judge Jones had to say before acting on their own platform to remove it from science classes. Only Bryan Rehm was firm, saying that if it was up to him, the policy would be gone immediately. He wanted to put the science teachers back in the business of writing the curriculum for the board to approve, not vice versa. "They know what they're doing," he said, a not-so-subtle jab at the outgoing board.

The election had two immediate impacts. The first was a crushing blow to Richard Thompson and the Thomas More Law Center, whose rationale for taking on the case was not necessarily to win at the trial level but to reach the Supreme Court, where they might permanently change the law. Now if the school district lost in court and the ID policy was thrown out, the new board would never appeal. The policy and the case would be dead.

But if the board won and Judge Jones said that the policy was constitutional, Thompson would still lose. The plaintiffs wouldn't appeal, either. The ruling would just mean that the ID policy could stay in place, not that it *must* stay in place. The new board would then simply kill it.

No matter what happened, intelligent design couldn't win.

This time.

The other impact was a change in Dover's image. It was no longer the darling of the religious right—Dover had gone to he dark side. The Reverend Pat Robertson, on his television show *700 Club*, had a message for the township. "I'd like to say to the good citizens of Dover, if there is a disaster in your area, don't turn to God. You just rejected him from your city, and don't wonder why he hasn't helped you when problems begin."

The appalling un-Christian arrogance of this announcement rankled people across the political spectrum in Dover. The pastor of a local Lutheran church called Robertson a "warmonger," and at Mountain Grove Chapel, the pastor said that Robertson's idea of God was "warped." He also noted that Robertson seemed afflicted with foot-in-mouth disease, having recently caused a diplomatic incident by calling for the assassination of the president of Venezuela.

One of the new members of the school board, Larry Gurreri, exclaimed, "This is a man of the cloth judging people and judging the whole town? I would expect a whole lot better from someone in his position."

Later, Robertson responded to the furor his comments had created by making matters worse, accusing Dover of poking a finger in God's eye. "If they have future problems in Dover, I recommend they call on Charles Darwin. . . . Maybe he can help them."

Jeff Brown drew the most obvious conclusion from Robertson's comments: "According to sworn testimony, intelligent design has nothing to do with God. Then Pat Robertson says if you don't support it, God will hate you. These clowns want it both ways."

A week later, the lame-duck school board met in its regular session. It resolutely avoided the topic of intelligent design and instead spent an hour talking about replacement air conditioners. Finally, at the end of the meeting, one member, David Napierski, suggested that it was time to throw in the towel with regard to ID, admit the voters had spoken, and avoid squandering any more resources on it. No one seconded his motion.

"I know that some of the board members are very personally attached, but why continue to defend it?" Napierski said afterward. "We can't defend it. It's gone."

On December 20, 2005, Judge John E. Jones III rendered his opinion in *Kitzmiller v. Dover.* The 139-page ruling spelled out in detail his legal reasoning and the evidence to support it, and then he made it available on the Internet to anyone who wishes to read it anywhere in the world.

He ruled the school board's ID policy unconstitutional, a violation of the First Amendment's prohibition against a government establishment of religion.

He found that the school board had sought to inject religion into the classroom while undermining the teaching of evolution and science.

He found that school board members lied under oath.

He found that intelligent design was a religious proposition, and not science.

He found the scientific arguments for intelligent design unpersuasive and the scientific evidence nonexistent.

It was a triumph for the plaintiffs, and a complete defeat for the intelligent design movement. Vic Walczak exulted, "It's a victory beyond our wildest dreams."

Jones's eloquent opinion painstakingly reviewed the case for evolution ("overwhelmingly accepted by the scientific community"); the question whether intelligent design is religious ("not one defense expert was able to explain how the supernatural action suggested by ID could be anything other than an inherently religious proposition"); the value of "irreducible complexity" to science ("meaningless"); and whether the school board's ID policy had promoted critical thinking and improved science education ("the opposite of such purposes occurred").

As for *Of Pandas and People,* Jones credited Barbara Forrest's "meticulous" analysis and endorsed her conclusions that the book's provenance as a work on intelligent design was a ploy to dodge legal precedent: "The systemic change from 'creation' to 'intelligent design' occurred sometime in 1987, *after* the Supreme Court's important *Edwards* decision. This compelling evidence strongly supports Plaintiffs' assertion that ID is creationism re-labeled."

Jones's conclusions immediately became required reading across the world, as advocates of ID parsed each word and prepared their attacks, the science bloggers posted excerpts and rejoiced, and other school districts and their critics—particularly in Ohio and Kansas—considered Jones's logic and their own uncertain future. For better or worse, *Kitzmiller v. Dover* was the most complete test thus far of evolution versus intelligent design. Jones's decision, as he had anticipated, was going to reverberate far beyond the township of Dover, with or without an appeal.

At the end of his long review of evidence, testimony, and legal analysis, Jones concluded:

> The facts of this case make it abundantly clear that the Board's ID Policy violates the Establishment Clause. In making this determination, we have addressed the seminal question of whether ID is science. We have concluded that it is not, and moreover that ID cannot uncouple itself from its creationist, and thus religious, antecedents.
>
> Both Defendants and many of the leading proponents of ID make a bedrock assumption which is utterly false. Their presupposition is that evolutionary theory is antithetical to a belief in the existence of a supreme being and to religion in general. Repeatedly in this trial, Plaintiffs' scientific experts testified that the theory of evolution represents good science, is overwhelmingly accepted by the scientific community, and that it in no way conflicts with, nor does it deny, the existence of a divine creator.
>
> To be sure, Darwin's theory of evolution is imperfect. However, the fact that a scientific theory cannot yet render an explanation on every point should not be used as a pretext to thrust an untestable alternative hypothesis grounded in religion

into the science classroom or to misrepresent well-established scientific propositions.

Jones also made a pointed analysis of the conduct of the Dover school board members:

> Remarkably, the 6–3 vote at the October 18, 2004, meeting to approve the curriculum change occurred with absolutely no discussion of the concept of ID, no discussion of how presenting it to students would improve science education, and no justification was offered by any Board member for the curriculum change.
>
> In fact, one unfortunate theme in this case is the striking ignorance concerning the concept of ID amongst Board members.
>
> The citizens of the Dover area were poorly served by the members of the Board who voted for the ID Policy. It is ironic that several of these individuals, who so staunchly and proudly touted their religious convictions in public, would time and again lie to cover their tracks and disguise the real purpose behind the ID Policy.
>
> The inescapable truth is that both Bonsell and Buckingham lied at their January 3, 2005, depositions about their knowledge of the source of the donation for *Pandas*, which likely contributed to Plaintiffs' election not to seek a temporary restraining order at that time based upon a conflicting and incomplete factual record. This mendacity was a clear and deliberate attempt to hide the source of the donations by the Board President and the Chair of the Curriculum Committee to further ensure that Dover students received a creationist alternative to Darwin's theory of evolution. We are accordingly presented with further compelling evidence that Bonsell and Buckingham sought to conceal the blatantly religious purpose behind the ID Policy.

Finally, the judge anticipated the criticism that he knew his decision would receive—the usual arguments about liberal judges defeating the will of the people—and he chose to tackle this head-on:

Those who disagree with our holding will likely mark it as the product of an activist judge. If so, they will have erred as this is manifestly not an activist Court. Rather, this case came to us as the result of the activism of an ill-informed faction on a school board, aided by a national public interest law firm eager to find a constitutional test case on ID, who in combination drove the Board to adopt an imprudent and ultimately unconstitutional policy.

The breathtaking inanity of the Board's decision is evident when considered against the factual backdrop which has now been fully revealed through this trial. The students, parents, and teachers of the Dover Area School District deserved better than to be dragged into this legal maelstrom, with its resulting utter waste of monetary and personal resources.

With that, Dover's intelligent design policy was dead. Judge Jones ordered it rescinded. There would be no more statements read to students in high school biology classes in Dover, no more referring kids to an altered creationist textbook, no more limitation on the teaching of evolution to fit a creationist board member's agenda.

The ruling was eloquent and well reasoned and, despite the inevitable criticisms, there is really no doubt that Jones merely applied existing law as opposed to making new law. The Supreme Court had mandated a strict method for analyzing violations of the establishment clause—the "Lemon test"—and there can be no doubt that this test showed Dover's policy to be unconstitutional. Anyone who sat through the trial and watched, anyone who took the time to read the transcripts, would see a devastating case against the ousted school board. The testimony that the board had a religious agenda and tried to hide it was overwhelming and credible, while board members' denials were not—their credibility was savaged by their own inconsistencies and their conveniently lost and rediscovered memories. The board had adopted an agenda that, as Richard Thompson said, was "harmonious with Christian faith." Then the Thomas More Law Center gave board members information on *Pandas*, legal advice, and free representation. Finally, there was the enticing image of Dover as an impetus for bringing God back into the schools, taking the fight all the way up to the Supreme Court and the waiting embrace of Justice Scalia. That was the vision that Thomas More had been trying, unsuccessfully, to sell to school districts for five years, until finally Bill Buckingham and Alan Bonsell responded, only to find that the "sword and shield" couldn't deliver.

Anyone who watched the trial or read the transcripts would also discern a scientific and legal imbalance, with the weight of talent, expertise, and evidence overwhelmingly on the side of the plaintiffs. Thomas More, the depleted ranks of the Discovery Institute's experts, and ID itself were badly outmatched. Intelligent design, so alluring in the Discovery Institute's slick brochures and CDs and PowerPoint presentations, just couldn't hold up under the questions and relentless scrutiny of a six-week trial.

Science must be tested, criticized, poked, and prodded. It took forty years before the big bang was taken seriously, before the accumulation of evidence made this theory acceptable to and accepted by the scientific community. Intelligent design has not done this hard work, has not proved itself—has not proved anything at all, as Michael Behe admitted when he said that there was a test for irreducible complexity but that he would not invest the time or effort to perform it. Nick Matzke put it this way in the science blog PandasThumb, while he continued to labor long after the trial, archiving the voluminous evidence collected in the case:

> If the ID movement were intellectually serious, they would withdraw completely from interfering with public education, realizing that introductory science classes simply have to educate students in the basics of accepted science, and are not the right places to try getting recruits for fringe science. They would stop trying to make their case in the media, and instead take the only legitimate route to academic respectability— winning the scientific battle, in the scientific community. IDists have made much of comparing ID to the Big Bang model—but did Big Bang proponents kick off their model in a high school textbook? Did they go around the country mucking with kiddies' science standards to promote their view? Did they ever lobby legislators? I don't think so.

The reaction from other quarters ran true to form and was just as Judge Jones predicted. The word "activist" figured prominently in the criticism. The Discovery Institute, in particular, angrily attacked the ruling for straying from the narrow question of whether the board had acted unconstitutionally. "Judge Jones found that the Dover board violated the Establishment Clause because it acted from religious motives. That should have been the end to the case," said John West, associate director of the Discovery Institute's Center for Science and Culture. "Instead, Judge Jones got on his soapbox to offer his own views of science, religion, and evolution. He makes it clear that he wants his place in history as the judge who issued a definitive decision about intelligent design. This is an activist judge who has delusions of grandeur."

Bill Buckingham had a much more direct response: "If the judge called me a liar, then he's a liar." He added, "I'm still waiting for a judge or anyone to show me anywhere in the Constitution where there's a separation of church and state. We didn't lose; we were robbed."

"A thousand opinions by a court that a particular scientific theory is invalid will not make that scientific theory invalid," Richard Thompson said. "It is going to be up to the scientists who are going to continue to do research in their labs that will ultimately determine that."

Spokespeople for the religious right, once again showing that ID's most ardent supporters considered it a religious cause, condemned Jones and predicted that he had destroyed his career. "This decision is a poster child for a half-century secularist reign of terror that's coming to a rapid end with Justice Roberts and soon-to-be Justice Alito," Richard Land, president of the Southern Baptist Convention's Ethics and Religious Liberty Commission and a political ally of White House adviser Karl Rove, told the *Washington Post*. "This was an extremely injudicious judge who went way, way beyond his boundaries—if he had any eyes on advancing up the judicial ladder, he just sawed off the bottom rung."

In Dover, the overwhelming reaction was relief that the case was over and the town was no longer in the media's spotlight. The plaintiffs, the lawyers, and many of the experts congregated at the Pepper Hamilton offices in Philadelphia for a national teleconference, a last press conference. Reporters around the world tuned in to hear these men and women express their pride and relief—and their hope that other communities could follow the example of Jones's opinion and avoid similar debacles.

One after another took a turn to speak, but it was Christy Rehm who closed it out, the woman who first saw a problem on the horizon when her husband came home distressed at the school board's odd talk about creationism, and who had seen the case through to its end. She hadn't seen a monkey dance since the school board election, a sign that, perhaps, the healing had begun. She hugged her husband and spoke into the microphone:

> I'm obviously ecstatic about this decision. We plaintiffs are
> here because we care extremely about our community. This is
> a win for our community, this is a win for science. This is a win
> for science education . . . and as a teacher, I'm well satisfied.

That evening, Judge Jones was basking in the glow of having finished his opinion before the holidays. It had been hanging over his head, and he was glad it was done, glad that the students of Dover would not have to endure yet another round of one-minute statements after the holiday break. It was time for his life to return to normal.

He flicked on the television just in time to hear the Fox News host Bill O'Reilly and O'Reilly's guest, Judge Andrew Napolitano, eviscerate him. As Jones recalls, "O'Reilly looked at Napolitano, and he said, 'Oh, judge, five days before Christmas and here we have another judge taking away your civil liberties.' And I just had to gasp. I had never been taken apart on national television before."

Things took an ugly turn after that, when e-mailed death threats led the U.S. Marshal's Office to place Judge Jones under twenty-four-hour guard.

And so Jones spent December with bodyguards, sitting with his family and hearing himself discussed on the evening news. "That's an experience you don't usually have in your everyday life," he says. Jones had a thick hide from his years in politics, but he worried about his kids. He also fretted that there was so much more to the case than was being discussed in the media, so much great material that no one would ever see, so much evidence showing that this had not been a close case at all. He found himself wishing, not for the last time, that he had resisted the memory of Judge Lance Ito's unfortunate experience with the O. J. Simpson trial and said yes to CourtTV. But that, he says, is his only regret in the case of *Kitzmiller v. Dover.*

"There's a lot of things you can do in life that embarrass your children," John Jones said. "I hope I made them proud in this."

Later, his daughter, Meghan, offered a thought that cheered him. "When I go to law school, just think, I might be talking about your opinion in class."

EPILOGUE

It is humanity's unique blessing and peculiar curse to be the only species on Earth, as far as we know, that worries so obsessively and at such great expense about where we came from and why we're here. In an age in which DNA analyses reveal a 96 percent similarity in sequences between human and mouse genomes, it is precisely the ability to ask such abstract questions that seems to distinguish us most from other creatures, and that drives such diverse and uniquely human enterprises as science, poetry, art, and religion. Unfortunately, this list of by-products also includes jihads, inquisitions, racism, genocide, and a 5,000-year history of war against the different, the godless, and the infidels. Recent genetic evidence suggests that this beautiful and damning quality arose at least in part because humans are genetically disposed to believe in mysteries, miracles, God, and faith—that we evolved such a capacity because natural selection favors the unity and behavioral codes of the faithful. Ironically, given the clash of worldviews represented by the Dover case, nature itself may have fashioned us to believe in supernatural causes.

Of course, just as belief in evolution does not require disbelief in God, accepting a scientific theory that there may be a "God gene," to use a phrase coined by the molecular biologist and author Dean Hamer,[1] does not answer the underlying question of how that gene came to exist in the first place. Did it evolve through random mutation and the powerful but unconscious force of natural selection, or was it intelligently designed?

This is the essence of the questions that animated the Kitzmiller case, and that Judge John E. Jones sought not so much to answer as to compartmentalize. Taking pains to separate the scientific from the spiritual in his criticism of the intelligent design movement, he wrote: "While ID arguments may be true, a proposition on which the Court takes no position, ID is not science."

He did this to resist overreaching (a truly activist judge could just as well have concluded, as one of the witnesses did, that ID was both pseudoscience and a form of blasphemy), and to avoid offending the heartfelt beliefs of many, if not most, Americans. For Jones, the question came down to this clear formulation: Science deals with the natural world; religion deals with the spiritual world. One world can be explained by physical law; the other is beyond the reach or control of the physical. Both can exist without ever occupying the same space and without contradicting each other. Maybe the proponents of intelligent design are right about design in life and in the universe, Jones says; but theirs is a religious idea, not a scientific one, no matter how hard they argue the contrary, and so their ideas do not belong in high school biology classes. For Jones, this was a simple, rational distinction, and one he fervently hoped would help lower the heat in the national culture wars.

The death threats he received as a Christmas present suggest that his ruling did not succeed on this score.

In the end, Jones's opinion in *Kitzmiller v. Dover* was both a rousing achievement and a failure. It succeeds brilliantly in the arenas of science and law, which arguably is all that could be expected of a judge. Yet as a vehicle for turning the case into a national "teachable moment," it seems to have had the opposite effect. Jones concluded—correctly—that the evidence in favor of evolution is convincing and compelling, and that the the counterarguments are far less so. The idea has been supported time and again in living nature, in the lab, and in the fossil record, this powerful and elegant notion that there are naturally occurring variations in any species—bigger brains, sharper teeth, more nimble fingers—and that some of those changes can confer an edge in the struggle to survive and thrive. Within four months of the decision in *Kitzmiller*, yet another "transitional fossil," a species with qualities that combined those of finned fish and a four-legged creature, the *Tiktaalik rosae*, was discovered in 375-million-year-old arctic sediments in Canada. The creature had a head and neck like a crocodile, and fins that show primitive arm and wrist bones, enabling it to prop itself up, head out of the water, and to drag itself through shallow wetlands in search of prey. The fossils were found exactly where evolutionary theory predicted they should be found—paleontologists actually pinpointed the place and period, using their understanding of evolution and the geologic past. In fact, the expedition to find this fossil was undertaken specifically to prove the predictive power of evolution and modern paleontology. If Judge Jones's opinion stands for anything, it is this: Only those unwilling to consider such powerful evidence with an open mind can continue to assert that the theory of evolution is a fairy tale.

But this is not a welcome message in America at this juncture in history and politics. Jones's opinion has had the unintended effect of reinforcing an already pervasive belief that the scientific community and the judiciary are "against" God and faith. The false notion that belief in a scientific principle such as evolution and in a constitutional principle such as the separation of church and state is hostile to religion, family values, and public morality has become a central issue in the culture wars. Since the conclusion of *Kitzmiller*, Judge Jones has spoken out publicly about the importance of judicial independence and the rule of law, arguing that the demonizing of judges and scientists represents a dangerous direction for the country, even as some national leaders cynically embrace it for political gain, using the Dover case as Exhibit 1 in their diatribes. Sadly, from Jones's point of view, the chief promoters of the assault on science and judicial independence have been in his own Republican Party.

"I'm ashamed of some folks of the party I came from and the way they've demagogued," he said in an interview with the author. "It's pandering. It dumbs down the public."

In the aftermath of *Kitzmiller*, a wounded Discovery Institute has seemed willing, even eager, to contribute to the continuing societal polarization and hostility to science that Jones denounces.

The institute has sought to foster this polarization in a variety of ways both strange and disingenuous. Displeased with the trial and its results, Discovery joined in an odd sort of doppelgänger event at a Christian college in Los Angeles, Biola University, a few months afterward.

The event was called "Intelligent Design Under Fire," and it purported to pit the leading lights of the ID movement against a panel of experts who could cross-examine them. The main difference from the trial in Dover was that this event was rigged in favor of ID. Stephen Meyer of the Discovery Institute, who had pulled out of testifying at the trial in Pennsylvania, took part, along with Michael Behe and several other leaders of the ID movement. On the other side sat a network news show correspondent who kept apologizing for his ignorance; an animal rights activist who was also an author of self-help works; and a group of earnest but outgunned professors from a local campus of California State University, whose "cross-examination" was often verbose and incomprehensible. There was no Eric Rothschild on hand to produce with a flourish some damning document or inconsistency. The panel was accompanied by the elderly British philosopher Anthony Flew, who renounced a lifetime of atheism to endorse ID but then promptly withdrew that endorsement. This was a surreal event, a kind of show trial played out before a partisan audience that cheered the ID group at every turn and catcalled the evolutionists. The proponents of ID were able to make all the claims that had been subjected to withering cross-examination in *Kitzmiller*, this time without serious challenge. No one spoke of the obvious inconsistency of proponents of ID asserting that there was nothing religious about ID before a cheering audience at a Bible college—an audience intermittently shouting "Amen." Flew, who looked frail and was clearly exhausted by his transatlantic flight, slept onstage throughout the proceedings. To no one's surprise, intelligent design "won."

In a similar vein, within three months of Jones's decision, the Discovery Institute Press published a 123-page book on the case.[2] It consisted primarily of an adaptation of angry Internet postings from the Discovery Institute's website and its affiliates. The title, *Traipsing into Evolution*, is a mocking reference to a phrase used by Jones in his opinion, in which he stated that the extensive testimony in *Kitzmiller* had left him in a good position to "traipse into this controversial area" of whether or not ID is science, so that other courts and governmental bodies could benefit from his review.[3]

The authors of Discovery's book object to this approach and suggest it shows that Jones has an oversize ego and an inflated opinion of his trial court's reach and influence. The book's first page features a dictionary definition of *traipse* as "To walk about idly or intrusively," and that is how the authors attempt to characterize Jones's ruling and personal conduct—as a self-aggrandizing, unnecessary journey, which is both legally and scientifically suspect.

But the 123-page book achieves this signature "traipsing" metaphor through quite a stretch of its own. It resorts to quoting a 1982 edition of the *American Heritage Dictionary* for its mocking definition. More current editions of the dictionary give the following principal definition: "To walk or tramp about; gad: *traipsed from one picnic site to another.*" This is a small point, yet it occurs on the very first page of the introduction of the book, and it is emblematic of many of the substantive arguments that follow. Few, if any, hold up to scrutiny.

The first substantive complaint about Jones in the book follows this pattern. It concerns what the authors describe as the judge's "partisan history of Intelligent Design." Jones is attacked for "conflating ID with fundamentalism," and after making this accusation, the book excoriates him by offering extensive information about how the intelligent design movement has nothing at all to do with Christian fundamentalism.

Perhaps so. But the accusation against Jones is a complete fabrication. An examination of the pages in Jones's written opinion that are specifically footnoted in *Traipsing*—pages that are supposed to support this attack on the judge—shows that they contain nothing like what the Discovery Institute claims. Rather, Jones was writing in those passages about the history of federal litigation on the teaching of evolution in public schools, not the history of intelligent design. And in those passages, he drew the uncontroversial and undeniable conclusion that the challenges to teaching evolution and attempts to balance it with alternative theories have in the past been led by Christian fundamentalists— including members of the Dover school board. This is not a partisan history; it is a fact. Discovery's book sets up a straw man; but the truth is that nowhere in Jones's opinion does he "conflate" intelligent design with fundamentalism. The Discovery Institute just made this up.

Likewise, one of the book's central accusations—that Jones "misrepresented the facts" in his key ruling that intelligent design is not science— also fails to stand under scrutiny. The key to this question is whether intelligent design requires a supernatural designer (the designer that proponents of ID steadfastly decline to identify, but that the vast majority of their supporters assume to be God). *Traipsing* says, no, Jones is wrong; it is possible to scientifically detect design in nature, on the basis of all our experience observing man-made intelligent designs. (E.g., we can tell Mount Rushmore from a rock and software code from gibberish, and therefore we can detect design in nature, too.) The book quotes various passages of testimony in *Kitzmiller* to support its position and flails at Jones for not accepting them. But *Traipsing* fails to address some of the key pieces of testimony Jones cited in his opinion to support his finding that ID is a supernatural, religious idea and not science: the very words of the Discovery Institute's own experts. Jones based his opinion in part on this information:

- Michael Behe has written that it is "implausible that the designer is a natural entity," and that the phrase *intelligent design* means "not designed by the laws of nature." Behe also admitted on the witness stand that a definition of science that included intelligent design would also include astrology, a supernatural belief that is essentially magic.
- Scott Minnich, another defense expert at the trial and a fellow of the Discovery Institute, said that the ground rules of science had to be broadened to include supernatural forces in order for ID to be considered "science."
- The defense expert Steve Fuller said that the mission of the ID movement was in part to change the ground rules of science to include the supernatural.
- Discovery's own leading academic, William Dembski, has argued that methodological naturalism—the scientific method—must be overturned if ID is to be accepted as science.
- *Of Pandas and People* describes the designer as a "master intellect," a phrase that, Jones asserts, strongly suggests a deity.

In the end, *Traipsing into Evolution* is not a refutation of Judge Jones's opinion so much as a rehash of the arguments that lost at the trial. It is the rant of a sore loser, whose experts backed out of the fight at the last minute and then complained that their opinions were given short shrift. It is worth noting that proponents of ID praised Judge Jones's credentials at the start of the case; the administrator of William Dembski's website, UncommonDescent, even intimated then that a victory for ID was all but ensured—not for legal reasons but for partisan ones. The administrator, whose online identity is DaveScot, wrote:

> Judge John E. Jones . . . is a good old boy brought up through the conservative ranks. He was state attorney for D.A.R.E., an Assistant Scout Master extensively involved with local and national Boy Scouts of America, political buddy of Governor Tom Ridge (who in turn is deep in George W. Bush's circle of power), and finally was appointed by GW himself. Senator Rick Santorum is a Pennsylvanian in the same circles (author of the "Santorum Language" that encourages schools to teach the controversy) and last but far from least, George W. Bush himself drove a stake in the ground saying teach the controversy. Unless Judge Jones wants to cut his career off at the knees he isn't going to rule against the wishes of his political allies. Of course the ACLU will appeal. This won't be over until it gets to the Supreme Court. But now we own that too.

Since the trial, of course, Jones has become just another liberal activist judge, a dogmatist entranced by the godless Darwin, according to the leaders of the intelligent design movement. Now he's the type of judge that the religious right regularly denounces. He's the type targeted by the outrageously mislabeled Constitution Restoration Act, which right-wing extremists in Congress hope will strip federal judges of the power to decide cases like *Kitzmiller*. And now Jones is the type of judge who gets death threats from the zealots of the "culture of life."

The false accusations emanating from the Discovery Institute have consequences. Discovery and its leaders have a loyal following, and their words influence discourse, as well as ideologues with larger megaphones. The extremist author Ann Coulter has taken up Discovery's cause in her fact-challenged, hate-filled best seller, *Godless*, which foists on a credulous readership one distortion after another about evolution, intelligent design, Dover, and Judge Jones's ruling. Coulter has the ability to influence millions of Americans; she could have educated her readers about evolution, about the separation of church and state, about ways that science and religion can coexist. Instead, she wrote a work laden with falsehood that makes Dr. Dino sound positively scholarly, and she did it with the help of the intelligent design movement's intellectual leader, William Dembski.

Coulter provides a nice summary of her views on evolution in *Godless*. Every significant statement in it is a lie or an egregious error:

> Liberals' creation myth is Charles Darwin's theory of evolution, which is about one notch above Scientology in scientific rigor. It's a make-believe story, based on a theory that is a tautology, with no proof in the scientist's laboratory or the fossil record—and that's after 150 years of very determined looking. We wouldn't still be talking about it but for the fact that liberals think evolution disproves God.

There are only three sentences and sixty-nine words in that passage, yet it contains five lies and one ludicrous error:

It is a lie to say evolution is a "creation myth," as it does not attempt to explain the origins of life, the earth, the universe, or any other act of creation—nor is it "myth" in any sense of the word.

It is a lie to characterize the modern science of evolution as "Darwin's theory," as it also now encompasses genetics, DNA analysis, microbiology, embryology, artificial life experiments, and a host of other findings, methods, and scientific disciplines that Darwin (and apparently Coulter) never heard of.

It is a lie to rank evolution one notch above scientology in terms of rigor. Scientology is not science but evolution is, and while *rigor* is a somewhat loose term, there is no question that evolutionary theory (unlike intelligent design) has been rigorously tested in the field, in laboratory experiments, and in computer simulations. Arguably, evolution has been more rigorously tested, and enjoys more evidence in its support, than any other theory in the history of science. As a religion, scientology is no more or less rigorous than any other religious belief, including Christianity.

It is a lie to say evolution lacks proof. Proof that evolution has occurred (and is occurring) has been observed in the laboratory and detected in the fossil record many times. No species could survive beyond a few generations without evolution: Germs evolve all the time to evade our immune systems, and our immune systems evolve across generations to cope with that threat.

Perhaps the most outrageous lie contained in this three-sentence passage is Coulter's claim that liberals think evolution disproves God. In truth, the exact opposite is true: It is conservatives who think this way. Religious conservatives, not liberals, have tried to ban evolution from public schools for decades because it contradicts their literal reading of the Bible. Recent data from the Harris Poll show that Democrats and Republicans believe in God in roughly equal numbers in America (90 percent of Democrats, 93 percent of Republicans); yet the percentage of liberals (primarily Democrats) who accept evolution is twice the percentage of conservatives (primarily Republicans) who accept it.[4]

Finally, it is a ludicrous error—and a standard, discredited creationist argument—to claim that evolution is based on a tautology. Coulter is referring to the popular (but incomplete and oversimplified) notion that natural selection is analogous to "survival of the fittest." The tautology is supposedly this: The fittest creatures survive, and we know which creatures are fittest because those are the ones that survive. This is actually an example of circular logic, not a tautology, which is defined as a statement that needlessly repeats an idea. (Here's a true tautology: Ann Coulter gets so many things wrong because she gets so few things right.) In any case, the criticism is mistaken. Even a leading creationist organization, Answers in Genesis, labels the tautology argument against evolution as "doubtful, hence inadvisable to use," while describing natural selection as a "useful explanatory tool." Natural selection is not a matter of individual creatures surviving because of their fitness; it's an explanation of how, across an entire population, traits (or variations or mutations) that enhance the ability to survive in a particular environment will be passed on in greater numbers to subsequent generations, while traits that do not enhance survival in that environment tend not to be passed on in such great numbers. In time, this process leads to changes in a population or an entire species—a process that is observed in laboratories every day. The traits that determine this "fitness" constantly change as an environment changes, but the alterations in a species are very gradual, which is why many supremely "fit" species of dinosaurs went extinct when the world climate and food supply took a drastic turn 65 million years ago (and why some others continued to evolve until they became creatures we now call birds).

Nowhere in Coulter's book will readers find a discussion of why one-third of the U.S. population insists on a literal interpretation of a religious document, the Bible, that is rife with proven geographic, scientific, and historical inaccuracies; that endorses immoral acts including slavery, stoning, capital punishment of the children of sinners, genocide, and mass murder. (God is supposed to have ordered the Israelites to commit genocide against seven nations, specifically ordering the execution of women and children; he caused the extermination of the firstborn infants of Egypt during the first Passover and even caused bears to slay forty-two children for calling the prophet Elisha "old baldy.") Religious conservatives who insist there is no proof that evolution has occurred are silent on the fact that there is no historical proof outside of the scriptures that Jesus of Nazareth ever lived; those obsessive record keepers the Romans left no unambiguous evidence that he existed, and even the Bible contains no eyewitness accounts of Jesus—the gospels were written long after the fact.

After her book's publication, Coulter was listed as a "featured expert" on a broadcast entitled "Darwin's Deadly Legacy," presented by the radical right preacher D. James Kennedy of Florida's Coral Ridge Ministries. The broadcast attempted to make a case that Nazism, the rise of Adolf Hitler, and the Holocaust are direct results of Darwin and his theory of evolution. According to Coulter, quoted on the Coral Ridge website, Hitler initiated the extermination of Jews because he "was applying Darwinism. He thought the Aryans were the fittest and he was just hurrying natural selection along." Adds Kennedy: "To put it simply, no Darwin, no Hitler."

Aside from the fact that this bit of revisionist history is dismissed by virtually every serious historian of the Holocaust and the Nazis, as well as by the Anti-Defamation League, the claim makes no sense even in scientific terms, as Hitler's notions of racial purity and his desire to exterminate Jews have nothing to do with Darwin's ideas of common descent and natural selection. A philosophical bastardization of evolutionary theory known as "social Darwinism"—a dark view of human societies that Charles Darwin, a pacifist, neither considered nor accepted—was used in the nineteenth century as an argument in favor of social inequality, laissez-faire capitalism, totalitarianism, and racism. It may have had some influence on Hitler, who employed many ideas as covers for his hatred, bigotry, and desire for power. But blaming Charles Darwin and such a misuse of his theory of evolution for Hitler's Final Solution is tantamount to blaming Lincoln and the Emancipation Proclamation for the Watts riots, or blaming Jesus of Nazareth for the horrors of the Spanish Inquisition.

Kennedy and Coulter never mention a much more tangible and widely accepted historical influence on Hitler and the Holocaust: a religious tract that formed the basis of modern anti-Semitism, that was quoted by the Nazis many times, and that the publisher of the Nazi newspaper *Der Stürmer*, in testifying at the Nuremberg Trials, identified as a major justification for the Final Solution. The tract, called "On the Jews and Their Lies," advocated persecution of all Jewish people (described as "poisonous worms" in the tract) through burning their synagogues, homes, schools, and prayer books; placing them into forced-labor and concentration camps; seizing their money and valuables; and denying them both mercy and legal rights.

"On the Jews and Their Lies" was written in 1543 by none other than Martin Luther—the most influential theologian in German history, who began the Reformation and Protestantism, in which the roots of both Hitler's Germany and modern American fundamentalism are firmly planted.

Unfortunately, Ann Coulter and her absurd book and TV appearances will be the only source of information many of her readers will have on evolution, intelligent design, and *Kitzmiller*. (A typical lie by Coulter: "They didn't win on science, persuasion, or the evidence. They won the way liberals always win: by finding a court to hand them everything they want on a silver platter.") Misinformation is everywhere in her book, yet many will believe it. And few will take the time to go online and read Judge Jones's opinion for themselves and learn firsthand what really happened. They will believe the lies because that is what they want to believe, as America is transformed into a nation that prefers opinion over fact.

"Ann Coulter foments a kind of civic stupidity, in my opinion," Jones lamented.

Although the propaganda war continues unabated, there have been some tangible results of *Kitzmiller*, and these do not bode well for the intelligent design movement. Several states that had been considering a "teach the controversy" approach have dropped it. Much of the pending legislation in states around the country has gone nowhere. Ohio, which had been the Discovery Institute's greatest victory, dumped a curriculum requirement for the "critical analysis" of evolution, a huge defeat for ID directly linked to *Kitzmiller* and to Ohioans' fear of a similarly costly and embarrassing lawsuit. That left only the state of Kansas in play, where the creationist majority on the state school board faced a tough reelection battle and anti-evolution state standards were on the line. In a bitter May 2006 election, several incumbents were voted out of office and the 6–4 creationist majority became a 6–4 pro-evolution majority, guaranteeing a repeal of the anti-evolution standards that the Discovery Institute had so strongly supported. The vote, in one of the nation's most politically conservative and Republican states, seemed an even tougher blow to the ID movement than *Kitzmiller*. This time, there was no "activist judge" to blame.

The Republican Party's loss of control of Congress in the fall 2006 elections diminished, for the time being, the Discovery Institute's influence in Washington. It also doomed federal legislative proposals to promote intelligent design and to limit future judicial involvement in cases such as *Kitzmiller*. Less clear is how the shift in national leadership will affect the culture wars outside the Beltway, where America remains as polarized as ever. There is little to suggest that the election will lead to new compromise or attempts at mutual understanding, and considerable evidence that it could drive the cultural wedge even deeper. In one telling postelection moment, the president-elect of the powerful Christian Coalition political organization, Reverend Joel C. Hunter, resigned over his failed efforts to broaden the group's mission beyond its traditional stances against abortion, evolution, and gay marriage. He wanted the coalition to tackle such traditionally liberal causes as environmentalism, global warming, and poverty, too—precisely the sorts of initiatives that could build common ground between conservative and liberal, religious and secular.

The group's board rebuffed Hunter, who recalled being told, "It won't speak to our base, so we just can't go there."

In Dover, the plaintiffs have returned to something like normality, but the rift in the community has not healed. Bill Buckingham has returned to York County, and the ousted members of the school board in Dover, including Alan Bonsell, have begun to attend meetings and to criticize the new board members, seeking their own political revival or perhaps just simple revenge. The new board's decision against renewing the contracts of the superintendent and assistant superintendent who had served the previous board majority has stirred new controversy in Dover, leading even Casey Brown to publicly break with the newly elected board members she had supported. It is not a happy place to be a school board member for now.

To the students, however, the controversy continues to perplex more than anything else, and perhaps that is a good thing. The taunts of "monkey girl" have subsided, and evolution, as far as Dover's classrooms are concerned, has returned to the back burner, where everyone seems to want it. In some ways, the children have been the most sensible players in this entire drama.

"I still don't know why people got so upset about it," one thoughtful student at Dover High said a few months after Jones's decision. "People are going to believe whatever they want when it comes to Darwin and God and coming from monkeys. That's why they call it belief. Facts have nothing to do with it. So why get all upset about it?"

NOTES

Prologue: One Thing's for Sure, We Didn't Come from Any Monkey

1. According to the testimony in *Kitzmiller v. Dover* of Casey Brown and other board members present at the retreat, including the former board member Aralene "Barrie" Callahan, as well as notes taken at the retreat by Dover Area School superintendent Richard Nilsen, which were entered as exhibits in *Kitzmiller*. Bonsell has said he does not remember making these statements, but conceded that he must have made them if they were in Nilsen's notes.

2. As at the retreat in 2002, Bonsell's interest in creationism was reflected in Superintendent Richard Nilsen's notes, though Bonsell could not remember making any such comments. According to the testimony and recollections of Casey Brown, Jeff Brown, and Barrie Callahan, all board members present at the retreat, Bonsell was adamant in expressing his belief that creationism and evolution ought to be taught fifty-fifty at Dover High School.

3. According to numerous eyewitnesses and accounts in two competing newspapers, Buckingham made this statement during the discussion of balancing evolution and creationism in school textbooks and lessons. Buckingham would later say he made such a remark, but at a meeting months earlier, in connection with a discussion of the phrase "under God" in the Pledge of Allegiance.

Chapter 1: Balancing Act

1. In an interview and testimony, Bertha Spahr said that Baksa used the word "creationism." Spahr reported Baksa's remarks to Dover High's principal at that time, who produced a memo summarizing Baksa's remarks, again using the word "creationism." Baksa admitted in court that he did not attempt to correct the principal after the memo was circulated.

Chapter 2: What Lies Beneath

1. The full results of Gallup's demographic analysis of who accepts evolutionary theory are fascinating. As reported by Gallup News Service on November 19, 2004, these percentages of the population "believe that Darwin's theory of evolution is a scientific theory well supported by the evidence":

Postgraduate education	65%	**Sample average**	35
Liberal	56	Nearly weekly church attendance	35
College graduate	52	30- to 49-year-olds	34
West	47	Some college	32
Seldom, never attend church	46	Women	30
Catholics	46	Republican	29
50- to 64-year-olds	44	Midwest	29
East	42	Protestant	28
Men	42	South	27
18- to 29-year-olds	41	Conservative	26
Independent	40	Weekly church attendance	22
Democrat	38	Age 65+	21
Moderate	36	High school or less	20

Chapter 3: I'd Rather Take a Beating Than Back Down

1. Members of the audience at that June school board meeting, as well as teachers, three board members, and two reporters for competing newspapers, heard Bill Buckingham and Alan Bonsell say they wanted a balance between creationism and evolution in the curriculum and in any textbook the high school purchased. Buckingham and Bonsell deny this, and their denials have been supported by another board member allied with them, along with the district superintendent. The official audiotapes of the meeting, which could have resolved exactly what was said, were destroyed. However, in a letter to the editor, in notes taken by district officials at other meetings, and through Buckingham's statements captured on news video, it is clear that Buckingham, Bonsell, and at least one other board member all indicated at one time or another that they wanted to introduce "creationism" into the curriculum. Buckingham and Bonsell assert that board members only used the term "intelligent design," and that reporters mistakenly (or maliciously) substituted the term "creationism." However, Buckingham concedes that he did not really know much about intelligent design at that point, that his memory was impaired about many events in that time frame because he had become in-

creasingly dependent on prescription painkillers that made it hard for him to concentrate or even drive a car, and that neither he nor anyone else in the school district ever lodged an official complaint or formal request for a retraction with any news organization about the alleged errors in reporting about creationism.

2. Bonsell and Buckingham have remained adamant throughout the controversy that they did not advocate creationism in the classroom at this or any other board meeting, just intelligent design. However, on the basis of accounts in two separate newspapers, and on accounts by a majority of eyewitnesses. U.S. District Court judge John E. Jones III found that Bonsell and Buckingham had made statements about creationism at board meetings.

3. Buckingham admits making such a statement—and believing it to be true—but asserts, contrary to many others' recollections, that he made the statement months earlier at a meeting in which the board addressed the phrase "under God" in the Pledge of Allegiance.

4. Judith Grabiner and Peter Miller, "Effects of the Scopes Trial," *Science*, September 6, 1974. Grabiner and Miller write: "It is easy to identify a text published in the decade following 1925. Merely look up the word 'evolution' in the index or glossary; you almost certainly will not find it."

5. According to the findings of U.S. District Court judge John E. Jones, despite Buckingham's denial.

Chapter 4: Darwin's Nemesis

1. *Scientific American,* July 1992. In his review, Gould extensively catalogued his objections to *Darwin on Trial,* including the central theme that evolution is hostile to religion, which he said was "the oldest canard and non sequitur in the debater's book." Gould wrote: "Science simply cannot (by its legitimate methods) adjudicate the issue of God's possible superintendence of nature. We neither affirm nor deny it; we simply can't comment on it as scientists. . . . Science can work only with naturalistic explanations; it can neither affirm nor deny other types of actors (like God) in other spheres (the moral realm, for example)." Gould went on to identify numerous scientific errors Johnson made in the book, including confusing or omitting the different roles of mutation, sexual selection, and polyploidy in evolutionary change; quoting out-of-date research; and even getting the basic definition of science wrong. "Johnson then upholds the narrow and blinkered caricature of science as experiment and immediate observation only. Doesn't he realize that all historical science, not just evolution, would disappear by his silly restriction?"

2. Barbara Forrest and Paul R. Gross, *Creationism's Trojan Hourse: The Wedge of Intelligent Design*, Oxford University Press, 2004. The book is an outsider's history of the intelligent design movement and the wedge strategy; Forrest was an expert witness in *Kitzmiller v. Dover* and has been the target of vigorous criticism from leading figures at the Discovery Institute.

3. Excerpts from the "Wedge Strategy," from exhibits in *Kitzmiller v. Dover*, follow.

THE WEDGE STRATEGY
INTRODUCTION

The proposition that human beings are created in the image of God is one of the bedrock principles on which Western civilization was built. Its influence can be detected in most, if not all, of the West's greatest achievements, including representative democracy, human rights, free enterprise, and progress in the arts and sciences.

Yet a little over a century ago, this cardinal idea came under wholesale attack by intellectuals drawing on the discoveries of modern science. Debunking the traditional conceptions of both God and man, thinkers such as Charles Darwin, Karl Marx, and Sigmund Freud portrayed humans not as moral and spiritual beings, but as animals or machines who inhabited a universe ruled by purely impersonal forces and whose behavior and very thoughts were dictated by the unbending forces of biology, chemistry, and environment. This materialistic conception of reality eventually infected virtually every area of our culture, from politics and economics to literature and art.

The cultural consequences of this triumph of materialism were devastating. Materialists denied the existence of objective moral standards, claiming that environment dictates our behavior and beliefs. Such moral relativism was uncritically adopted by much of the social sciences, and it still undergirds much of modern economics, political science, psychology and sociology.

Materialists also undermined personal responsibility by asserting that human thoughts and behaviors are dictated by our biology and environment. The results can be seen in modern approaches to criminal justice, product liability, and welfare. In the materialist scheme of things, everyone is a victim and no one can be held accountable for his or her actions.

Finally, materialism spawned a virulent strain of utopianism. Thinking they could engineer the perfect society through the application of scientific knowledge, materialist reformers advocated coercive government programs that falsely promised to create heaven on earth.

Discovery Institute's Center for the Renewal of Science and Culture seeks nothing less than the overthrow of materialism and its cultural legacies. Bringing together leading scholars from the natural sciences and those from the humanities and social sciences, the Center explores how new developments in biology, physics and cognitive science raise serious doubts about scientific materialism and have re-opened the case for a broadly theistic understanding of nature. The Center awards fellowships for original research, holds conferences, and briefs policymakers about the opportunities for life after materialism.

THE WEDGE STRATEGY
PHASE I: Scientific Research, Writing & Publicity
PHASE II: Publicity & Opinion-Making
PHASE III: Cultural Confrontation & Renewal

THE WEDGE PROJECTS
PHASE I: SCIENTIFIC RESEARCH, WRITING & PUBLICATION
    Individual Research Fellowship Program
    Paleontology Research program (Dr. Paul Chien et al.)
    Molecular Biology Research Program (Dr. Douglas Axe et al.)

PHASE II: PUBLICITY & OPINION-MAKING
    Book Publicity
    Opinion-Maker Conferences
    Apologetics Seminars
    Teacher Training Program
    Op-Ed Fellow
    PBS (or other TV) Co-Production
    Publicity Materials / Publications

PHASE III: CULTURAL CONFRONTATION & RENEWAL
    Academic and Scientific Challenge Conferences
    Potential Legal Action for Teacher Training
    Research Fellowship Program: Shift to social sciences and humanities

4. From "Statement by Seth L. Cooper Concerning Discovery Institute and the Decision in Kitzmiller v. Dover Area School Board Intelligent Design Case," Discovery Institute website, http://www.evolutionnews.org/2005/12/statement_by_seth_l1_coope_con.html (as of November 2006).

Chapter 5: Class Acts

1. Karen Muller, "Faithful Divided on Book," *York Daily Record,* June 23, 2004.
2. Jeff Brown related this conversation in testimony and in an interview with the author. Harkins would later testify that she did not recall the conversation, and that she gave *Pandas* only a cursory one-hour review before Jeff Brown came over to pick it up.
3. This quote attributed to Heather Geesey and the surrounding discussion is drawn from "Intelligent Design Voted In," by Joseph Maldonado, *York Daily Record,* October 19, 2004, and from interviews and testimony of Bertha Spahr, Barrie Callahan, and Jeff and Casey Brown. In testimony in *Kitzmiller v. Dover,* Geesey confirmed that she uttered these words, but claims she was misinterpreted and that she was actually referring to firing the board's solicitor, Stock and Leader, not the teachers. She explained this by testifying that the solicitor had approved the board's actions in regard to intelligent design. A memo on the issue from the solicitor, filed as evidence in *Kitzmiller,* contradicts Geesey on this point.

Chapter 6: Broken Watches

1. In fish the proportion of brain weight to body weight is 1:5,000. By comparision, the proportion is 1:180 for the average mammal and 1:50 for humans.
2. At the same time, natural selection would also preserve other advantageous traits—bipedalism, the ability to run upright, the ability to vocalize. One fascinating but disputed theory suggests that many human traits were selected because, unlike other primates, the ancestors of humans lived for extended periods in semiaquatic environments, wading and swimming in order to obtain marine life as a principal food source. Natural selection for success in an aquatic environment would explain some characteristics humans have that other primates lack. Bipedalism is rare among mammals and not particularly efficient or stable for pursuing (or fleeing) prey, though it is ideal for wading in shallow water. Further support for the "aquatic ape premise" of human evolution: Outside a watery environment, the upright posture leaves humans uniquely prone to back problems, varicose veins, hemorrhoids, hernias, and difficulty birthing—which are hardly hallmarks of "intelligent" design. The human ability to control and hold breathing, rare among non-marine mammals and absent in other primates, is a necessity for living in and around water. Sweating and tearing, which no other primates do, remove salt and excessive water from the body (though their other purpose, temperature regulation, does not fit the aquatic ape theory). The waterproof coating on a newborn hu-

man's skin, vernix caseosa, is unique among primates (but present in newborn seals). Finally, sparse body hair, while a common trait in marine mammals, is not known among other primates. Fetal chimpanzees and fetal humans both form a fine coat of hair all over the body as they develop in utero. Humans shed this coat, called lanugo, before birth; chimps, obviously, keep theirs. The gene regulating a furry coat, present in both animals, has been "switched off" in humans, apparently through the power of natural selection. The continued presence of that gene in human DNA is cited as evidence of both common descent and the continued adaptations of natural selection, and as an argument against intelligent design.

3. A single-cell water creature, euglena, has such a primitive light-sensitive sensory patch that lets it navigate using its flagellum. Other water creatures, such as the simple, multicell planaria, elaborate on this arrangement with a pit that is lined with light-sensitive cells, allowing them to determine the direction of light. More advanced visual arrangements include covering the pit and leaving only a small hole to admit light, as in a pinhole camera. The nautilus has such an eye, which has a retina but no lens or cornea. Other creatures have progressively more advanced eyes, and the mammalian eye has a cameralike design. Darwin used the analogy with living species to suggest that natural selection could have generated a similar progression long ago to create the first eyes and their successors, so long as each small change from the first photosensitive cells conferred a survival advantage.

Chapter 7: The Watchmaker Returns

1. Behe's assertion that evolutionary theory lacks "proof," a complaint he has repeated many times, goes beyond Denton's argument that evolutionary theory fails to explain the emergence of new organs and creatures, and it is a curious complaint for a scientist to make. Scientific theories are rarely if ever "proved," as Behe should know. Rather, either a theory succeeds in explaining the available evidence, or the evidence disproves, or (to use the scientific term) falsifies, the theory. Although creationists believe that evolutionary theory has been falsified by their Bible-based research, which purports to indicate a 6,000-year-old Earth and a world in which men and dinosaurs coexisted, few mainstream scientists agree. For the mainstream, Darwin's theory provides a very effective explanation for the diversity of life and the genetic relationships between diverse life-forms. This is not to say that the theory may not be falsified someday—science is designed to be provisional, given its history of new discoveries, such as the round Earth and quantum physics, that have shattered old assumptions. But by asserting that aspects of evolutionary theory are

not proved, rather than claiming that they have been falsified, Behe creates a potent-sounding argument which resonates with a lay readership, though his critics say it has no scientific merit. On the other hand, he does speak of his own theory of irreducible complexity as a falsification of evolutionary theory.

2. The argument that there would be no mate for the first member of a new species falsely assumes that the first creature in a new species would simply pop into existence, completely different from its ancestors. In fact, whole populations evolve, not individual creatures, and the path toward a new species is very gradual. Recent evidence in the human and chimpanzee genomes, for instance, suggest that the distant ancestors of the two species separated at some point 6 million or 7 million years ago, but that the two groups of primates remained so closely related that they may have continued to interbreed until the species finally diverged completely 5 million years ago.

Henry Morris came up with the calculation that cosmic dust would have buried the entire earth under a 182-foot layer if the planet really were billions of years old. Morris's calculations are wrong; they are based on estimates of dust accumulation that fail to exclude the dust generated on the planet's surface. An accurate calculation shows that an inconsequential 66 centimeters of cosmic dust would have fallen to earth over 4.5 billion years (0.000000015 centimeter a year).

The rate of the moon's movement away from Earth has not been constant across billions of years. Paleological evidence recovered from fossil coral formations, which provide a kind of natural calendar of the tides, the movement of the moon, and the length of the seasons, has shown that the ancient Earth had a much longer year, and that the moon was moving away at a much slower rate. The best data on the movement of the Earth and the moon disproves a "young Earth" and supports an age of about 4.5 billion years.

Five exquisitely preserved archaeopteryx fossils showing obvious feathers were discovered in different places on different occasions under well-documented conditions. The original was obtained in 1863 by the curator of the British Museum of Natural History, Sir Richard Owen, who was a creationist and an implacable opponent of Darwin and the theory of common descent.

A complete list of creationist claims and evolutionist responses can be found at the website of talkorigins.org at http://www.talkorigins.org/indexcc/list.html.

3. The list of discredited claims that Answers in Genesis, a young-Earth creationist organization, asks creationists not to use includes these: the thickness of moon dust proves a young universe; NASA discovered a missing day; woolly mammoths were flash-frozen during Noah's flood; if we evolved from

apes, then apes shouldn't exist today; archaeopteryx is a fraud; no new species have ever been produced. The complete list can be found at http://www.answersingenesis.org/home/area/faq/dont_use.asp.

4. Clinton had served one two-year term as governor, then was narrowly defeated by Frank White, a banker who would come to be viewed by many Arkansans as a religious zealot before he was voted out of office after one term—defeated by a resurgent Clinton. The creation-science law White had championed while in office (which he later admitted having signed without actually reading) was an embarrassment, but the main reason White was voted out by a wide margin was due more to his close ties to the utility industry and his role in passing an enormous increase in electric rates. It would have been politically unwise for any candidate to rule out the teaching of alternatives to evolution in public schools, and so Clinton instead cautioned against attempting to introduce overtly religions teachings, while remaining open to scientific alternatives.

5. From Lawrence S. Lerner, "Good Science, Bad Science: Teaching Evolution in the States," Thomas B. Fordham Foundation, September 2000.

Chapter 8: The Waters of Kansas Part

1. The conclusion of "Science and Creationism: A View from the National Academy of Sciences," 2nd ed., a paper published by the National Academy of Sciences in 1999, explains why the academy believes incorporating creationism or intelligent design into science courses would undermine public education:

> Science is not the only way of acquiring knowledge about ourselves and the world around us. Humans gain understanding in many other ways, such as through literature, the arts, philosophical reflection, and religious experience. Scientific knowledge may enrich aesthetic and moral perceptions, but these subjects extend beyond science's realm, which is to obtain a better understanding of the natural world.
>
> The claim that equity demands balanced treatment of evolutionary theory and special creation in science classrooms reflects a misunderstanding of what science is and how it is conducted. Scientific investigators seek to understand natural phenomena by observation and experimentation. Scientific interpretations of facts and the explanations that account for them therefore must be testable by observation and experimentation.
>
> Creationism, intelligent design, and other claims of supernatural intervention in the origin of life or of species are not science because they are not testable by the methods of science. These claims subordinate observed data to statements based on authority, revelation, or religious belief.

Documentation offered in support of these claims is typically limited to the special publications of their advocates. These publications do not offer hypotheses subject to change in light of new data, new interpretations, or demonstration of error. This contrasts with science, where any hypothesis or theory always remains subject to the possibility of rejection or modification in the light of new knowledge.

No body of beliefs that has its origin in doctrinal material rather than scientific observation, interpretation, and experimentation should be admissible as science in any science course. Incorporating the teaching of such doctrines into a science curriculum compromises the objectives of public education. Science has been greatly successful at explaining natural processes, and this has led not only to increased understanding of the universe but also to major improvements in technology and public health and welfare. The growing role that science plays in modern life requires that science, and not religion, be taught in science classes.

Chapter 9: *What Will We Tell the Children?*

1. The chemists Stanley Miller and Harold Urey conducted the now famous Miller-Urey experiment in 1953 at the University of Chicago, simulating what was then believed to be the early environment on Earth before life appeared. They put hydrogen, methane, ammonia, and water inside a sealed sterile beaker. The water was then heated to vapor and allowed to condense, simulating the water cycle, and periodic electrical sparks simulating lightning were introduced. After one week, organic compounds had formed inside the beaker, including thirteen of the twenty-one amino acids used to make proteins in living cells.

2. The statement Meyer mentioned, signed by various scientists, many of them with degrees in engineering, computers and other nonbiological fields, is titled, "A Scientific Dissent from Darwinism," and reads: "We are skeptical of claims for the ability of random mutation and natural selection to account for the complexity of life. Careful examination of the evidence for Darwinian theory should be encouraged." The list of signers may be found at http://www.discovery.org/scripts/viewDB/filesDB-download.php?id=302.

3. The list has been criticized by the National Center for Science Education (NCSE), among others, on a number of points, including its incorrect description of the evolutionary process. Natural selection acts on biological mechanisms other than just random mutation—such as genetic drift and gene duplication—to create variation in organisms, but the statement attached to the list mentions only random mutation. Some scientists who believe in evolution might have signed the statement because according to one school of

thought these other factors may at times be more powerful than random mutation as a force for evolution. The NCSE also questions the credentials of the signers, who include many scientists in fields so far removed from evolutionary biology that their opinions may be no more significant than a layperson's. The NCSE has parodied the Discovery Institute's list by compiling a longer and more prestigious list of supporters of evolution limited to scientists whose first name is "Steve." Because scientists named Steve represent about 1 percent of all scientists, the NCSE argues, its list represents tens of thousands of scientists who support evolution. More information on Project Steve can be found at http://www.ncseweb.org/resources/articles/3541_project_steve_2_16_2003.asp.

Chapter 11: Monkey Suit

1. There is nothing inherently wrong with critical analysis of any subject, according to the ACLU, the National Center for Science Education, and other scientific organizations. In fact, critical analysis in the generic sense ought to be part of the study of every subject—it's just good teaching. But the term "critical analysis" in recent years has been appropriated by opponents of evolution to describe attacks on the theory of evolution and no other school subject. As the term has been used by creationist groups, by the Discovery Institute, and by various school districts and states, notably Ohio, it has incorporated erroneous and historically creationist attacks that single out evolutionary theory alone for criticism. This type of critical analysis could be viewed as a violation of the separation of church and state.

2. The account of this last-minute meeting was provided by Mark Ryland and Richard Thompson during a recorded panel discussion at a conference entitled "Science Wars," convened in Washington, D.C., by the American Enterprise Institute for Public Policy Research on October 21, 2005. During a sharp exchange, Ryland and Thompson revealed both the advice received by the board from Discovery and the severe rift that developed between the Discovery Institute and the Thomas More Law Center. The most remarkable aspect of the panel discussion, which also featured Ken Miller, author of the *Biology* textbook at the center of controversy in Dover, was that it took place in the middle of the *Kitzmiller v. Dover* trial.

Chapter 12: Sword and Shield, Shock and Awe

1. From *Hamlet,* by William Shakespeare:
   HAMLET: Do you see yonder cloud that's almost in shape of a camel?
   POLONIUS: By the mass, and 'tis like a camel, indeed.
   HAMLET: Methinks it is like a weasel.

POLONIUS: It is backed like a weasel.

HAMLET: Or like a whale?

POLONIUS: Very like a whale.

2. Richard Dawkins, *The Blind Watchmaker*, Norton, New York, 1986. Also see the Weasel Program at http://home.pacbell.net/s-max/scott/weasel.html.

3. From the sworn affidavit filed in *Kitzmiller v. Dover* by Stephen Meyer, dated June 27, 2005.

Chapter 13: Paleozoic Roadkill, Kentucky Fried Chicken, and Bad Frog Beer

1. Linda Heuman, "The Evolution of Ken Miller," *Brown University Alumni Magazine*, November–December 2005.

2. Gordy Slack, "Intelligent Designer: The Chief Defender of Intelligent Design in the Dover Evolution Trial Insists He Has Science and God on His Side," Salon.com, October 20, 2005.

3. One example Miller seemed to consider almost a personal affront is a passage in *Of Pandas and People* that cites biochemical analyses showing that, on the evolutionary ladder, the bullfrog and the horse are at the same distance from one another as from the carp. This, *Pandas* maintains, is a problem for evolution, because the amphibious frog should be more closely related in its biochemistry to the fish than to the horse, which is so much more dissimilar. But this is a layperson's mistake, an appeal to common sense that shows how little the authors of *Pandas* know about evolution, or suggests that they are deliberately misrepresenting it, Miller said. Evolutionary relationships are best represented as a tree, not a ladder; and carp, frogs, and horses have all been evolving along the branches of this tree for the same amount of time from the point on the tree's trunk where they had a common ancestor. This is counterintuitive rather than commonsensical, but it falls into place with a simple analogy: Imagine three people starting a walk at the same point—call it the common ancestor point. They walk off in three different directions for exactly the same amount of time and distance. They get to three separate, different places—call those places frog, horse, and carp—but all have traveled the same distance to get to their respective positions, and all are the same distance from their point of origin. So rather than disproving evolution, the biochemical data cited by *Pandas* actually confirm it. "But students using *Pandas*," Miller said, "would misunderstand this point completely."

Chapter 14: Of Panders and People

1. In her testimony, Forrest also pointed out six major themes that intelligent design and creationism share—the same "playbook," as she called it. Creationists and proponents of ID both:

- Make the same arguments that naturalistic and "materialistic" theories are insufficient to explain life and the universe, leaving only supernatural explanations.
- Make the same arguments that evolution poses a threat to culture, morality, and society—evolution portrays "human beings not as moral beings but as animals and machines."
- Claim that life made an "abrupt appearance" on Earth and that this implies a supernatural creator.
- Focus on "gaps in the fossil record," with particular complaints (increasingly bogus in light of new discoveries) that the Cambrian Explosion cannot be explained by evolution.
- Assert that only the supernatural can account for biochemical complexities, citing DNA as evidence of a designer at work. (Creationists also cited the bacterial flagellum as evidence of molecular structures that could not evolve naturally—in an article published two years before Michael Behe's book *Darwin's Black Box*.)
- Both advocate a "teach the controversy" approach—presenting alternative theories and the "strengths and weaknesses" of evolutionary theory in public schools.

Chapter 15: Under the Microscope, Deer in the Headlights

1. For example, Behe said that showing how some parts of the flagellum could serve other purposes, such as the "toxic dart" on the plague bacillus, as Miller had done, was insufficient. But to refute Miller, Behe had to modify his original definition of irreducible complexity to say that his theory could be proved wrong only by the discovery of simpler precursors of the flagellum that *served the same purpose*, in this case, propulsion. That last condition was new, and Behe's critics immediately accused him of changing the rules when it appeared he had lost. Furthermore, his restatement of this theory, according to Miller, rendered the whole concept of irreducible complexity meaningless. To take the example of a bird's wing, evolutionary theory dictates that it would have had to evolve originally to serve purposes other than flight, as when flightless dinosaurs brooded their eggs with feathers and arm flaps. Behe's redefinition would make wing evolution as impossible as it made the evolution of the flagellum.

2. Note to the author from Bill Buckingham, April 20, 2006 (excerpt).

Epilogue

1. Dean Hamer, *The God Gene: How Faith Is Hardwired into Our Genes*, Double-day, Garden City, N.Y., 2004.

2. David K. DeWolf, John G. West, Casey Luskin, and Jonathan Witt, *Traipsing into Evolution*, Discovery Institute Press, Seattle, 2006.

3. Here is the passage from Jones's opinion in *Kitzmiller v. Dover*: "We find it incumbent upon the Court to further address an additional issue raised by Plaintiffs, which is whether ID is science. . . . While answering this question compels us to revisit evidence that is entirely complex, if not obtuse, after a six-week trial that spanned twenty-one days and included countless hours of detailed expert witness presentations, the Court is confident that no other tribunal in the United States is in a better position than are we to traipse into this controversial area. Finally, we will offer our conclusion on whether ID is science not just because it is essential to our holding that an Establishment Clause violation has occurred in this case, but also in the hope that it may prevent the obvious waste of judicial and other resources which would be occasioned by a subsequent trial involving the precise question which is before us."

4. Harris Poll 52, July 6, 2005, "Nearly Two-thirds of U.S. Adults Believe Human Beings Were Created by God." Harris Poll 11, February 26, 2003, "The Religious and Other Beliefs of Americans."

INDEX

abiogenesis, 14, 95, 277
abortion issue, 143, 231
Abrams, Steve, 150–52, 154–56, 158–59,
    162–63, 165, 171–72, 174–76
Adams, John, 310
adaptation concept, 116, 118
Afghanistan, 81
agnosticism, 134, 237
Ahmanson, Howard, 70, 139, 222
Ahmanson, Roberta, 139
AIDS, 145, 293
Alabama, 26, 136, 140–41, 243
Alaska, 141
Alberts, Bruce, 190, 242
Alito, Justice, 336
Alters, Brian, 270
ambulocetus, 255
American Association for the
    Advancement of Science, 301
American Civil Liberties Union (ACLU),
    xii, xiv, 23, 49, 51, 53, 55, 73, 100–101,
    104–5, 169, 178, 181–91, 193, 201–2,
    210, 220–21, 228, 282, 287, 328–29
American Museum of Natural History,
    190, 237
Americans United for Separation of
    Church and State, xiv, 62, 101, 193,
    202, 290
"Ancestor, The" (computer program), 271
animal breeding, 119–20, 129
Answers in Genesis, 31, 135, 190, 347
Anti-Defamation League, 348
Anti-Evolution League of America, 50

anti-evolution standards, Kansas and,
    152–57. *See also* evolution
anti-evolution statutes, 51, 53–56
Apollo astronauts, 32
Aquinas, Thomas, 113, 115
archaeopteryx fossil, 114–15, 134
archetypes, 115
Argento, Mike, 81–82
Aristotle, 112–13, 145
Arkansas, 54–57, 135
    Supreme Court, 55
Armageddon, 29
arthropods, 17–18
Associated Press, 280
astrology, 301, 344
Atchison, Michael, 302–3
atheism, 33–34, 66, 105, 134, 137, 155,
    162–63, 169, 175, 202, 206–7, 226, 237,
    248, 252, 286–88, 296
atomic theory, 93, 177, 275
atomists, 111–12, 121
Australia, 121
Australopithecus afarensis, 42, 123, 124
autocatalytic reaction, 205

bacteria, 293, 304–5
bacterial flagellum, 138, 145, 174, 199, 234,
    256, 263–64, 267, 274, 280, 298, 306
Baird, John Logie, 109
Baksa, Mike, 9–10, 12, 14, 83–85, 91–93, 95,
    100, 102–3, 215–16, 221, 324
"balanced treatment" issue, 55–58, 205
Baltimore Sun, 49

Barton, David, 100
Basin and Range (McPhee), 147
bats, 118
Baylor University, 238
Beagle, 92, 117, 121
Behe, Michael, 31, 69–70, 74, 128–29,
        131–33, 138–40, 144–45, 156, 166,
        170–71, 173–74, 190, 199, 233–34, 240,
        245–46, 251, 263–64, 269, 271, 274–76,
        280, 297–306, 309, 335, 342, 344
Berlinski, David, 245, 246
Bernhard-Bubb, Heidi, 316–17
Bible
    inerrancy doctrines and, 48, 114
    as literal, xii–xiii, xviii, 15, 20–23, 27,
        29–32, 36, 38, 52, 70, 133–34, 196, 271,
        322, 347, 348
    as metaphor, 114, 271
big bang theory, 32, 93, 112, 142, 153, 263,
        277, 281, 299, 335
Big Story with John Gibson, The (TV show),
        245
Bill of Rights, 184–87
biochemistry, 70, 122, 128–29, 131, 140, 171,
        250, 269
biodiversity, 92
biogeography, 196
Biola University, 341–42
Biological Sciences Curriculum Study
        (BSCS), 54
Biology (Miller and Levine, "Dragonfly
        Book") 39–44, 61, 78, 83–90, 183, 190,
        206, 237, 263, 266, 277, 308
Biology: The Dynamics of Life (McGraw-
        Hill textbook), 84
bipedalism, 204
birds, 114–15, 121, 126, 129–31, 255–56,
        348
Blind Watchmaker, The (Dawkins), 66, 286
blood-clotting cascade, 70, 274, 280
Bob Jones University, 83
Bonsell, Alan, xv–xvi, xvii, xviii, 9–10,
        12–15, 39–44, 56, 60–61, 87, 89–91,
        93–96, 98–100, 102, 183, 212–13,
        216–19, 226, 249, 294–95, 313–14, 318,
        320, 322–24, 327–29, 333, 335, 350
Bonsell, Donald, 314, 323–24
Booher, Larry, 243–44, 247
Boyd, Ralph F., Jr., 143
Bragg, William Lawrence, 291
brain size, 123–24
Brazil, 117

Bristol, Virginia, 243–44, 246–47
Brown, Carol Honor "Casey," xiv–xviii, 39,
        60–62, 85, 87, 89, 93, 96, 99–101, 212,
        350
Brown, Jeff, xv, xvii, 39, 43, 61, 85–87, 96,
        99–100, 212, 218, 330
Brown, John, 140
Brownback, Sam, 140, 243
Brown University, 265
Brown v. Board of Education, 54, 185
Bryan, William Jennings, 45–53, 60, 66, 73,
        158
bubonic plague, 274
Buckingham, Betty, 311
Buckingham, Charlotte, 38, 61, 322
Buckingham, Walter, 311
Buckingham, William "Bill," xvii, xviii,
        xix, 12, 14–15, 24, 39–44, 56, 60–62,
        76–78, 80–91, 93–96, 99–102, 104,
        164, 206, 184, 208–10, 212–14, 216–20,
        225–26, 248–49, 333, 262, 294, 307–16,
        320, 322–25, 327, 333, 335–36, 350
Buckley, William F., 72, 265–66
Buell, Jon A., 240, 241–42
Burlington-Edison school district, 73
Burr, Richard, 243
Burton, Dan, 145
Bush, George H.W. (Bush I), 192
Bush, George W. (Bush II), 26–27, 83, 142,
        143, 150, 155, 206, 245–46, 258, 345
Bush, Jeb, 231
Butler, John Washington, 44–46
Butler Act (1925), 45–46, 48–49, 51–54

Cable News Network (CNN), 211, 224
California, 26, 136, 206
Callahan, Aralene "Barrie," 42, 90–91, 95,
        97, 203, 294, 295
Callahan, Frederick, 203, 295–96
Calvert, John, 157–59, 161–62, 165–66,
        169–71, 175–76
Cambrian Explosion, 17, 156, 197–99, 237
Campbell, John Angus, 233, 240, 242
carbon dating, 56
Carter, Jimmy, 23
Cashman, James, 222–23
Cashman, Rebecca, 223
Catholics, 29, 35–36, 137, 143, 188, 192, 265,
        294
cell biology, 132, 269
    protein sequences and, 274, 276,
        304–5

Center for the Renewal of Science and Culture (*later* Center for Science and Culture), 69, 75–76, 211

Chapman, Bruce, 69

Charles Darwin's Dangerous Idea (Dennett), 132

Chautauqua circuit, 46, 49

Cheney, Dick, 22

chimpanzees, 198, 254–55, 267–69, 272–73, 326

Christian Coalition, 137

"Christian nation" concept, 44, 159, 169, 195, 294, 310

Christian Rapture books, 24, 30

"Christian Reconstructionism," 70

Christians, "war against," 23, 133, 145, 187–88, 232–33

Christian symbols in public spaces, 231

chromosomes, 268–69

Cicero, 116

Civic Biology (Hunter), 49, 53

Clay Center, 159

Cleaver, Jane, 99, 317, 327

Clinton, Bill, 135, 222

Cobb County, Georgia, 26, 146

coccyx, 19

Cold War, 7

Columbine school shootings, 33–34

Columbus, 48

Comedy Channel, xv

common descent concept, 23, 41, 62, 66, 116, 122, 129–30, 132, 152–53, 164–65, 168–70, 267–68, 272–73, 299, 326, 348–49

communism, 184, 186, 230

Compass International, 23

"complex specified information," 234–35

"Confronting the Judicial War on Faith" conference (2005), 242

Constitutional Restoration Act, proposed, 243, 345

"Contract With America," 150

controversy, as wedge strategy, 211, 233, 245–47. *See also* "teach the controversy" strategy

Cooper, Seth, 76–78, 209–10

Copernicus, 3, 48

Coral Ridge Ministries, 348

Coulter, Ann, 236, 345–50

courts (judiciary), 7, 10, 23, 135–36, 187, 243, 341–42

creation, definition of, in *Pandas*, 285–86

Creation Battles Evolution (Booher compendium), 243

Creation Biology (early draft of *Pandas*), 285. *See also* *Of Pandas and People*

creationism and creation "science." *See also* intelligent design (ID)

  ACLU and early cases on, 188–89

  birth of, 56–57

  claims vs. evolution and, 134–35

  court rulings vs., 25, 57–59, 135–36

  Darwin and, 11

  Dover school board, and push to teach, xv–xvi, xvii, xviii, 9–10, 43–44, 61–62, 77, 81

  Dover school board, and use of term, 81, 183, 217–18, 223, 225–27, 320, 308–10, 313–14, 316–18, 320–22, 324–26

  Dover science faculty opposes teaching of, 9–10

  ID and, 70–71, 134, 140–41, 195, 201, 209, 253, 299–300, 307

  Kansas Board of Education and, 148, 152–54, 157–59, 167–68, 177

  Pandas book linked with, 200, 241–42, 284–86

  parochial schools and, 26

  push to teach, in public schools, 27–28, 56

  scientific arguments vs., 196–200

Creationism's Trojan Horse (Forrest), 75, 286

"Creation MegaConference," 23

Cretaceous period, 148

Crick, Francis, 277

"critical analysis" approach, 210

"Critical Analysis of Evolution" curriculum, 166

Cro-Magnon man, 42

"Culture of Life" campaign, 143

Daily Show, The (TV show), xv

"Dangers of Evolution, The" (DVD), 31

Daniel v. Waters, 56–58

Darrow, Clarence, xii, 49–54, 162

Darwin, Annie, 121

Darwin, Charles, xii, 29, 80, 179, 258

  background and early life of, 116–17

  biology curriculum and textbook controversy on, 40–47, 53, 91–92, 103, 136

  blamed for racism, 34

  Catholics and, 36, 137

Darwin, Charles (*continued*)
  critiques of, 4–5, 8, 22, 66, 75, 129–31,
    140, 230, 233, 236–37, 349–50
  development of theory and *Origin of
    Species* of, 113, 116–27
  fame and influence of, 110–11
  religious beliefs of, 120–21
Darwin, Erasmus, 116
Darwin exhibit (American Museum of
  Natural History), 190
"Darwinism," as term, 153, 252
Darwin on Trial (Johnson), 64, 67, 69, 139
Darwin's Black Box (Behe), 131–32, 139,
  166, 302–4
"Darwin's Deadly Legacy" broadcast, 348
Davis, Cynthia, 206–7
Dawkins, Richard, 66–67, 234–35, 237,
  286–87
Dayton, Tennessee, 48–50, 82
death penalty, 143
death threats, 144, 337, 340
Declaration of Independence, 310
DeFever, Val, 154–55
DeHart, Roger, 72–75, 77–78, 169, 173
deists, 3
Delay, Tom, 23–24, 143–44
Delone Catholic School of Eastern York, 84
Dembski, William, 190, 233–42, 247–48,
  270, 288–92, 300, 344–46
Democratic Party, 23, 149, 249, 347
Democritus, 111
Dennett, Daniel C., 132
Denton, Michael, 66, 129–31, 138
"descent with modification," 119, 123, 219
Design of Life, The (new edition of *Pandas*),
  239–40, 300. See also *Of Pandas and
  People*
developmental biology, 246
Dini, Michael, 142–43
"Dino, Dr." *See* Hovind, Ken
Dinosaur Adventureland Park, 31, 34
dinosaurs, 152, 188, 196, 204
  fossils, 17, 18–19, 114–15, 254
  -to-bird evolution, 126, 131, 134, 190,
    255–56, 348
"Dinosaurs in the Bible" (DVD lecture), 31
Discovery Institute, xiv–xv, xvii, 12, 140,
  205–6, 222, 242
  background and wedge strategy of,
    63–76, 84, 152
  Design of Life edition of *Pandas* and,
    239–40

Dover board and, 61–62, 76–78, 81, 101,
    189, 209–11, 224–25, 233–34, 238–41,
    244–46, 252, 270
  Kansas and, 155–58, 161, 163–64, 166,
    169–70, 173, 178
  NCSE vs., 190, 198–200
  polarization and accusations by, 341–43,
    345–46
  "ten questions" of, 155–56
  Traipsing book on trial and, 342–45
  trial and, 283–84, 288, 290, 298–99,
    304–5, 314, 325, 335–36
disease organisms, resistance and, 14
DNA, 27, 157, 162, 197–99, 234, 346. *See
  also* humans and, 122–23, 165,
  267–68, 339
dolphins, 123–24, 274
Domino's Pizza, xv, 230
Dover Area School Board, xii, xiii–xiv. See
  also *Kitzmiller et. al v. Dover Area
  School District;* and specific members
  and issues
  annual retreat of 2002 and evolution
    controversy origin, xv–xvii, 9
  audiotapes erased by, 81, 213–14, 316
  biology curriculum including ID voted
    in by, xvii–xviii, 91–102
  Christian evangelical movement and, 24
  conduct of, and trial rulings, 194, 333–34
  decision on creationism vs. evolution
    and pressure on science teachers, 9–12
  demand that members be "born again,"
    xvii
  depositions of, 216–20, 225–27
  Discovery Institute and, 62, 76–78, 81,
    169–70, 210–11, 240–41
  elections of November 2005 and Dover
    C.A.R.E.S. slate voted in, 328–31,
    350–51
  elections of spring 2005 and, 248–52
  evolution as conspiracy theory and, 133
  ID statement proposed by, 102–4, 215
  lawsuit and, 89–90, 104–6, 181–93
  lawsuit as test case for ID and, 78, 101–2,
    229–30
  lawsuit filing response of, 212–15
  meetings of June 2004, 43–44, 59–62,
    308
  meetings of July 2004, 84–85
  meetings of August 2004, 87–89
  meetings of October 18, 2004, 91, 213,
    333

news coverage of, 79–86, 100–101
newsletter on ID, 251–53, 294
Pandas book and, 78, 84–91, 95, 97, 101
press release by, on ID, 102–4
religious intent and, 42–44, 223–25
science teachers oppose teaching creationism or ID and, 12–14
test case for ID, 101–2
textbook controversy and, 14, 39–44, 81–91, 95, 97, 101
trial testimony and, 295–96, 307–27
Dover Assembly of God Church, 80, 223
Dover Citizens Actively Reviewing Educational/Economic Strategies (Dover C.A.R.E.S.), 248–49, 251, 328–31
Dover school controversy, xi–xiii, xvii–xix. *See also Kitzmiller et. al v. Dover Area School District;* and specific issues, organizations, and people
development of, xv–xvii, 12–14, 39–44
history of schools and, 5–9
ID and evolution statement read to students, 102–4, 195, 202, 214–17, 220–21, 223, 252–53, 262, 270, 274–76, 325, 337
mission statement of, 13
polarization and, xiv, xix, 29, 80–83, 350–51
Dover science faculty, 87, 202
board meetings with, 12–15
board vote on curriculum change and, 94–105
DVDs sent to, 73
evolution not taught by, 293–94
opposes teaching of ID, xviii
pressured to teach creationism, 8–12
refuses to read ID statement, 214–16, 221
textbooks controversy and, 39–44, 61–62
trial and, 262, 295
"Dragonfly Book." See *Biology* (Miller and Levine)

Earth. *See also* young-Earth creationists
age of, 3, 23, 114, 134, 142, 145, 153, 156, 162, 166–68, 170
as globe orbiting around sun, 8, 93, 134
Eckhardt, Robert, 204–5
education levels, 28–29

Edwards case, 44, 57–60, 67–68, 71, 76, 135–36, 153, 200, 229, 232, 261, 284–85, 326, 332
Einstein, Albert, 110, 277, 291
Eisenhower, Dwight D., 6
Elsberry, Wes, 238
Eldridge, Niles, 190
elections of 2004, 27
embryology, 231, 164, 346
Emig family, 6
end of the world, 20–21, 29–30
Enlightenment, 3, 263
"Enlisting Science to Find the Fingerprints of a Creator" (article), 72
Eocene era, 255
Epicurus, 111
Epperson case, 44, 54–56
Eschbach, Robert, 215, 218
Eschbach, Warren, 218
establishment clause, 56, 194, 229, 331–32, 334, 336
Ethics and Religious Liberty Commission, 336
Eusthenopteron, 256
evangelicals, 9, 20, 22, 24, 27, 76, 84, 206, 232–33
Eveland, Beth, 203
evolution and evolutionary theory. *See also* creationism; Dover Area School Board; intelligent design; *Kitzmiller et al. v. Dover Area School District*
attack on, nationwide, 7–8, 25–27, 53–56
belief in God and, 347 (*see also* atheism)
Biblical literalists vs., 20–25
biology curriculum and, 91–92
Catholic Church and, 137
conspiracy theory and, 133–34
"controversy" in scientific community argument, 204–5
controversy over teaching of, 137–38
Coulter's attack on, 345–50
Darwin's theory of, defined, 118–27
Discovery Institute's attack on, 63–76, 145
discredited claims vs., 134–35
Dover board's lack of understanding of, xix
Dover teaching on, before pressure begins, 10–12
early controversy over, 4–5
"five fundamentals" and, 48

evolution and evolutionary theory
(*continued*)
fossil record and, 16–19, 23, 32–33, 42,
48, 56–57, 113–15, 123, 126, 133–34,
139–40, 147–49, 197–99, 204, 255–56,
281, 340–41
"gaps" in theory of, 26, 71, 92–93, 98,
102–3, 140, 183, 198, 276
genetics and, 19, 119, 122, 152, 165, 171,
197–99, 234–35, 267–69, 272–74,
346
history of, in science, 111–27
ID challenge to, 128–33, 138–45, 156–57,
209
Jones's ruling and, 332–33
Kansas hearings of May 2005 and,
159–78
Kansas school board controversy over,
146–60
Miller's testimony on, as tested and
verified, 264–69
NCSE defense of, 190–92, 195–200
origin of species issue and, 15
public opinion on, xiii, 27, 29, 140, 149,
186–87, 211–12, 247, 348
Republican Party and, 142–45
scientific community support of, 27,
204–5
Scopes trial and, 44–54
"Steeling the Mind" conference against,
22–24, 27–34
tautology issue, 347–48
"teach the controversy" strategy, xiv–xv,
xix, 63, 71–72, 74, 77, 166, 172–73, 211,
247, 345, 350
term, derivation of, 29, 112
testing of, 346–47
evolutionary biologists, 18, 33, 66, 84, 123
evolutionary scientists, xiv, 73–76. *See also*
specific disciplines and people
Kansas hearings boycotted by, 159–60,
176–77
Evolution: A Theory in Crisis (Denton), 66,
129–31
"Evolution is a theory, not a fact" sticker,
146
"Evolution Is Stupid" presentation, 134
extinction, 68, 114
eye, 19, 124–25, 130–32

Fall, 21
falsification, 93, 273

Farrell, Steve, 207, 223
fascism, 47, 73, 134
"faunal succession" principle, 113
Fenimore, Deborah, 202
finches, 14, 118–19, 126
Finding Darwin's God (Miller), 265
"fine-tuning" argument, 163, 244
Firing Line (TV show), 72, 266, 290
First Amendment, 26, 186, 194, 222, 229,
261, 296
fish, 121, 124, 255, 256, 304
"five fundamentals," 48, 151
Flew, Anthony, 342
flight, 125–26, 255, 256
flood and flood geology, 21, 23, 56–57, 133,
141, 152
Florida, 206, 231
Food and Drug Administration, 245
Forrest, Barbara, 284–91, 294, 297,
299–300, 326, 331–32
Fortas, Abe, 55
fossil record, 17–19, 281
Behe and, 139–40
creationists on, 32–33, 56–57
flight and, 126
flood geology and, 23, 133–34
human evolution and, 42, 48, 123, 204
Kansas and, 147–49
new genetic information issue, 197–99
Smith and, 113–15
transitional, 255–56, 340–41
Foundation for Thought and Ethics
(*Pandas* publisher), 200, 239–41, 310
Founding Fathers, 3, 44, 100, 187, 231
Fox 43 news video (June 2004), 309–10,
313–15
Fox News, 245, 337
Franklin, Benjamin, 3, 270, 273
Frazier, Gary, 30
freedom of press, 186
freedom of public assembly, 186
freedom of religion, 186–88, 193
freedom of speech, 186
freedom to raise grievances, 186–87
Freud, Sigmund, 75
Frist, Dr. Bill, 144
Fuller, Steve, 306–7, 344
fundamentalists, 38, 48, 55, 133–35, 137–38,
143, 232–33, 343, 349

Galápagos archipelago, 14, 45, 117–19
Galileo, 8, 48

"gaps" in evolution theory, 26, 71, 92–93, 98, 140, 183, 198, 276

Geesey, Heather, 61, 81, 96–97, 99, 101, 318–23, 327

genes and genetics, 19, 119, 122, 152, 165, 171, 267–69, 272–73, 346
   new information issue, 197–99, 234–35, 274

Genesis, xii, 13, 15, 21, 27, 32, 38, 45, 52, 56, 134, 194, 208, 219, 232, 307

Genesis Flood, The (Morris), 133–34, 153, 265

genocide, 188, 348

geocentrism, 5

Geological Society of London, 114

geology, 3, 16–19, 46, 56, 113–15, 117, 133, 147–48, 151–53. *See also* flood and flood geology

Germany, 47

germ theory, 10, 275, 347

giant sloths, 118

Gibson, Mel, 142

Gideon case, 185

Gilder, George, 69

Gillen, Pat, 217, 262–64, 315, 322–25, 327

Gitlow case, 185

global warming, 30, 144

God, belief in, and evolution, 347. *See also* atheism; religion

God Delusion, The (Dawkins), 237

"God gene," 339

Godless (Coulter), 236, 345–48

Gonzalez, Guillermo, 244

Gonzalez-Bravo, Jill, 172–74

Gore, Al, 23

Gould, Gordon, 109

Gould, Stephen Jay, 68–69, 198, 281

Grabiner, Judith, 53

Grand Canyon, 141, 153, 243

Grantsburg, Wisconsin, school board, 206

Great Plains, 147

Greeks, ancient, 93, 111–12

Grove, Rev. Jim, xi–xiii, xvi, 205, 277–83

Gurreri, Larry, 330

Hamer, Dean, 339

Hamlet, 234

Harkins, Sheila, 85–86, 93, 97, 99, 212, 216–17, 220, 327, 329

Harmony Grove Community Church, 38

Harris, William H., 161, 163

Harrisburg conference on evolution and creationism (2004), 84

Harris Poll, 347

Harvey, Steve, 192–93, 195, 217, 296, 306–10, 313–15, 323

hate mail, 248

Haught, John, 271, 272

Hays, Kansas, 147

Heinz, John, 259

heliocentric solar system, 263

Hell and the High Schools (Martin), 50

Henry VIII, King of England, 230

hippocampus, 115

Hippocrates, 270, 273

"historical" science, 151

Hitler, Adolf, 34, 111, 348–49

Home Savings banks, 70

hominids, ancient, 123–24, 164, 168, 204

Homo erectus, 42, 123

Homo habilis, 123, 124

Homo sapiens, 123, 198

Hovind, Kent ("Dr. Dino"), xi, 24, 31–34, 47, 145, 203–4, 279, 281–82, 346

humanism, 20–23, 25, 232, 172, 287

humans, 4–5
   brain, 19, 115
   creation of, in Genesis, 21, 23
   descent of, 13–15, 41–42, 123–24, 164, 168, 204
   genome of, 19, 165, 254–55, 267–69, 272–73, 326

Humburg, Burt, 283

Hunter biology textbook, 49

hypothesis, defined, 11

Icons of Evolution (DVD, video), 12, 73, 77

Icons of Evolution (Wells), 164, 166, 172

IMAX theaters, 26

immune system, 306, 347

information theory, 234

Inherit the Wind (film), 53, 135–36

Inhofe, James, 144

Inquisition, 8

Institute for Creation Research (Creationist U), 27, 30, 133–36, 141

intelligent design (ID), 26, 30–31, 59. *See also* creationism
   Behe's definition and defense of, 298–307
   creationism and, 201
   defeats for, after *Kitzmiller*, 350

intelligent design (ID) (*continued*)
 Discovery Institute leads movement for,
  63–76, 210–11
 Dover board includes in curriculum,
  xi–xiv, xvii–xviii, 12–14, 39–44, 61,
  93–102, 181–85, 189–93, 214–17, 219
 Dover school district newsletter on,
  251–53
 Dover school statement on, 102–4
 expert witnesses for, 232–42
 Idaho and, 243
 Jones's ruling and implications for,
  331–38, 345
 Kansas and, 155–57, 161–78
 Kitzmiller and, 181–210, 232–42,
  298–307, 331–38, 345
 Matzke as opponent of, 196–200
 Paley as intellectual progenitor of, 115
 Pandas text and, 239, 72–73, 78, 85–91,
  95, 97, 101–3, 107, 199–200, 209–10,
  215–16, 219, 227, 231, 239–42, 255, 257,
  274, 284–87, 290, 295, 299–300,
  314–15, 318–19, 323–26, 331–32, 333,
  344
 polarization and, 341–42
 publicity campaign and, 72
 religion vs. science issue and creationism
  and, xviii, 84, 129–35, 138–45, 194–95,
  219, 223–27, 262–63, 265, 271, 284–89,
  290–92, 299–304, 325, 326, 331, 340,
  344
 research and peer-reviewed articles
  lacked by, 74
 Thomas More Center and test case for,
  xviii, 78
Intelligent Design Network, 157–58, 161
"Intelligent Design Under Fire" event, 342
Internet, 196
Iowa, 141
Iraq Survey Group, 233
Irigonegaray, Pedro, 159–60, 162, 165–68,
  170, 174–75
"irreducible complexity" concept, 69–70,
  74, 138–40, 145, 233, 251, 263–64, 271,
  274, 276, 302, 304–7, 331, 335
IRS, 34
Ito, Lance, 338

James, Forrest Hood "Fob," 64
Jefferson, Thomas, 3
Jenner, Edward, 109
Jesus, 48, 348

Jews, 29, 70, 188, 192
John, gospel of, 71, 194, 289, 290
John Paul II, Pope, 137, 143
John S. Battle High School, 243–44
Johnson, Phillip E., 64–73, 139–40, 144–45,
  156, 158, 178, 179, 210–11, 233, 288, 325
Jonah, 52
Jones, Beth Ann, 260, 290
Jones, Jeri Lee, 16–20, 22
Jones, John E., III, 258–59, 262, 264–65,
  268, 272, 275, 278, 284–87, 293–95,
  299–300, 305, 307, 309, 315, 317,
  323–24, 327, 329–34, 336–38, 340–45,
  349–50
Jones, Meghan, 338
Jones, Paula, 222
Joshua, 134
Jurassic Park (film), 255
Jurassic period, 16
Justice Department, 26, 142–44, 192
"Justice Sunday," 23

Kansas, 72, 140–42, 145–60, 197, 206,
  209–10, 332, 350
Kansas Board of Education, 146, 148–56,
  350
 hearings of May 2005, 158–78, 206, 242,
  304–5
 "Majority Report," 156–58, 162, 168–69,
  174
 "Minority Report," 156–58, 161, 163, 168,
  170, 177
Kansas Citizens for Science, 151, 159,
  162–63, 165–66
Kansas Republican Assembly, 155
Kansas: The Priarie Spirit Lives (textbook),
  148
Katskee, Richard, 193
Keller, Helen, 46
Kennedy, D. James, 348–49
Kentucky Mountain Bible College, 154
Kenyon, Dean, 284
Kevorkian, Dr. Jack, 231
Kilby, Jack, 109
Kitzmiller, Megan, 104, 294
Kitzmiller, Tammy, 6, 104–5, 192, 203, 222,
  248, 250–52, 280, 294
*Kitzmiller et al. v. Dover Area School
  District*, xii, xviii, xiv–xv, 44, 222, 239,
  254–327, 331–38, 340–43
 ACLU develops case, 104–6, 181–86,
  189–200

aftermath of, 340–45, 350–51
announced, and public reactions, 201–10
board's response to, 212–14
closing arguments of trial, 324–28
Coulter on, 345–50
decision against TRO, 219–21
defense case built, 228–42
defense case presented, 269–73
depositions of ID experts and, 240–42
depositions of school board members
    and, 216–20, 225–27, 320–21, 323–26
Discovery Institute and, 209–11, 342–45
ID expert testimony, 269–70, 297–307
judge chosen for, 258–61
judge's ruling on, 331–38, 340–45,
    349–50
Matzke and scientific issues in, 195–200
news media and, 279–81
opening arguments and, 261–63
Panda publisher and, 241–42
parents petition to intervene in, 222–23
parents vs. school issue in trial and,
    292–96
public opinion and, 247–48
religious intent issue, 223–25
reporters testimony and, 316–18
residents' reaction to, 211–12
school board testimony and, 307–15,
    318–24
scientific expert testimony and, 254–58,
    261–78, 283–89
spring 2005 board elections and, 248–53
test case for DI, 178, 209
Krebs, Jack, 159, 163, 165, 168

Laboratory of Comparative Morphology
    and Mechanics (Penn State), 204
LaHaye, Tim, 24, 30
Lamb and Lion Ministries, 20
Lancaster, California, school district, 26
Land, Richard, 336
Larson, Edward, 84
"Left Behind" seminars, 30
"Left Behind" series (LaHaye), 24
"legislative intent" issue, 194. *See also*
    religious intent issue
Lehigh University, 301
Leib, Joel, 202
Leonard, Bryan, 166–68
Levine, Joe, 39
Lewontin, Richard, 190
Liberal Christians, 70

liberals, 23, 347
"Lies in the Textbooks" (DVD lecture), 31
lightning, 263, 270
Limbaugh, Rush, 40, 292
Linker, Robert, 11–12
Little Conewego Creek, 17
Little Rock Central High School, 54–55
Los Angeles County, 231
Los Angeles Times, 72
Loser, Rev. Michael, 79–80
Louisiana, 57, 141
Louisiana Balanced Treatment Act,
    57–58
Louisville, Kentucky, schools, 141
Lucretius, 112
Luther, Martin, 349
Lutherans, 5–6, 260, 330
Lyell, Charles, 5, 114, 117
Lynn, Rev. Barry, 202, 290

Machiavelli, 291
macroevolution, 132, 140, 161–62
Maldonado, Joseph, 316–17
Malone, Dudley Field, 51–52
Man for All Seasons, A (film), 230
marsupials, 121
Martin, Kathy, 159, 164, 168–69, 176
Martin, T. T., 50
Marx, Karl, 75, 111, 230
materialist naturalism, 66, 73, 137, 230, 232,
    238
mathematicians, 190
Matzke, Nick, 195–200, 201, 211, 238, 241,
    270–73, 284–86, 302, 335
McLean v. Arkansas, 44, 57, 135, 260
McPhee, John, 147
medicine, 26, 27
"Menace of Darwinism, The" (Bryan), 46
Mencken, H. L., 49, 52, 81–82
Mendel, Gregor, 122
"methodological naturalism," 156, 264, 270,
    344. *See also* scientific method
metoposaur fossils, 17
Meyer, Stephen C., 67–70, 170, 190,
    197–99, 209, 233, 240–42, 245, 290,
    342
"Meyer's Hopeless Monster" (Matzke
    internet posting), 198
Michigan, 90, 206, 231
microbiology, 346
microevolution, 14, 129, 140, 152
Miller, Jen, 11, 102

Miller, Kenneth, 39–40, 85, 107, 190, 237–38, 263–70, 272, 274–78, 280, 290, 294, 298–99, 308, 326
Miller, Peter, 53
Miller, Stanley, 164
Minnich, Scott, 233, 240, 256–57, 306, 344
Miranda case, 185
"missing day" concept, 134
Missouri, 26, 141, 206–7, 243
"mock the monkeys" case, 188–89
Modern Biology (Holt, Rinehart, and Winston textbook), 84
molecular biology, 122, 164–65, 269
Monaghan, Thomas, 230
"monkey dance" man, 248
"monkey girl" branding, 183, 222, 351
moon, 32–33, 134
Moon, Sun Myung, 165
morality issue, 28, 33, 48, 73, 341
More, St. Thomas, 230
Morris, Connie, 158–59, 162, 164, 174, 176
Morris, Henry, 133–36, 153, 265
mosasaurs, 147, 148
Mountain Grove Chapel, 330
mouse genome, 339
Muise, Robert, 275–77, 299
Muller, Hermann J., 54
"Mustard Seeds" (Atchison), 302–3
Myers, P. Z., 245–46
Myth of Separation, The (Barton), 100

Napierski, David, xviii–xix, 331
Napolitano, Andrew, 337
NASA, 134, 245
National Academy of Sciences, 74, 150, 153–54, 172, 242, 300–301
National Association of Biology Teachers, 137–38
National Center for Science Education (NCSE), xiii, 169, 177, 190–92, 195–200, 201, 237–38, 272–73, 301
National Geographic, 226
National Institutes of Health, 128
National Intelligent Design Movement, 175
National Merit scholars, 153
National Public Radio, 154
National Science Foundation, 54
National Science Teachers Association, 28, 242
naturalism, 162

natural selection, 14, 40, 198, 339
   applied to human society, 47
   Behe's critique of, and mousetrap analogy, 171, 263–64
   blood-clotting cascade and, 70
   Dawkins on, 66
   Denton's critique of, 129
   developed and defined by Darwin, 118–27
   Dover schools and, 80
   Kansas and, 152
   nonrandom nature of, 29, 130–31
   pseudogenes and, 268
   "tautology" argument and, 347–48
Natural Theology (Paley), 115–16
Nature, 74, 267
"nature's contrivance" concept, 116
Nazis, 34, 184–85, 348–49
Neal, Brad, 100
Nelson, Paul, 288–89
neo-Darwinian evolutionary theory, 122, 199
news media, 162–63, 166, 181, 201, 204–5, 207, 212–15, 218, 220, 245–47, 251, 261, 272, 277, 279–80, 316–18
new species issue, 152, 234, 274. *See also* genetics, new information issue
New Testament, 21, 36, 71
Newton, Isaac, 10, 110
New York, 243
New York Times, 154, 161, 206, 207, 211, 245, 246
Nightline (TV show), 279
Nilsen, Richard, 86–88, 91, 215–16, 219–20
Niobrara Chalk, 147–49, 154
Noah, 23, 57, 133, 141, 149, 152
No Child Left Behind Act (NCLB), 140
"Nuremberg files" case, 231
Nuremberg war crimes trials, 185
Nye, Bill, 190

Ocean Hills Community Church, 24
Of Pandas and People (intelligent design textbook), 72–73, 78, 85–91, 95, 97, 101–3, 107, 209–10, 215–16, 219, 227, 231, 255, 257, 274, 299–300, 318–19, 344
   Dembski and new edition *Design of Life*, 239–41, 325
   donation of, 314–15, 323–24, 333
   early drafts and creationism in, 199–200, 241–42, 284–87, 290, 295, 325–26

Jones's ruling on, 331–32
  publisher of, asks to intervene, 241–42
Ohio, 72, 146, 166, 197, 210, 332, 350
Oklahoma, 141, 206
Old Testament, 21, 36, 134
"On the Jews and Their Lies" (Luther), 349
On the Nature of Things (Lucretius), 112
open-meeting laws, 226
Oregon, 141
O'Reilly, Bill, 211, 224–25, 337
"organizing energy" concept, 115
origin of life
  confused with origins of species, 15, 94,
    95, 250, 326
  Dover science faculty vs. Bonsell and,
    14–15
  Hovind on, 32
  Kansas and, 163–64
  Pandas and, 97
origin of species, 15, 94–95, 152, 250, 326
Origin of Species, The (Darwin), 5, 29, 47,
  54, 111, 118–27, 152, 233, 258
Overton, William R., 260, 261
oviraptor, 255
Owen, Richard, 114, 115
OxyContin, 38–40, 248, 308, 311–12, 325

Padian, Kevin, 110, 190, 255–56, 270, 281,
  283–84
Pajaro Dunes conference of 1993, 69–70, 139
paleoanthropologists, 203
paleomagnetism, 17, 19
paleontology, 3, 17–18, 33, 68, 110, 148, 190,
  255, 341
paleozoic era, 17–18
Paley, William, 115–16, 118, 127, 138, 145,
  298
PandasThumb.org (website), 198–99, 335–36
Pangaea, 16
panspermia, 277
parasites and hosts, 121
parochial schools, 83–84
Passion of the Christ, The (film), 142
PBS, 72
Pell, Max, 43
Penn, William, 5–6
Pennock, Robert, 237, 238, 270–72, 280, 281
Pennsylvania, 146
  geological history of, 16
  "Little Bible belt," 8, 9
  Republicans, old school, 259
  state education standards, 10, 14

Pennsylvania state legislature, 6, 91–92
People magazine, 226
peppered moths, 164, 172–74
Pepper Hamilton law firm, xv, 191–93,
  336–37
pesticide-resistance insects, 14, 129
Pharyngula.org, 246
physicists, 33
phytosaur fossils, 17
Pittsburg schools, 188
Plano, Texas, 142
plate tectonics theory, 93, 177
Plato, 3, 112
Poland, 185–86
Pontifical Academy of Sciences, 137
"prebiotic soup" concept, 163–64
Privileged Planet (film), 244
*Proceedings of the Biological Society of
  Washington*, 197
prokaryote, 304–5
pseudogenes, 267–68, 273
public opinion, xiii, 27, 140, 149, 186–87,
  211–12, 247, 348
public school law (1834), 6
pythons, 118

Quakers, 5–6
quantum theory, 93

racism, 34, 47, 73
radioactive decay, 142
radiometric dating techniques, 30
random mutation concept, 119, 121,
  130–31
Raup, David, 68
Reagan, Dave, 20–23, 25, 31
Reagan, Ronald, 69
"Reclaiming America for Christ
  Conference" (Florida, 1999), 71
RedStateRabble blog, 176
Reed, Ralph, 137
Reformation, 349
Rehm, Bryan, 9–15, 82, 95, 97–98, 105–6,
  202, 213, 248, 252, 262, 295, 329
Rehm, Christy, 8–9, 12–15, 82–83, 97–98,
  105–6, 202, 248, 252, 295, 337
relativity theory, 93, 177
religion
  belief in scientific and constitutional
    principles seen as hostile to, 341
  defined, vs. science, 271, 340, 344
  reconciliation of science and, 265–66

religious intent issue, xviii, 11–12, 23–24, 42–44, 69–70, 192, 194–95, 208–9, 209, 218, 223–25, 225–30, 257, 262, 334–35, 336

religious right, 7, 27, 137, 330, 336

Rendell, Ed, 259

Republican Party, 34, 135, 140–41, 143–45, 149, 153, 155, 243, 249, 259, 329, 341, 347, 350

Revolutionary War, 5

Riddle, Eric, 97, 213

Riddle, Mike, 31

Ridge, Tom, 259, 345

Riedel, 216

"right to receive information" issue, 222–23

Roberts, Justice John G., 336

Robertson, Rev. Pat, 330

Robert's Rules of Order, 98

rock strata "laid down quickly," 152–53

Roe v. Wade, 23, 185, 231

Rohler family, 6

Romans, 112, 116

Rothschild, Eric, 191–93, 200, 217–19, 225–27, 241–42, 261–62, 288–89, 297–306, 309, 313, 317–18, 324–27, 342

Rove, Karl, 336

Rowand, Ed, 213, 329

Rowand, John, 223

Rowand, Joshua, 88

Rupp, Matthew, 80

Ruse, Michael, 190

Russell, Bertrand, 179

Russell, Stephen, 89–90

Rutherford Institute, 222–23

Ryland, Mark, 210

Sahelanthropus tchadensis, 123

Salon (web magazine), 206, 273

Salvation Army, 142

San Diego, 27, 137

San Francisco Chronicle, 211, 223

San Jose Mercury News, 64

Santorum, Rick, 140, 251

Satan, evolutionists as servants of, 20–22, 48, 133, 141, 144, 243, 282

Scalia, Antonin, 58–61, 76, 95, 229, 335

Schiavo, Terri, 143–44, 231, 242

school prayer, xvi, 55, 187–88

science
    assault on, 341–42
    Bush administration and, 245
    defined, vs. religion, 340, 344

defined at trial, 194–96, 264–65, 306

ID and, 194–96

meaning of, altered by Kansas Board, 148–49, 156

natural vs. supernatural phenomena and, 67, 266–67, 273

NCSE and defense of, 190

replicable research and, 74

real vs. junk, 11

as star of *Kitzmiller* trial, 254–58, 263–69

Science, 139

science, philosophers of, 190

science blogs, 245–246

science education, Sputnik and, 6–7

science teachers, pressured to avoid evolution, 26, 242

Scientific American, 69, 242

scientific method, 156, 257, 264–65, 270–71, 344

scientific revolution, 3

scientists, 27–28
    biblical literalists oppose, 23
    dissenting 400, 170
    vilification of, 144

scientology, 346

Scopes, John T., 45, 49–53, 73

Scopes Monkey Trial (1925), xi, xii, xv, 7, 44, 49–53, 55, 59–60, 62, 80, 82, 84, 136, 158, 185, 193, 258

Scott, Eugenie, 176–77, 191–92, 195, 237, 273

second coming, 20, 48

separation of church and state, xvi, 3, 20, 23, 43, 52, 58, 60–61, 95, 100, 187–88, 192, 229, 247, 308, 310, 336, 346

September 11, 2001, 207

700 Club (TV show), 330

Shakespeare, 234–35

Shallit, Jeffrey, 234

sharks, 147

Shelby, Richard, 243

Short, Ron, 328

"Signs of Intelligence" (Dembski), 289

Simpson, O. J., 338

Slack, Gordy, 273

Smith, Julie, 203, 293–94

Smith, Katherine, 293

Smith, William, 113–14

Smithsonian Institute, 197

Smithsonian Museum of Natural History, 244

Smoky Hills, 147
Sneath, Cynthia, 202–3
Snoke, David, 171, 304
social Darwinism, 47, 349
socialism, 134
Solidarity movement, 185–86
soul, 4–5, 137
South Carolina, 206
South Dakota, 231
Soviet Union, 6
Spahr, Bertha "Bert," 8, 9, 41–42, 85, 90, 96–97, 184, 227, 295
"specified complexity," 234–35. *See also* "irreducible complexity"
Springsteen, Bruce, 185
Sproull, David, 223
Sproull, Yvette, 80
Sputnik, 6, 135
"staged drawings" controversy, 173–74
Stalin, Joseph, 34
standards and standardized tests, 148, 150–51, 156–57
Star Trek (TV show), 245
"Steeling the Mind" conferences, 22–24, 27–34
stem-cell research, 143, 245
Sternberg Museum of Natural History, 147–48
Stock and Leader, 96
Stough, Steven, 104, 203, 250, 252
Sullivan, Anne, 46
Summer for the Gods (Larson), 84
supernatural vs. natural causes, 67, 157, 263–64, 266–67, 270, 273, 277, 288–89, 291, 299, 306, 325–26, 331, 339, 343–44
survival of the fittest, 119, 347–49
Swatara State Park, 17–18
symbiosis, 121

Talbot, Margaret, 260
tautology, 347
"teach the controversy" strategy, xiv–xv, xix, 63, 71–72, 74, 77, 166, 172–73, 211, 247, 345, 350
teeth, 121
teleological argument, 113
Temple of Nature (Erasmus Darwin), 116–17
Ten Commandments, 23, 242–43, 271
Tennessee, xii, 45–46, 52–54, 56, 206
Tennessee Supreme Court, 52–53

termites, 121
Tesla, Nikola, 109
Texas, 26, 136, 141, 206
Texas Tech University, 142–43
textbooks
    Alabama and, 140–41
    California and, 136
    Dover schools and *Pandas* to supplement *Dragonfly*, 14, 78, 83–91, 95, 97, 183
    evolution in, 7, 53–56, 136
    Kansas and, 148
    Missouri and, 206–7
    Texas and, 136
Thaxton, Charles, 163–64
theistic evolution, 54, 175, 192, 277
theistic science, 270
theocracy, 20, 144
theory, scientific, definition of, 11, 149, 270, 275–76, 300–301
thermodynamics
    first law of, 281, 331
    second law of, 11, 32, 33
Thomas Fordham Foundation, 141, 146, 153
Thomas More Law Center, xv, xvii, xviii, 78, 81, 84, 89–90, 95, 187–88, 208, 210, 214, 217, 220–21, 223–24, 230–32, 237, 240–42, 249, 251–53, 262, 269, 272–73, 275, 283–84, 286, 288, 298, 307, 314, 316, 327, 329, 334–35
Thompson, Richard, 78, 84, 89, 187–88, 208, 210, 214–15, 220–21, 224–25, 228–332, 240, 270–73, 277, 284, 287–90, 329–30, 334, 336
Three Mile Island, 191–92
Tiktaalik rosae, 340
Traipsing into Evolution (Discovery Institute Press), 342–45
Treaty of Tripoli (1796), 310
Triassic period, 16, 20
trilobite, 17–19, 147
TRO (temporary restraining order) decision, 181–83, 195, 217, 219–21, 232, 333
Twain, Mark, 107
Tylosaurus, 147

UFOs, 29, 31, 277
UncommonDescent blog and website, 236, 345
Unification Church, 165
United Kingdom, 47

U.S. Congress, 27, 140, 345
U.S. Constitution, 23, 44, 184–85
U.S. House Government Reform
   Committee, 145
U.S. Park Service, 153, 243
U.S. Senate, 140, 144, 310
U.S. Sixth Circuit Court of Appeals, 56
U.S. Supreme Court, xvi, xviii, 25, 51, 53,
   55, 57–60, 67–68, 71, 84, 95, 100, 135–36,
   146, 178, 188, 200, 210, 229, 231,
   260–61, 284–85, 329–30, 332, 334–35
universe, 21, 23, 142, 156
University of California, 206
University of Kansas, 153
USA Today, 242
Utopia (More), 230

vaccinations, 144–45
Van Meter, Iris, 155
Vardiman, Larry, 30
Vatican, 137, 175
Victoria, Queen of England, 114
virgin birth, 48
Virginia, education standards, 26
"Vise Strategy, The" (Internet posting),
   236–37
Vista, California, school board, 137

Walczak, Witold "Vic, 181–91, 201–2, 217,
   220–21, 264, 266, 280, 318–20, 328
"War against Christians, The" conference,
   23
Warren, Earl, 55, 65
"Was Darwin Wrong?" (article), 226
Washington, George, 3
Washington County, Virginia, 244
Washington Post, 206, 336

"watchmaker analogy," 115–16, 118,
   138–39, 298
Waters case, 44
"Wedge Document: So What?, The"
   (website article), 75
wedge strategy, 63–64, 152, 156, 201, 211,
   225, 238, 286, 288, 325
   document, 75–76, 288–89
Wedgwood, Josiah, 116
Wells, H. G., 49
Wells, Jonathan, 164–65, 166
Wenrich, Noel, 98–100, 212, 226
West, John G., 161, 211, 336
West Virginia, 90
"What Is Science?" essay assignment, 11
"Why Evolution Is Stupid" (DVD), xi
   279–82
Wilson, Woodrow, 45
Wired, 206
Wisconsin, 26
World War I, 47, 50
World War II, 35

Yagodich, Danielle, 221
Yingling, Angie, 6, 88–89, 99, 205–6, 208,
   212, 222, 225–27
York Catholic School, 84
York College, 208, 223
York County, Pennsylvania, 5, 249
York Daily Herald, 81–82, 217
York Daily Record, 40, 208, 213–14, 217, 308,
   313, 316
York Dispatch, 208, 308, 313, 316, 329
young-Earth creationism, 13, 23, 30–31, 83,
   137, 141, 150–51, 153, 169, 188, 196

Z-DNA, 128